Plant Tissue Culture and Plant Somatic Embryogenesis

Plant Tissue Culture and Plant Somatic Embryogenesis

Guest Editors

Justyna Lema-Rumińska
Danuta Kulpa
Alina Trejgell

Basel • Beijing • Wuhan • Barcelona • Belgrade • Novi Sad • Cluj • Manchester

Guest Editors

Justyna Lema-Rumińska
Department of
Environmental Biology
Kazimierz Wielki University
in Bydgoszcz
Bydgoszcz
Poland

Danuta Kulpa
Department of Genetic, Plant
Breeding and Biotechnology
West Pomeranian University
of Technology
Szczecin
Poland

Alina Trejgell
Department of Plant
Physiology and
Biotechnology
Nicolaus Copernicus
University
Toruń
Poland

Editorial Office
MDPI AG
Grosspeteranlage 5
4052 Basel, Switzerland

This is a reprint of the Special Issue, published open access by the journal *Agronomy* (ISSN 2073-4395), freely accessible at: www.mdpi.com/journal/agronomy/special_issues/64O04LMG41.

For citation purposes, cite each article independently as indicated on the article page online and using the guide below:

Lastname, A.A.; Lastname, B.B. Article Title. *Journal Name* **Year**, *Volume Number*, Page Range.

ISBN 978-3-7258-3682-6 (Hbk)
ISBN 978-3-7258-3681-9 (PDF)
https://doi.org/10.3390/books978-3-7258-3681-9

© 2025 by the authors. Articles in this book are Open Access and distributed under the Creative Commons Attribution (CC BY) license. The book as a whole is distributed by MDPI under the terms and conditions of the Creative Commons Attribution-NonCommercial-NoDerivs (CC BY-NC-ND) license (https://creativecommons.org/licenses/by-nc-nd/4.0/).

Contents

About the Editors vii

Preface ix

Małgorzata Malik, Ewelina Tomiak and Bożena Pawłowska
Effect of Liquid Culture Systems (Temporary Immersion Bioreactor and Rotary Shaker) Used During Multiplication and Differentiation on Efficiency of Repetitive Somatic Embryogenesis of *Narcissus* L. 'Carlton'
Reprinted from: *Agronomy* 2024, 15, 85, https://doi.org/10.3390/agronomy15010085 1

Danuta Kulpa and Marcelina Krupa-Małkiewicz
Propagation of Clematis 'Warszawska Nike' in In Vitro Cultures
Reprinted from: *Agronomy* 2023, 13, 3056, https://doi.org/10.3390/agronomy13123056 13

María Guadalupe Barrera Núñez, Mónica Bueno, Miguel Ángel Molina-Montiel, Lorena Reyes-Vaquero, Elena Ibáñez and Alma Angélica Del Villar-Martínez
Chemical Profile of Cell Cultures of *Kalanchoë gastonis-bonnieri* Transformed by *Agrobacterium rhizogenes*
Reprinted from: *Agronomy* 2024, 14, 189, https://doi.org/10.3390/agronomy14010189 23

Piotr Licznerski, Justyna Lema-Rumińska, Emilia Michałowska, Alicja Tymoszuk and Janusz Winiecki
Effect of X-rays on Seedling Pigment, Biochemical Profile, and Molecular Variability in *Astrophytum* spp.
Reprinted from: *Agronomy* 2023, 13, 2732, https://doi.org/10.3390/agronomy13112732 43

Ahmed E. Higazy, Mohammed E. El-Mahrouk, Antar N. El-Banna, Mosaad K. Maamoun, Hassan El-Ramady and Neama Abdalla et al.
Production of Black Cumin *via* Somatic Embryogenesis, Chemical Profile of Active Compounds in Callus Cultures and Somatic Embryos at Different Auxin Supplementations
Reprinted from: *Agronomy* 2023, 13, 2633, https://doi.org/10.3390/agronomy13102633 62

Marzena Warchoł, Edyta Skrzypek, Katarzyna Juzoń-Sikora and Dragana Jakovljević
Oat (*Avena sativa* L.) In Vitro Cultures: Prospects and Challenges for Breeding
Reprinted from: *Agronomy* 2023, 13, 2604, https://doi.org/10.3390/agronomy13102604 78

Jarosław Tyburski and Natalia Mucha
Antioxidant Response in the Salt-Acclimated Red Beet (*Beta vulgaris*) Callus
Reprinted from: *Agronomy* 2023, 13, 2284, https://doi.org/10.3390/agronomy13092284 102

Wantian Yao, Diya Lei, Xuan Zhou, Haiyan Wang, Jiayu Lu and Yuanxiu Lin et al.
Effect of Different Culture Conditions on Anthocyanins and Related Genes in Red Pear Callus
Reprinted from: *Agronomy* 2023, 13, 2032, https://doi.org/10.3390/agronomy13082032 120

Rabia Koçak, Melih Okcu, Kamil Haliloğlu, Aras Türkoğlu, Alireza Pour-Aboughadareh and Bita Jamshidi et al.
Magnesium Oxide Nanoparticles: An Influential Element in Cowpea (*Vigna unguiculata* L. Walp.) Tissue Culture
Reprinted from: *Agronomy* 2023, 13, 1646, https://doi.org/10.3390/agronomy13061646 144

Piotr Pałka, Bożena Muszyńska, Agnieszka Szewczyk and Bożena Pawłowska
Elicitation and Enhancement of Phenolics Synthesis with Zinc Oxide Nanoparticles and LED Light in *Lilium candidum* L. Cultures In Vitro
Reprinted from: *Agronomy* 2023, 13, 1437, https://doi.org/10.3390/agronomy13061437 157

About the Editors

Justyna Lema-Rumińska

Professor (associate) Justyna Lema-Rumińska, PhD., DSc., deals with plant biotechnology, particularly the application of biotechnological methods in the breeding and cultivation of ornamental and medicinal plants in vitro. She is interested in somatic embryogenesis, the genetic stability of plants propagated in vitro, secondary metabolites, and molecular markers.

Danuta Kulpa

Professor (associate) Danuta Kulpa, PhD. DSc., deals with biotechnology and plant breeding. Her research interests focus on developing methods for the micropropagation of horticultural plants—especially ornamental plants and herbs. Currently, she is particularly interested in studying the effect of metal nanoparticles on the development and physiology of plants in in vitro cultures.

Alina Trejgell

Professor (associate) Alina Trejgell, PhD., DSc., deals with plant physiology and biotechnology. Her research interests focus on developing micropropagation systems for protected species and the genetic stability of plants propagated in vitro, as well as slow-growth cultures and cold adaptation mechanisms.

Preface

In vitro plant cultures and somatic embryogenesis are the focus of this Special Issue. Scientists worldwide are developing and improving plant propagation and regeneration methods in in vitro cultures. Micropropagation is used on a large scale to produce high-quality cuttings of ornamental plants and, to a lesser extent, vegetable or agricultural plants. In addition, the technique is used in gene banks or to produce important secondary metabolites. The most efficient plant regeneration methods include somatic embryogenesis (SE). As a result of SE, somatic embryos can potentially be produced from any living somatic cell.

The genetic plant stability can be disturbed during the propagation and regeneration of plants in in vitro cultures. To be sure that the plants obtained this way are true-to-type, their genetic stability must be confirmed. Molecular markers that can be used at any stage of plant development work best here. The genetic variability created in this way and mutagenesis or genetic transformation in in vitro cultures facilitate the breeding of new cultivars of crops.

Justyna Lema-Rumińska, Danuta Kulpa, and Alina Trejgell
Guest Editors

Article

Effect of Liquid Culture Systems (Temporary Immersion Bioreactor and Rotary Shaker) Used During Multiplication and Differentiation on Efficiency of Repetitive Somatic Embryogenesis of *Narcissus* L. 'Carlton'

Małgorzata Malik, Ewelina Tomiak * and Bożena Pawłowska

Department of Ornamental Plants and Garden Art, University of Agriculture in Krakow, 29-Listopada 54, 31-425 Kraków, Poland; malgorzata.malik@urk.edu.pl (M.M.); bozena.pawlowska@urk.edu.pl (B.P.)
* Correspondence: ewelina.tomiak@urk.edu.pl

Abstract: Liquid culture systems, including bioreactors, are valuable tools for the scaling up of production. Their involvement leads to the automation of the highly efficient, reproducible somatic embryogenesis of *Narcissus* L. 'Carlton'. Alternative procedures for efficient embryogenic tissue and early somatic embryo multiplication have been developed. The long-term embryogenic callus of narcissus 'Carlton', obtained by repetitive somatic embryogenesis, was multiplicated and differentiated in different liquid culture systems. For multiplication, the Rita® temporary immersion bioreactor and the rotary shaker at 60 rpm and 100 rpm were used, and, for differentiation, the rotary shaker at 60 rpm and solid cultures were investigated. Cultures immersed with a frequency of 15 min every 24 h during multiplication were characterized by the greatest increase in biomass (1.3), and the greatest number of embryos (152.6 embryos per 1 g of inoculum) was seen during differentiation. Higher immersion frequencies (15 min every 8 and 12 h) decreased the tissue quality and yield. The use of a bioreactor during multiplication promoted the number of embryos obtained during differentiation. In turn, cultivation in a rotary shaker during differentiation, regardless of the multiplication system, stimulated the multiplication of embryogenic tissue. The liquid medium used for the multiplication and differentiation of somatic embryos improved the synchronization of their development, which reached up to 95–99% depending on the system.

Keywords: daffodil; immersion frequency; in vitro; liquid medium; Rita®; temporary immersion system

1. Introduction

Narcissus L. is a very important long-lived perennial geophyte from a botanical, horticultural, folkloric, and medicinal perspective [1–3]. These bulbous plants are widely used as ornamental plants in urban landscaping and home gardens and as a cut flower. Bulbs are produced on a large scale throughout Europe, especially in the Netherlands, and globally [4,5].

In recent years, daffodils have gained enormous popularity due to the content of biologically active compounds in their bulbs, and they are a rich source of specialized metabolites. Alkaloids such as galanthamine (Gal), lycorine, and narciclasine are known for their pharmaceutical properties; they are used during therapy in Alzheimer's [6] and cancer [7]. *Narcissus pseudonarcissus* L. 'Carlton' was chosen as a source due to the relatively high concentration of Gal in the bulbs [1,3,4,8,9]. Plant tissue culture is a promising alternative to optimize alkaloid extraction [6,7]. Ferdausi's [10,11] studies also confirmed the presence of galantamine in bulbs, shoots, and calli obtained in vitro from the 'Carlton' cultivar at high auxin concentrations.

In vitro cultures are used for the mass production of plants that are difficult to propagate traditionally. They provide an opportunity to obtain elite propagation material and are also helpful in the propagation of plants with industrial and pharmacological potential, as well as the propagation of endangered species and their cryopreservation [12–16]. Under field conditions, the natural vegetative propagation rate of narcissus when separating adventitious bulbs is very slow, at nearly 1.6 bulbs/year [17,18]. Chipping and twin-scaling are two traditional propagation techniques for rapid reproduction, which are more efficient but are insufficient and do not meet the current needs of mass production [19,20].

In vitro propagation is a technique that includes somatic embryogenesis, which is one of the basic tools that is widely used in the high-yield reproduction of many species. In the genus *Narcissus*, somatic embryogenesis has been achieved in *Narcissus confusus* [21]; *Narcissus pseudonarcissus* cv. 'Golden Harvest', 'St. Keverne' [22], and 'Carlton' [23]; *Narcissus tazzeta* [24,25]; and *Narcissus papyraceus* cv. Shirazi [26].

Repetitive somatic embryogenesis (RSE) offers even greater opportunities for the automation and commercialization of production due to the scale-up of production through the continuous multiplication of somatic embryos [27,28]. New embryos can proliferate continuously through consecutive cycles of secondary somatic embryogenesis. Embryos are formed from embryos that were previously formed [27,29]. The advantages of long-term embryogenic cultures obtained in this way are adaptability to different liquid culture systems, efficient handling, and the independence of this process from the original explant source [30]. High-frequency secondary and repetitive somatic embryogenesis has been developed for *Coffea arabica* [31], *Camellia assamica* [32], *Quercus robur* [33], *Camellia sinensis* [30], and *Hepatica nobilis* [34]. High-yielding RSE for narcissus cv. Carlton for solid cultures has also been developed [29].

The use of liquid culture systems for propagation by somatic embryogenesis provides many benefits, including increased efficiency, increased nutrient uptake, more uniform culture conditions, more homogeneous material, and reduced production costs [31,35]. A special system consists of a bioreactor designed for intensive plant propagation in liquid media under controlled environmental conditions. In the Rita® temporary immersion bioreactor, these include the time of tissue contact with the media, aeration, and air exchange [36]. The method of operation of temporary immersion systems involves periodically immersing the cultured tissue in a liquid medium, followed by draining and exposing the plant tissue to a sterile gaseous environment [35]. Temporary immersion and gas exchange in the bioreactor vessel allow one to overcome the occurrence of hyperhydricity or asphyxia, which is often observed in cultures continuously immersed in a liquid medium [37].

To date, for narcissus 'Carlton', a protocol involving liquid media for the initiation of somatic embryogenesis has been developed. Somatic embryogenesis was induced in ovary cultures by alternating the use of liquid media and solid media [38]. The aim of the present study was to check whether the use of different liquid systems (rotary shaker and TIS technology) during multiplication and differentiation would increase the efficiency of narcissus RSE, the synchronization of somatic embryos, and biomass production.

2. Materials and Methods

2.1. Plant Material

A long-term (seven-year-old) embryogenic callus obtained by RSE from ovary explants isolated from *Narcissus* L. 'Carlton' flower buds originating from bulbs (12 cm in circumference) chilled for 3 weeks at 5 °C was taken for the experiments [29]. The callus was initiated and grown on solid Murashige and Skoog (MS) medium [39] with 25 μM Picloram and 5 μM 6-benzylaminopurine (BA) and 3% sucrose. The medium was adjusted to pH 5.5 before autoclaving and was gelled with 0.5% agar (Lab-Agar™, Biocorp, Warsaw, Poland). The callus was maintained at 20 ± 2 °C in darkness. Cultures were transferred to fresh medium every 4–6 weeks.

2.2. Effect of Immersion Frequency—Experiment 1

A two-stage experiment was assumed: six-week multiplication followed by six-week differentiation stages. For multiplication, clusters of embryogenic tissue (callus with differentiating primary and secondary embryos at the globular stage) were placed in the vessels of a Rita® bioreactor (CIRAD, Montpellier, France) with a temporary immersion system filled with 200 mL of multiplication liquid medium. Experiment 1 was designed as a single-factor experiment (immersion frequency used during multiplication) and investigated the influence of the factor on the RSE performance during the multiplication and differentiation stages. The effect of the following immersion frequencies was investigated: (i) 15 min every 24 h (TIS 1 × 15), (ii) 15 min every 8 h (TIS 3 × 15), and (iii) 15 min every 2 h (TIS 12 × 15). The control involved cultures in Petri dishes (diameter 9 cm, height 2.5 cm) filled with 25 mL of multiplication agar-solidified medium (0.5%, Lab-Agar™, Biocorp, Warsaw, Poland). The multiplication medium contained macro- and micronutrients and vitamins, as described by Murashige and Skoog [39], as well as 25 μM Picloram, 5 μM BA, and 3% sucrose. The medium was adjusted to pH 5.5 before autoclaving. Each Petri dish was filled with 2 g of embryogenic tissue and each Rita® vessel was filled with 5 g. Cultures were maintained at 20 ± 2 °C in darkness.

For the differentiation stage, embryogenic tissue (2 g per dish) obtained in all combinations of the multiplication stage was transferred to a Petri dish (diameter 9 cm, height 2.5 cm) on solid MS medium with 5 μM BA and 0.5 μM naphthalene-1-acetic acid (NAA) and 3% sucrose. The medium was adjusted to pH 5.8 before autoclaving and was gelled with 0.5% agar (Lab-Agar™, Biocorp, Warsaw, Poland). The callus was maintained at 20 ± 2 °C in darkness.

2.3. Effect of Liquid Culture System—Experiment 2

Experiment 2 was designed as a two-factor experiment (culture system used during multiplication × culture system used during differentiation) and investigated the influence of both factors on the RSE performance during the differentiation stage. Embryogenic tissue multiplied for six weeks in three liquid systems was used for experiment 2: (i) continuous cultivation in liquid medium on a rotary shaker at 60 rpm (RS 60), (ii) continuous cultivation in liquid medium on a rotary shaker at 100 rpm (RS 100), (iii) temporary immersion system at an immersion frequency of 15 min every 24 h (TIS 1 × 15).

Clusters of embryogenic callus obtained during multiplication (on multiplication medium) for regeneration were transferred to MS regeneration medium containing 5 μM BA and 0.5 μM NAA and 3% sucrose. Calli were cultured for six weeks in two ways: continuously in liquid medium on a rotary shaker at 60 rpm or on solid medium and then for six weeks on solid medium of the same composition. Solid media were gelled with 0.5% agar (Lab-Agar™, Biocorp, Warsaw, Poland). Control cultures were maintained on solid media during multiplication and differentiation.

The following cultivation vessels were used: 100 mL Erlenmeyer flasks with 20 mL of medium and 1 g of embryogenic tissue for cultivation on a rotary shaker and Petri dishes (diameter 9 cm, height 2.5 cm) with 20 mL of solid medium and 2 g of tissue for cultivation on solid medium. Cultures were maintained at 20 ± 2 °C in darkness.

2.4. Growth Evaluation and Statistical Analysis

After six and 12 weeks, the biomass growth index and the total number of somatic embryos per 1 g of inoculum tissue were calculated. For the biomass growth index, the following formula was used: (FFW − IFW)/IFW, where FFW = final fresh weight and IFW = initial fresh weight. The percentage shares of the individual developmental stages (early, mature and cotyledonary) in the total number of embryos obtained were also determined. The percentage of early embryos in the total number of embryos was defined as synchronization.

The results of the observations were evaluated by analysis of variance. Means that differed significantly were identified using Tukey's multiple test at a significance level of $p \leq 0.05$ (Statistica version 10, StatSoft, Kraków, Poland).

3. Results and Discussion

3.1. Biomass Growth

The scale-up of production resulting from the involvement of RSE can be achieved due to faster tissue growth and easy adaptation to liquid culture systems [32,40,41]. The direct contact of plant cells with the liquid medium, i.e., the better availability of nutrients and growth regulators, allows for increased efficiency [42] but also influences the synchronization of production [31].

In the narcissus 'Carlton' cultures, the multiplication of the embryogenic tissue and the differentiation of somatic embryos were observed in all culture systems tested. The decisive parameters were the state of the medium and the time of contact with the liquid medium. The long-term callus representing RSE, continuously multiplied on solid medium containing 25 µM Picloram and 5 µM BA after transfer to the Rita® bioreactor in liquid multiplication medium, proliferated best when the frequency of immersion was 15 min every 24 h (TIS 1 × 15). The multiplication rate was 1.3, which was higher compared to that obtained on solid media, i.e., TIS 3 × 15 and TIS 12 × 15 (Table 1 and Figure 1a). However, the lowest biomass increase was observed in cultures immersed for 15 min every 8 h (0.2, TIS 3 × 15). In the cultures that were immersed the most frequently, the dedifferentiation of the embryogenic tissue was observed more often, as well as its faster ageing (Figure 1b).

Table 1. Effect of immersion frequency on efficiency of *Narcissus* L. 'Carlton' repetitive somatic embryogenesis in Rita® bioreactor during multiplication and differentiation stages.

Immersion Frequency [A]	Multiplication		Differentiation	
	Biomass Growth (6 Weeks)	No. of Somatic Embryos (6 Weeks)	Biomass Growth (6 Weeks)	No. of Somatic Embryos (6 Weeks)
TIS 1 × 15	1.3 ± 0.2 d [B]	51.7 ± 5.2 ab	1.6 ± 0.7 ab	152.6 ± 8.4 b
TIS 3 × 15	0.2 ± 0.0 a	61.9 ± 6.9 b	0.7 ± 0.2 a	117.6 ± 5.9 a
TIS 12 × 15	0.4 ± 0.1 b	41.8 ± 8.1 a	1.6 ± 0.4 ab	110.4 ± 27.2 a
Solid	0.8 ± 0.0 c	49.2 ± 7.3 a	2.1 ± 0.7 b	100.2 ± 8.5 a
Significant effects [C]	***	***	***	***

[A] Immersion frequency: Solid—solid medium (control); TIS 1 × 15—cultures in bioreactor Rita® with immersion frequency of 15 min every 24 h; TIS 3 × 15—cultures in bioreactor Rita® with immersion frequency of 15 min every 8 h; TIS 12 × 15—cultures in bioreactor Rita® with immersion frequency of 15 min every 2 h. [B] Mean values ± SD followed by different lowercase letters are significantly different at $p \leq 0.05$ according to Tukey's multiple range test. [C] Significant effects: ***—at $p \leq 0.01$.

Figure 1. Repetitive somatic embryogenesis of *Narcissus* L. 'Carlton' in different liquid culture systems. (**a**) Embryogenic tissue during multiplication in Rita® bioreactor immersed with frequency of 15 min every 24 h. Bar = 1 cm. (**b**) Embryogenic tissue during multiplication in Rita® bioreactor immersed with frequency of 15 min every 2 h. Bar = 1 cm. (**c**) Differentiation on solid medium (control cultures). (**d**) Early stages (globular and torpedo) of somatic embryos and bioreactor cultures (TIS 1 × 15) after transfer from rotary shaker at 60 rpm to solid medium. (**e**) Mature somatic embryo. (**f**) Cluster of embryogenic tissue with cotyledonary embryos. (**g**) Bioreactor cultures (TIS 1 × 15) after transfer to solid medium. Bar = 1 cm. (**h**) Embryo conversion in light—20th week of culture. Bar = 1 cm.

When transferring the callus cultures from the bioreactor to a solid regeneration medium with 5 µM BA and 0.5 µM NAA, the biomass growth in the TIS 3 × 15 cultures (0.7) was still the lowest, while intensified growth was observed in the control on solid medium (2.1). This resulted not only from the proliferation of the embryogenic tissue but also from the development of somatic embryos and their conversion (Figure 1c).

The efficiency of embryogenic callus proliferation and somatic embryo differentiation in the Rita® temporary immersion bioreactor has been repeatedly reported to be highly dependent on the immersion frequency. As a rule, more frequent immersions promoted an increase in fresh weight, which was not observed in the case of the embryogenic callus of narcissus. In cork oak [43], an immersion frequency of 1 min every 6 or 4 h increased the fresh weight compared to a frequency of 1 min every 12 h and with semi-solid media. Frequencies of 1 min every 6, 8 and 12 h increased the biomass growth of *Quercus robur* embryo clusters compared to a solid medium [33]. The immersion frequency of 5 min every 4 h, compared to 1, 10 and 15 min every 4 h and with a semi-solid medium, promoted the multiplication of secondary somatic embryos of *Eurycoma longifolia*. Increased productivity was also observed when the immersion frequency was changed from every 4 h to every 8 h [44].

An increase in biomass when using liquid systems can also be achieved at the differentiation stage. After a differentiation cycle in various culture systems, i.e., solid and rotary shakers at 60 rpm, significant differences in narcissus biomass growth were observed. Regardless of the culture system used during the multiplication of the narcissus embryogenic tissue (RS 60, RS 100, TIS 1 × 15, solid medium—control), a higher or similar biomass growth index was obtained in shaking cultures compared to solid cultures (Table 2).

Table 2. Effect of the liquid culture system used during multiplication and differentiation on the efficiency of *Narcissus* L. 'Carlton' repetitive somatic embryogenesis.

Culture System During Multiplication (Mcs) [A]	Culture System During Differentiation (Dcs)	6th Week of Differentiation		12th Week of Differentiation	
		Biomass Growth	No. of Somatic Embryos	Biomass Growth	No. of Somatic Embryos
		Effect of Mcs and Dcs			
Solid	Solid	2.2 ± 0.5 a–c [B]	21.0 ± 1.3 ab	4.7 ± 1.7 a–c	38.5 ± 3.8 ab
Solid	RS 60	5.9 ± 0.8 d	14.1 ± 3.4 a	11.7 ± 1.3 d	41.0 ± 17.0 ab
RS 60	Solid	1.7 ± 0.2 ab	24.0 ± 5.1 ab	4.9 ± 1.0 a–c	29.8 ± 4.5 a
RS 60	RS 60	3.8 ± 0.8 c	16.8 ± 3.2 a	7.6 ± 0.6 c	29.9 ± 9.0 a
RS 100	Solid	1.0 ± 0.3 a	14.4 ± 10.2 a	2.3 ± 0.5 a	20.1 ± 12.1 a
RS 100	RS 60	3.2 ± 1.0 bc	17.0 ± 13.8 a	6.2 ± 1.5 bc	30.8 ± 17.2 a
TIS 1 × 15	Solid	1.5 ± 0.3 ab	42.1 ± 7.7 b	4.0 ± 0.2 ab	65.1 ± 7.1 b
TIS 1 × 15	RS 60	3.1 ± 0.3 bc	22.2 ± 11.3 ab	6.4 ± 0.6 bc	44.9 ± 11.7 ab
		Effect of Mcs			
Solid		4.0 ± 2.1 b	17.6 ± 4.4 a	8.2 ± 4.1 c	39.7 ± 11.1 ab
RS 60		2.7 ± 1.2 a	20.3 ± 5.5 ab	6.2 ± 1.7 a	29.8 ± 6.4 a
RS 100		2.1 ± 1.4 a	15.7 ± 10.9 a	4.3 ± 2.4 a	25.4 ± 14.5 a
TIS 1 × 15		2.3 ± 0.9 a	32.2 ± 13.9 b	5.2 ± 1.4 ab	55.0 ± 14.0 b
		Effect of Dcs			
	Solid	1.6 ± 0.5 a	25.4 ± 12.2 b	4.0 ± 1.4 a	38.3 ± 18.7 a
	RS 60	4.0 ± 1.3 b	17.5 ± 8.4 a	8.0 ± 2.5 b	36.6 ± 13.8 a
		Main effects [C]			
Mcs × Dcs		**	ns	***	ns
Mcs		***	**	***	***
Dcs		***	**	***	ns

[A] Culture system: solid—solid medium (control); TIS 1 × 15—cultures in Rita® bioreactor with immersion frequency of 15 min every 24 h; RS60/RS100—cultures in liquid medium on rotary shaker (60 and 100 rpm).
[B] Mean values ± SD followed by different lowercase letters are significantly different at $p \leq 0.05$ according to Tukey's multiple range test. [C] Main effects: ***—at $p \leq 0.01$; **—at $p \leq 0.05$; ns—not significant.

The shaking speed (60 or 100 rpm) used during multiplication only influenced biomass growth during the differentiation stage. After 12 weeks of differentiation, significantly more embryogenic calli were observed in cultures shaken at 60 rpm compared to 100 rpm (Table 2).

The highest biomass growth was noted in cultures grown on solid media, which were transferred to liquid media and shaken at 60 rpm (solid RS60) after both 6 and 12 weeks of cultivation (5.9-fold and 11.7-fold increase, respectively). The solid differentiation medium used for this material did not favor the multiplication of the embryogenic tissue. Biomass growth on the solid medium amounted to 2.2 and 4.7, respectively. Transferring the tissue grown in liquid media (systems: RS60, RS100, TIS 1 × 15) to solid media for differentiation inhibited multiplication. On the other hand, continuation in liquid media supported the multiplication of embryogenic tissue.

3.2. Somatic Embryo Differentiation

The differentiation of somatic embryos occurs when the auxin concentration in the medium decreases [38,45]. Reducing the auxin concentration allows the achievement of subsequent embryo developmental stages: globular, torpedo, mature, cotyledonary and converted plants [35]. In the 'Carlton' cultures maintained on a multiplication medium in a bioreactor, the largest number of embryos was formed when the embryogenic tissue had the least frequent contact with a medium with high auxin content, i.e., for an immersion frequency of 1 × 15 min or 3 × 15 min per day (51.7 embryos and 61.9 embryos, respectively; Table 1). In combinations where contact with the multiplication medium was longer, in solid and bioreactor cultures immersed for 12 × 15 min per day, the smallest number of embryos was formed (49.2 and 41.8, respectively). Further differentiation of the somatic embryos was observed on a solid regeneration medium with the reduced availability of auxin. Regarding the differentiation medium, the largest number of embryos was obtained in the cultures immersed the least frequently at the multiplication stage (1 × 15 min during the day)—152.6 embryos (Table 1).

Based on the results of experiment 2, it was found that the number of somatic embryos formed during differentiation depended on the culture system used at the multiplication stage (Table 2). Regardless of the other factors, the largest number of embryos was obtained in cultures multiplied in the Rita® bioreactor. The efficiency of embryo formation is also influenced by the culture system used immediately after multiplication. More embryos are observed when differentiation takes place on a solid medium. A similar relationship has been observed in the development of somatic embryos in many plant species, including Arabica coffee [31], peanut [45] and hybrid larch [46]. In the liquid medium (RS60) at the differentiation stage, the number of embryos was lower than that obtained on the solid medium in the case of cultures multiplied in the bioreactor (Table 2).

3.3. Somatic Embryo Maturation

The greatest diversity in the developmental stages among the narcissus embryos occurred in cultures grown on solid media, where all developmental stages of embryos appeared: globular, torpedo, mature, cotyledons and post-conversion (Figures 1d–f and 2).

On solid media, in contrast to liquid media, where the cultures are more homogeneous, a concentration gradient of growth regulators is created, which causes somatic embryos to develop asynchronously [47]. This is due to the gradual consumption of auxin from the medium and the formation of a concentration gradient of growth regulators in clusters of embryogenic tissue with RSE. The decrease in the stimulation of endogenous IAA synthesis induced by high concentrations of exogenous auxin releases the potential for embryo formation and maturation [48]. In turn, maintaining a high concentration of auxin is necessary to maintain the embryogenic nature of embryogenic cultures of monocotyledonous plants [49]. When the somatic embryos were continuously or frequently immersed in the liquid medium, embryo maturation was delayed. Somatic embryos were arrested in the globular stage. This phenomenon allows for the long-term (from several months to several years) multiplication of embryogenic tissue in the process of RSE and is beneficial at the stage of the highly efficient proliferation of embryogenic tissue.

During the multiplication of narcissus in the Rita® bioreactor, synchronization of 90–99% (early stages of embryos) was observed. Similar values were achieved after transferring the material from the bioreactor to the solid regeneration medium (94–98% of embryos in early stages; Figure 1g). In the control cultures on solid media in the multiplication and differentiation stages, further embryo stages appeared. Synchronization was 72 and 74%, respectively (Figure 2).

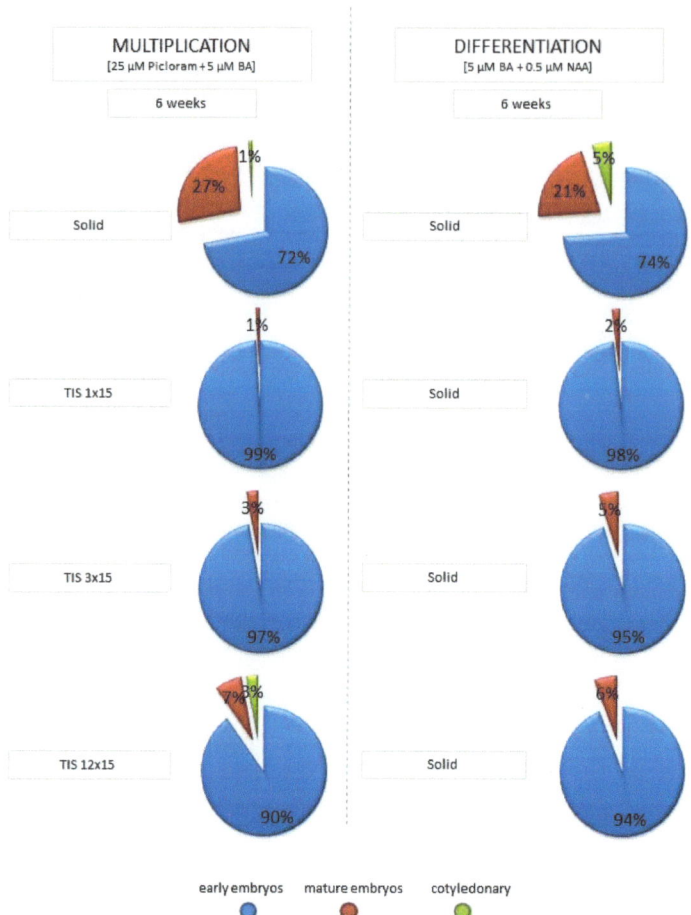

Figure 2. The effect of the immersion frequency on the percentage share of individual developmental stages of *Narcissus* L. 'Carlton' embryos (early, mature and cotyledonary) in the total number of embryos obtained during multiplication and differentiation (Solid—solid medium (control); TIS 1 × 15—cultures in Rita® bioreactor with immersion frequency of 15 min every 24 h; TIS 3 × 15—cultures in Rita® bioreactor with immersion frequency of 15 min every 8 h; TIS 12 × 15—cultures in Rita® bioreactor with immersion frequency of 15 min every 2 h).

Experiment 2 also confirmed that synchronization depended on the culture system used in the multiplication or differentiation stage. The narcissus 'Carlton' cultures that were propagated in the liquid culture systems (RS 60, RS 100 or TIS 1 × 15) after transfer to the regeneration medium, regardless of whether it was solid or liquid, maintained a high degree of synchronization for 6 weeks (77–95% of early-stage embryos). Further cultivation on solid regeneration medium led to the maturation of the embryos. The cotyledonary stage was observed in all combinations (1–20%), and the synchrony ranged from 37% in the control to 68–76% in cultures grown in the bioreactor (TIS 1 × 15) (Figure 3). Transferring the cultures to light for another 8 weeks stimulated further embryo development and conversion to plants (Figure 1h).

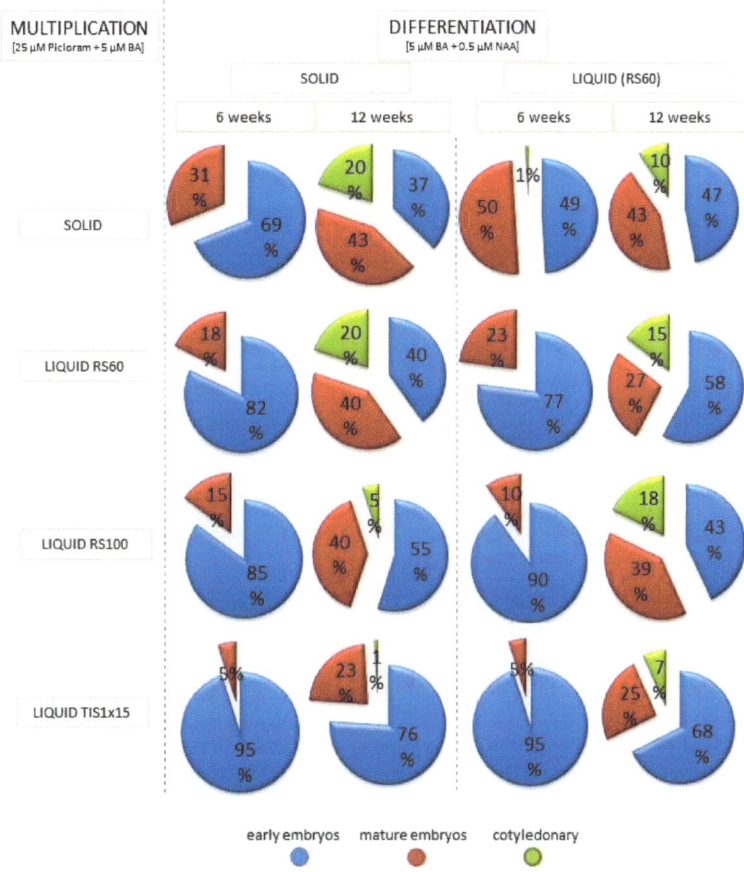

Figure 3. The effect of the liquid culture system on the percentage share of individual developmental stages of *Narcissus* L. 'Carlton' embryos (early, mature and cotyledonary) in the total number of embryos obtained during multiplication and differentiation (Solid—solid medium (control); TIS 1 × 15—cultures in Rita® bioreactor with immersion frequency of 15 min every 24 h; RS 60/RS 100—cultures in liquid medium on rotary shaker (60 and 100 rpm).

The commercial application of somatic embryogenesis has many limitations. One of the factors that makes it difficult to automate plant reproduction through somatic embryogenesis is the asynchronous development of embryos. Somatic embryogenesis consists of five stages: the initiation of embryogenesis, the proliferation of embryogenic cultures, differentiation—somatic embryo formation and early development, the maturation of somatic embryos and the conversion of embryos into plantlets [35,50]. These require the development of optimal factors (physical, chemical and biological) that enable the somatic embryos to achieve subsequent developmental stages synchronously at the same time. At the stage of embryo formation and its early development, in addition to influencing the composition of growth regulators in the medium, especially auxins, we can significantly influence the processes by selecting a liquid culture system. Temporary immersion systems that allow for the semi-automation or full automation of multi-stage regeneration processes are of particular importance. The use of the bioreactor for the RSE tissue of narcissus 'Carlton' allowed for the long-term multiplication of the embryogenic callus and early somatic embryos, effectively inhibiting embryo maturation above the globular stage. The goal for the future will be to develop a system or set of factors for the synchronous maturation and conversion of multiplied somatic embryos.

4. Conclusions

Based on the presented results, it can be stated that the highest efficiency of the RSE of narcissus 'Carlton' was obtained using the Rita bioreactor at the multiplication stage. After 6 weeks of immersion with a frequency of 15 min every 24 h, with a medium containing 25 µM Picloram and 5 µM BA, the biomass growth was 1.3. The highest efficiency of embryo differentiation (152.6 embryos per 1 g of inoculum) and, at the same time, the highest synchronization (98%) can be achieved by transferring the tissue multiplied in the bioreactor to a solid medium containing 5 µM BA and 0.5 µM NAA for the next 6 weeks. Carrying out differentiation in rotary shaker cultures stimulates multiplication while reducing the number of somatic embryos obtained.

Author Contributions: Conceptualization, M.M.; methodology, M.M. and E.T.; software, M.M.; validation, M.M. and B.P.; formal analysis, M.M. and E.T.; investigation, M.M. and E.T.; resources, M.M.; data curation, M.M. and E.T.; writing—original draft preparation, M.M. and E.T.; writing—review and editing, M.M. and E.T.; visualization, M.M. and E.T.; supervision, M.M.; funding acquisition, B.P.; project administration, M.M., B.P. and E.T. All authors have read and agreed to the published version of the manuscript.

Funding: This research was supported by the Ministry of Science and Higher Education of the Republic of Poland from subvention funds for the University of Agriculture in Krakow.

Data Availability Statement: The original contributions presented in the study are included in the article, further inquiries can be directed to the corresponding author.

Conflicts of Interest: The authors declare no conflicts of interest.

References

1. Křoustková, J.; Ritomská, A.; Al Mamun, A.; Hulcová, D.; Opletal, L.; Kuneš, J.; Bucar, F. Structural analysis of unusual alkaloids isolated from *Narcissus pseudonarcissus* cv. 'Carlton'. *Phytochemistry* **2022**, *204*, 113439. [CrossRef] [PubMed]
2. Jezdinský, A.; Jezdinská Slezák, A.; Vachůn, A.; Pokluda, R.; Uher, J. Effect of saline water on the vase life of *Narcissus poeticus* L. flowers. *Folia Hortic.* **2024**, *36*, 95–117. [CrossRef]
3. Boshra, Y.R.; Fahim, J.R.; Hamed, A.N.E.; Desoukey, S.Y. Phytochemical and biological attributes of *Narcissus pseudonarcissus* L. (Amaryllidaceae): A review. *S. Afr. J. Bot.* **2022**, *146*, 437–458. [CrossRef]
4. Berkov, S.; Georgieva, L.; Sidjimova, B.; Nikolova, M.; Stanilova, M.; Bastida, J. In vitro propagation and biosynthesis of Sceletium-type alkaloids in *Narcissus pallidulus* and *Narcissus* cv. 'Hawera'. *S. Afr. J. Bot.* **2021**, *136*, 190–194. [CrossRef]
5. The Daffodil Society. Bulb Suppliers. 2019. Available online: https://thedaffodilsociety.com/bulb-suppliers/ (accessed on 5 December 2019).
6. Santos, G.S.; Sinoti, S.B.P.; de Almeida, F.T.C.; Silveira, D.; Simeoni, L.A.; Gomes-Copeland, K.K.P. Use of galantamine in the treatment of Alzheimer's disease and strategies to optimize its biosynthesis using the in vitro culture technique. *Plant Cell Tissue Organ Cult. (PCTOC)* **2020**, *143*, 13–29. [CrossRef]
7. Sena, S.; Kaur, H.; Kumar, V. Lycorine as a lead molecule in the treatment of cancer and strategies for its biosynthesis using the in vitro culture technique. *Phytochem. Rev.* **2024**, *23*, 1861–1888. [CrossRef]
8. Lubbe, A.; Gude, H.; Verpoorte, R.; Choi, Y.H. Seasonal accumulation of major alkaloids in organs of pharmaceutical crop *Narcissus* 'Carlton'. *Phytochemistry* **2013**, *88*, 43–53. [CrossRef]
9. Georgiev, V.; Ivanov, I.; Pavlov, A. Recent Progress in Amaryllidaceae Biotechnology. *Molecules* **2020**, *25*, 4670. [CrossRef] [PubMed]
10. Ferdausi, A.; Chang, X.; Hall, A.; Jones, M. Galanthamine production in tissue culture and metabolomic study on Amaryllidaceae alkaloids in *Narcissus pseudonarcissus* cv. 'Carlton'. *Ind. Crops Prod.* **2020**, *144*, 112058. [CrossRef]
11. Ferdausi, A.; Chang, X.; Jones, M. Enhancement of galanthamine production through elicitation and NMR-based metabolite profiling in *Narcissus pseudonarcissus* cv. Carlton in vitro callus cultures. *Vitr. Cell. Dev. Biol. Plant* **2021**, *57*, 435–446. [CrossRef]

12. Juan Vicedo, J.; Pavlov, A.; Ríos, S.; Casas, J.L. Micropropagation of five endemic, rare and/or endangered *Narcissus* species from the Iberian Peninsula (Spain and Portugal). *Acta Biol. Crac. Ser. Bot.* **2021**, *63*, 55–61. [CrossRef]
13. Abu Zahra, H.M.F.; Oran, S.A. Micropropagation of the wild endangered daffodil *Narcissus tazetta* L. *Acta Hortic.* **2009**, *826*, 135–140. [CrossRef]
14. Sage, D.O. Propagation and protection of flower bulbs: Current approaches and future prospects, with special reference to *Narcissus*. *Acta Hortic.* **2005**, *673*, 323–334. [CrossRef]
15. Kocot, D.; Nowak, B.; Sitek, E. Long-term organogenic callus cultivation of *Ranunculus illyricus* L.: A blueprint for sustainable ex situ conservation of the species in urban greenery. *BMC Plant Biol.* **2024**, *24*, 212. [CrossRef] [PubMed]
16. Maślanka, M.; Panis, B.; Malik, M. Cryopreservation of *Narcissus* L. 'Carlton' somaticembryos by dropletvitrification. *Propag. Ornam. Plants* **2016**, *16*, 28–35.
17. Rees, A.R. The initiation and growth of *Narcissus* bulbs. *Ann. Bot.* **1969**, *33*, 277–288. [CrossRef]
18. Slezák, K.A.; Mazur, J.; Jezdinský, A.; Kapczyńska, A. Bulb size interacts with lifting term in determining the quality of *Narcissus poeticus* L. propagation material. *Agronomy* **2020**, *10*, 975. [CrossRef]
19. Rees, A.R. *Ornamental Bulbs, Corms and Tubers*; CABI International: Wallingford, UK, 1992; p. 220.
20. Hanks, G.R. *Narcissus and Daffodil: The Genus Narcissus*; CRC Press: London, UK, 2002; p. 452. [CrossRef]
21. Selles, M.; Viladomat, F.; Bastida, J.; Codina, C. Callus induction, somatic embryogenesis and organogenesis in *Narcissus confusus*: Correlation between the state of differentiation and the content of galanthamine and related alkaloids. *Plant Cell Rep.* **1999**, *18*, 646–651. [CrossRef]
22. Sage, D.O.; Lynn, J.; Hammatt, N. Somatic embryogenesis in *Narcissus pseudonarcissus* cvs. Golden Harvest and St. Keverne. *Plant Sci.* **2000**, *150*, 209–216. [CrossRef]
23. Malik, M.; Bach, A. Morphogenetic pathways from *Narcissus* L. 'Carlton' in vitro cultures of pcstage flower bud explants according to cytokinin and auxin ratios. *Acta Sci. Pol. Hortorum Cultus* **2016**, *15*, 101–111.
24. Chen, L.; Zhu, X.; Gu, L.; Wu, J. Efficient callus induction and plant regeneration from anther of Chinese narcissus (*Narcissustazetta* L. var. chinensis Roem). *Plant Cell Rep.* **2005**, *24*, 401–407. [CrossRef]
25. Al-Zghoul, T.M.; Shibli, R.A.; Qudah, T.S.; Tahtamouni, R.W.; Rawshdeh, N. Employment of Somatic Embryogenesis as a Tool for Rescuing Imperiled *Narcissus tazetta* L. Growing Wild in Jordanian Environment. *Jordan J. Biol. Sci.* **2021**, *14*, 559–564. [CrossRef]
26. Anbari, S.; Tohidfar, M.; Hosseini, R.; Haddad, R. Somatic embryogenesis induction in *Narcissus papyraceus* cv. Shirazi. *Plant Tissue Cult. Biotechnol.* **2007**, *17*, 37–46. [CrossRef]
27. Raemakers, C.J.J.M.; Jacobsen, E.A.; Visser, R.G.F. Secondary somatic embryogenesis and applications in plant breeding. *Euphytica* **1995**, *81*, 93–107. [CrossRef]
28. Vasic, D.; Alibert, G.; Skoric, D. Protocols for efficient repetitive and secondary somatic embryogenesis in *Helianthus maximiliani* (Schrader). *Plant Cell Rep.* **2001**, *20*, 121–125. [CrossRef]
29. Malik, M.; Bach, A. High-yielding repetitive somatic embryogenesis in cultures of *Narcissus* L. 'Carlton'. *Acta Sci. Pol. Hortorum Cultus* **2017**, *16*, 107–112.
30. Akula, A.; Becker, D.; Bedson, M. High-yielding repetitive somatic embryogenesis and plant recovery in a selected tea clone, TRI-2025, by temporary immersion. *Plant Cell Rep.* **2001**, *19*, 1140–1145. [CrossRef] [PubMed]
31. Aguilar, M.E.; Wang, X.-Y.; Escalona, M.; Yan, L.; Huang, L.-F. Somatic embryogenesis of Arabica coffee in temporary immersion culture: Advances, limitations, and perspectives for mass propagation of selected genotypes. *Front. Plant Sci.* **2022**, *13*, 994578. [CrossRef]
32. Bajpai, R.; Chaturvedi, R. Recurrent secondary embryogenesis in androgenic embryos and clonal fidelity assessment of haploid plants of Tea, *Camellia assamica* ssp. *assamica* and *Camellia assamica* ssp. *lasiocaylx*. *Plant Cell Tissue Organ Cult.* **2021**, *145*, 127–141. [CrossRef]
33. Mallón, R.; Covelo, P.; Vieitez, A.M. Improving secondary embryogenesis in *Quercus robur*: Application of temporary immersion for mass propagation. *Trees* **2012**, *26*, 731–741. [CrossRef]
34. Szewczyk-Taranek, B.; Pawłowska, B. Recurrent somatic embryogenesis and plant regeneration from seedlings of *Hepatica nobilis* Schreb. *Plant Cell Tissue Organ Cult.* **2015**, *120*, 1203–1207. [CrossRef]
35. Murthy, H.N.; Joseph, K.S.; Paek, K.Y.; Park, S.Y. Bioreactor systems for micropropagation of plants: Presentscenario and futureprospects. *Front. Plant Sci.* **2023**, *14*, 1159588. [CrossRef] [PubMed]
36. Berthouly, M.; Etienne, H. Temporary immersion system: A new concept for use liquid medium in mass propagation. In *Liquid Culture Systems for In Vitro Plant Propagation*; Springer: Dordrecht, The Netherlands, 2005; pp. 165–195. [CrossRef]
37. Etienne, H.; Berthouly, M. Temporary immersion systems in plant micropropagation. *Plant Cell Tissue Organ Cult.* **2002**, *69*, 215–231. [CrossRef]
38. Malik, M. Comparison of different liquid/solid culture systems in the production of somatic embryos from *Narcissus* L. ovary explants. *Plant Cell Tissue Organ Cult.* **2008**, *94*, 337–345. [CrossRef]

39. Murashige, T.; Skoog, F. A revised medium for rapid growth and bioassays with tobacco tissue cultures. *Physiol. Plant.* **1962**, *15*, 473–497. [CrossRef]
40. Burns, J.A.; Wetzstein, H.Y. Development and characterization of embryogenic suspension cultures of pecan. *Plant Cell Tissue Organ Cult.* **1997**, *48*, 93–102. [CrossRef]
41. Guan, Y.; Li, S.-G.; Fan, X.-F.; Su, Z.-H. Application of Somatic Embryogenesis in Woody Plants. *Front. Plant Sci.* **2016**, *7*, 938. [CrossRef] [PubMed]
42. Ziv, M. Simple bioreactors for mass propagation of plants. In *Liquid Culture Systems for In Vitro Plant Propagation*; Hvoslef-Eide, A.K., Preil, W., Eds.; Springer: Dordrecht, The Netherlands, 2005. [CrossRef]
43. Pérez, M.; Bueno, M.A.; Escalona, M.; Toorop, P.; Rodríguez, R.; Cañal, M.J. Temporary immersion systems (RITA®®) for the improvement of cork oak somatic embryogenic culture proliferation and somatic embryo production. *Trees* **2013**, *27*, 1277–1284. [CrossRef]
44. Mohd, N.M.; Ja'Afar, H.; Zawawi, D.D.; Alias, N. In Vitro Somatic Embryos Multiplication of *Eurycoma longifolia* Jack using Temporary Immersion System RITA®®. *Sains Malays.* **2017**, *46*, 897–902. [CrossRef]
45. Durham, R.E.; Parrott, W.A. Repetitive somatic embryogenesis from peanut cultures in liquid medium. *Plant Cell Rep.* **1992**, *11*, 122–125. [CrossRef] [PubMed]
46. Le, K.C.; Dedicova, B.; Johansson, S.; Lelu-Walter, M.A.; Egertsdotter, U. Temporary immersion bioreactor system for propagation by somatic embryogenesis of hybrid larch (*Larix × eurolepis* Henry). *Biotechnol. Rep.* **2021**, *32*, 684. [CrossRef] [PubMed]
47. Li, F.; Yao, J.; Hu, L.; Chen, J.; Shi, J. Multiple Methods Synergistically Promote the Synchronization of Somatic Embryogenesis Through Suspension Culture in the New Hybrid Between *Pinus elliottii* and *Pinus caribaea*. *Front. Plant Sci.* **2022**, *13*, 857972. [CrossRef] [PubMed]
48. Michalczuk, L.; Ribnicky, D.M.; Cooke, T.J.; Cohen, J.D. Regulation of indole-3-acetic acid biosynthetic pathways in carrot cell cultures. *Plant Physiol.* **1992**, *100*, 1346–1353. [CrossRef] [PubMed]
49. Krishna, R.S.; Vasil, I.K. Somatic embryogenesis in herbaceous monocots. In *In Vitro Embryogenesis in Plants*, 1st ed.; Thorpe, T.A., Ed.; Kluwer Academic Publishers: Dordrecht, The Netherlands, 1995; pp. 471–540. [CrossRef]
50. von Arnold, S.; Sabala, I.; Bozhkov, P.; Dyachok, J.; Filonova, L. Developmental pathways of somatic embryogenesis. *Plant Cell Tissue Organ Cult.* **2002**, *69*, 233–249. [CrossRef]

Disclaimer/Publisher's Note: The statements, opinions and data contained in all publications are solely those of the individual author(s) and contributor(s) and not of MDPI and/or the editor(s). MDPI and/or the editor(s) disclaim responsibility for any injury to people or property resulting from any ideas, methods, instructions or products referred to in the content.

Propagation of Clematis 'Warszawska Nike' in In Vitro Cultures

Danuta Kulpa * and Marcelina Krupa-Małkiewicz

Department of Plant Genetics, Breeding and Biotechnology, West Pomeranian University of Technology Szczecin, 71-434 Szczecin, Poland; mkrupa@zut.edu.pl
* Correspondence: danuta.kulpa@zut.edu.pl

Abstract: A micropropagation protocol for growing Clematis 'Warszawska Nike' was developed. The MS medium supplemented with 1 mg·dm^{-3} BAP showed good results in the case of microshoot initiation (80%). The addition of BAP to the medium at higher concentrations resulted in the formation of a large amount of callus tissue at the base of the explant. Of the explants growing on the medium with the lowest cytokinin concentration, 8% flowered. Very quickly, after just 14 days, the explants began to die: some of the leaves that developed in in vitro cultures began to turn yellow and wither. The propagation of shoots was performed in two steps. In the first step, cytokinin BAP and Kin in various concentrations (0.5–2 mg·dm^{-3}) were added to the MS medium. In the second step, MS medium with the combinations of BAP (0.5 and 1 mg·dm^{-3}) with IAA or GA$_3$ (1 and 2 mg·dm^{-3}) was used. The MS medium with 0.5 mg·dm^{-3} BAP and 2 mg·dm^{-3} GA$_3$ was the best medium for the multiplication stage of clematis. Plants growing on this medium had the largest number of leaves, shoots, and internodes, and were also heavier compared to plants propagated on other media. The proliferated clematis explants were rooted on MS medium with the addition of IAA or IBA in different concentrations (0.5 to 4 mg·dm^{-3}). Of the concentrations tested, 0.5 mg·dm^{-3} IAA was the most effective one for in vitro root induction. The highest percentage of acclimatized plants (75%) was observed when the shoots were rooted on MS medium with 0.5 mg·dm^{-3} IAA.

Keywords: climbers; auxins; cytokinin; gibberellins; in vitro; rooting

1. Introduction

One of the most well-known and widely distributed genera of climbers is the clematis (*Clematis* sp.). It has many representatives found in temperate climatic zones. *Clematis* belongs to the buttercup family (*Ranunculaceae* Juss.), and is a medicinally important genus with over 200 species and 400 varieties and hybrids [1,2]. Plants are a rich source of chemical compounds with a broad spectrum of applications in pharmacology, such as antibacterial, anti-inflammatory, anti-cancer, analgesic, and diuretic [3]. Due to their aesthetic value, many cultivars are grown as ornamental plants. However, the shelf life of cut clematis flowers is short. Therefore, in Europe, they are primarily planted outdoors, in contrast to the United States where clematis is cultivated for cut flowers [4].

The Polish breeder of clematis was a Jesuit Father Stefan Franczak (1917–2009), who named over sixty varieties. Several of its varieties, such as 'Blue Angel', 'Polish Spirit', and 'Warszawska Nike' have received the Award of Garden Merit, which is the highest distinction awarded by the Royal Horticulture Society in Great Britain for the best garden plants [5]. One of them, 'Warszawska Nike' (also known as Warsaw Nike) is a resistant, undemanding, very profusely flowering variety. It has velvety, dark purple-violet flowers with a diameter of 12–14 cm. Its values are emphasized by the light background of the wall, leaves, and flowers of other plants. It is recommended for growing in large containers on balconies and terraces. It reaches 2–3 m [6].

The traditional method is vegetative propagation, by cuttings, in the early summer, autumn, or winter. This method is more commonly employed than generative propagation due to its efficiency and the production of progeny plants that are genetically identical to the mother plant. Another notable propagation method is grafting, particularly when *Clematis vitalba* or *C. viticella* are utilized as rootstocks, as this leads to rapid plant growth. Grafting is seldom employed for perennial clematis. Although the method of propagation by grafting is occasionally used on an amateur scale [7], conventional propagation methods may not always be efficient for obtaining sufficient plant quantities as raw materials for the food and pharmaceutical industries. Moreover, the vegetative propagation of some species and their cultivars can be challenging, with cuttings proving difficult to root. This difficulty has prompted the development and commercial use of in vitro methods [1]. The essential objective is to develop and optimize a rapid in vitro culture production technology for generating plants with high economic and medicinal value, such as *Clematis*, all year round and in the desired quantities. In addition, the in vitro technique makes it possible to obtain plant material free of viruses, bacteria, and fungi, which allows for obtaining healthy plant material. This is particularly crucial for early large-flowered varieties [8].

The regenerative capacity depends on the genotype, nutrient solution composition, plant growth regulators (PGRs), and other organic substances [9]. Several studies have been conducted on *Clematis* micropropagation, e.g., Parzymies and Dąbski [10] describe the effect of cytokinin on the in vitro multiplication of *C. viticella* (L.) and *C. integrifolia* 'Petit Faucon'. Chavan et al. [1] achieved a high frequency of propagation of *C. heynei* through nodal bud segments. Izadi Sadeghabadi et al. [11] describe the effects of plant growth regulators on the rooting and growth of the *Clematis orientalis* L. in in vitro culture. Mitrofanova et al. [2] developed a propagation protocol for 13 cultivars of *Clematis* taken in collection plots of the Nikita Botanical Gardens using somatic embryogenesis and in vitro organogenesis. Although in vitro culture methods and conditions are similar for different *Clematis* genotypes, their requirements for growth regulators in culture media are different. A review of the literature showed that an in vitro regeneration protocol for this medicinal climber has not yet been standardized. Therefore, it is important to develop a highly efficient plant regeneration system for each genotype.

This research aims to develop a micropropagation method for *Clematis* 'Warszawska Nike'. The optimal content of plant growth regulators in the media at the initiation, multiplication, and rooting stages will be determined. The influence of auxin content in the media at the rooting stage on the adaptation of clematis to greenhouse conditions will also be determined.

2. Materials and Methods

2.1. Preparation of Plant Material

The materials used in this study were one-year-old *Clematis* 'Warszawska Nike' plants. This cultivar belongs to the *Viticella* section. To establish in vitro cultures, 2–3 cm stem fragments were used, from which the leaves were removed. The stem fragments were rinsed for 15 min under running water and then immersed for 30 s in 70% ethanol. The pre-decorated fragments were immersed in a 7% sodium hypochlorite (NaOCl) solution for 20 min. The next steps were carried out under sterile conditions in a laminar flow cabinet. The NaOCl was discarded and the shoots were rinsed three times with sterile distilled water. The disinfected shoot fragments were dried on sterile tissue paper. Then, the isolated fragments were placed individually in 15 mL test tubes containing 5 mL of mineral medium, according to Murashige and Skoog [12], with the addition of the 6-benzylaminopurine (BAP) (Duchefa Biochemie, Haarlem, The Netherlands) in a concentration from 1 to 3 mg·dm^{-3}.

A total of 50 explants were placed on each type of medium in three repetitions. After 21 days, the number of growth-initiating, dying, and infected cultures was assessed.

2.2. Multiplication and Rooting Stage

Shoot fragments with a length of 2 cm were used to induce organogenesis. The explants were placed on MS medium with the addition of plant growth regulators in different concentrations: 0.5 to 2 mg·dm^{-3} BAP, and 0.5 mg·dm^{-3} to 2 mg·dm^{-3} kinetin (Kin) (Duchefa Biochemie, Haarlem, The Netherlands). In the second experiment, combinations of MS medium with the addition of 0.5 or 1 mg·dm^{-3} BAP, used simultaneously with 1 or 2 mg·dm^{-3} IAA (indole-3-acetic acid) (Duchefa Biochemie, Haarlem, The Netherlands) or GA$_3$ (gibberellic acid A$_3$) (Duchefa Biochemie, Haarlem, The Netherlands) were used. As the control, the MS medium without plant growth regulators was used. After 6 weeks, the morphological characteristics were measured.

To induce in vitro rooting, the multiplied shoots were placed on a rooting MS medium with the addition of auxins, IAA or IBA (indole-3-butric-acid) (Duchefa Biochemie, Haarlem, The Netherlands), in concentrations from 0.5 to 4 mg·dm^{-3}. The MS medium without the addition of plant hormones was the control. After 6 weeks, the morphological characteristics were measured.

2.3. Culture Conditions

The experiment on the multiplication and rooting stage was carried out in 10 replicates of 6 explants in a 200 mL jar filled with 30 mL of the respective medium. At each stage, the MS medium was supplemented with 30 g·dm^{-3} sucrose, 0,1 g·dm^{-3} myo-inositol (Duchefa Biochemie), and 8 g·dm^{-3} agar. The pH of the medium was set at 5.7, using 0.1 M solutions of HCl and NaOH. The medium was sterilized in an autoclave at 121 °C and 1 MPa for 20 min. The cultures were placed in a phytotron at a temperature of 24 °C and relative humidity of 70–80%, under a 16 h photoperiod with a photosynthetic photon flux density (PPFD) of 40 µmol·m^{-2}·s^{-1} PAR (photosynthetically active radiation).

2.4. Adaptation to Ex Vivo Conditions

The rooting shoots were transferred to multiple pots filled with an organic-mineral substrate based on peat and perlite in a ratio of 3:1 at pH 6, and a polymer hydrogel additive of 2.5 g·dm^{-3} containing nitrogen (N) 0.64% (amide form 0.46%, nitrate form 0.18%) and potassium oxide (K$_2$O) 0.43% soluble in water.

After 2 weeks, mineral fertilization was applied with a liquid organic-mineral fertilizer with the following composition: total nitrogen (N) 1.00%, total phosphorus (P$_2$O$_5$) 0.50%, and total potassium (K$_2$O) 1.74%. Fertilization was applied at a rate of 10 mL·dm^{-3}. At the same time, Thiram Granuflo 80 WG protection products were applied at a rate of 0.6 g·dm^{-3} against grey mold.

After 4 weeks, the plantlets were transferred to the greenhouse, where they were transplanted into pots filled with a mixture of peat, sand, shredded pine bark, and compost in a ratio of 2:1:1:1 (pH of 6.0). ENTEC fertilizer (25 g·dm^{-3}) was added to the substrate with the following composition: total nitrogen (N) 14%, including nitrate nitrogen 5.5% and ammonium nitrogen 8.5%, total phosphorus (P$_2$O$_5$) 7%, total potassium (K$_2$O) 17%, magnesium (MgO) 2%, sulfur (S) 10%, and micronutrients such as boron (B) 0.02% and zinc (Zn) 0.01%.

From each rooting medium, 20 explants were planted in three repetitions. The survival rate (%) was evaluated 2 months after the beginning of the acclimatization.

2.5. Statistical Analysis

All statistical analyses were performed using Statistica 13.0 (StatSoft, Cracow, Poland). The statistical significance of the differences between means was determined by testing the homogeneity of variance and normality of distribution, followed by ANOVA with Tukey's post hoc test. The significance was set at $p < 0.05$. When effects were expressed as percentages, data were arcsin-square-root-transformed before analysis.

3. Results and Discussion

One of the factors influencing the initiation of explants under in vitro conditions is the type and concentration of plant growth regulators in the culture medium. Low concentrations of growth regulators are essential for the induction and regulation of key physiological and morphogenetic factors. For each species and cultivar, the optimal concentrations and combinations of plant hormones in the medium should be selected individually. According to several authors [2,3,13], the preferred medium for *Clematis* initiation is based on the mineral composition of Murashige and Skoog [12]. Raja Naika and Krishna [3] inoculated the stem explants on MS with the addition 2,4-D (2,4-dichlorophenoxyacetic acid) with FAP (6-furfurylamino purine). Still, the organogenic response was noticed only when the MS medium was augmented with 3 to 5 mg·dm^{-3} FAP. In contrast, Parzymies and Dąbski [10] concluded that KIN at a concentration of 10 mg·dm^{-3} or 5 mg·dm^{-3} 2iP (isopentenyl adenine) was the best for micropropagation of *Clematis viticella* shoot tips. In the present study, BAP was used as the main cytokinin at the initiation stage of clematis in vitro. Single microshoot formations were observed in the first week of culturing, on all media types. The largest number (80%) of explants initiated growth on MS medium supplemented with 1 mg·dm^{-3} BAP. The least (36%) of the plants initiated growth on MS medium with the addition of 3 mg·dm^{-3} BAP (Figure 1a,b). In explants placed on medium with the addition of 0.5 and 2 mg·dm^{-3} BAP, growth was initiated by 42 and 50%, respectively. Of the explants growing on the medium with the lowest cytokinin concentration, 8% flowered. Very quickly, after just 14 days, the explants began to die: some of the leaves that developed in in vitro cultures began to turn yellow and wither.

At the multiplication stage of clematis 'Warszawska Nike', the experiment was divided into two steps. In the first step, cytokinin BAP and Kin (Kinetin) in various concentrations (0.5–2 mg·dm^{-3}) were added to the MS medium (Table 1, Figure 1c,d).

Table 1. The influence of the cytokinins BAP and Kin in MS medium on the morphological traits of clematis 'Warszawska Nike'.

Morphological Traits	Control		Cytokinins					
			Kin (mg·dm^{-3})			BAP (mg·dm^{-3})		
			0.5	1	2	0.5	1	2
Number of leaves	4.75	abc *	3.83 bc	3.75 bc	2.75 c	7.75 a	6.50 ab	3.16 c
Shoot length (cm)	2.51	a	1.83 ab	1.29 b	1.54 ab	1.83 ab	2.16 ab	1.62 ab
Number of nodals	2.00	ab	1.41 abc	1.08 bc	1.08 bc	1.66 abc	2.25 a	0.83 c
Number of shoots	1.33	d	1.47 d	1.68 cd	2.02 bc	2.45 ab	2.88 a	2.03 bc
Fresh weight (g)	0.10	b	0.03 b	0.04 b	0.07 b	0.18 b	0.24 b	0.59 a

* Means followed by different letters in columns are significantly different at the 5% level according to Tukey's multiple ranges.

The addition of kinetin to MS medium regardless of the concentration used, had an inhibitory effect on the growth of plants (Table 1). The addition of BAP to the medium, especially at higher concentrations, resulted in the formation of a large amount of callus tissue at the base of the explant (Figure 1d). As the results of our research showed, plants growing on the medium with the addition of 1 mg·dm^{-3} BAP had the largest number of leaves, shoots, and internodes and were also heavier compared to plants propagated on other media. Only the length of the shoots was lower than in the case of plants propagated on a medium without plant growth regulators (Table 1). Shoots developed by explants growing on the medium with the highest BAP concentration (2mg·dm^{-3}) were distorted and vitrified.

Figure 1. Micropropagation of *Clematis* 'Warszawska Nike': (**a**) initiation of explant on the medium supplemented with 3 mg dm^{-3} BAP and (**b**) 0.5 mg dm^{-3} BAP; (**c**) explants propagated for 6 weeks on control medium (**d**) and medium with the addition of 3 mg·dm^{-3} BAP with a large amount of callus tissue; (**e**) explants propagated on MS medium with the addition of 0.5 mg·dm^{-3} BAP with IAA (0.5 mg·dm^{-3}) (**f**) and with GA$_3$ (2 mg·dm^{-3}); (**g**) roots of plants on rooting medium (0.5 mg·dm^{-3} IBA). Scale bar = 1 cm.

The effect of cytokinin on *Clematis* multiplication was also described by Kreen et al. [7]. They compared five cultivars of clematis belonging to the section *Atragene*. Based on their results, they concluded that the optimal medium for the propagation of this species was MS with the addition of 1 mg·dm^{-3} BAP. The efficacy of BAP and Kin during shoot induction and proliferation has been reported for *C. heynei* by Chaval et al. [1]. It was observed that the frequency of axillary shoot proliferation and the number of shoots per explant increased with increasing concentrations of BAP (4 mg·dm^{-3}) and Kin (3 mg·dm^{-3}) in the MS medium. The opposite results of our study were obtained by Raja Naika and Krishna [3]. They used MS medium supplemented with a 6-furfuryl amino purine (FAP) and BAP for the propagation of *Clematis gouriana* Roxb. Of the concentrations of FAP and BAP used, adventitious shoot organogenesis was observed only on the medium supplemented with FAP, whereas the use of low concentrations of BAP (0.5–1.5 mg·dm^{-3}) resulted in the development of callus tissue. Also, Khanbabaeva et al. [14] observed that BAP in a concentration of 0.3 mg·dm^{-3} and 0.4 mg·dm^{-3} for the propagation of 'Terry' varieties of *Clematis* plants allows for obtaining a high multiplication factor.

Several authors [1–3,13] recommend the addition of cytokinin in combination with gibberellin or lower auxin to the medium to induce organogenesis in *Clematis* plants. According to Gabryszewska [15], gibberellins and cytokinin play a crucial role in the activation of axillary buds (control of apical dominance) in lilacs (*Syringa vulgaris*). Gibberellins belong to a group of plant growth regulators with effects on the control of cell division and elongation, as well as on the control of apical dominance (para-dormancy). On the other hand, the use of higher auxin concentrations in the multiplication stage may potentially inhibit cytokinin, affecting shoot shortening.

During the multiplication stage on media with the addition of KIN and BAP, despite the growth of the shoots, we did not obtain satisfactory results due to their poor quality: a large amount of callus tissue and signs of vitrification were observed. In the second step of the multiplication stage, BAP in combination with IAA or GA$_3$ was added to the MS medium, in different concentrations (Table 2). It was observed that the number of leaves per plant increased with rising concentrations of BAP and IAA in the medium and decreased with increasing concentrations of BAP and GA$_3$ (Table 2). *Clematis* on MS medium supplemented with 1 mg·dm^{-3} BAP and 2 mg·dm^{-3} IAA, as well as on MS with 0.5 mg·dm^{-3} BAP + 2 mg·dm^{-3} GA$_3$ developed the largest number of leaves (114% more leaves compared to the control (4.75)) and shoots (3.21). On the other hand, elevated concentrations of cytokinin in combination with auxin or cytokinin with gibberellin in the MS medium exerted an inhibitory effect on the shoot length of *Clematis* explants. The shoot lengths were lower than the control (2.54 cm), from 33% to 42% in the MS medium with BAP + IAA, and from 49% to 12% in the MS medium with BAP + GA$_3$. Additionally, it was observed that a higher concentration of IAA or GA$_3$ relative to cytokinin BAP stimulated internode formation. The explants on MS medium supplemented with 0.5 mg·dm^{-3} BAP and 2 mg·dm^{-3} GA$_3$ developed the most internodes (2.83) and shoots (3.67). However, for both the shoot length and number of internodes, these differences were not statistically significant. The number of nodes is a commercially significant parameter that significantly influences the profitability of the conducted production. Only in two scientific reports published so far has the impact of adding plant growth regulators on its formation been determined for the *Clematis* species. Parzymięs and Dąbski [10] found that *Clematis viticella* (L.) propagated on MS medium had 2.2 nodes, while when cytokinins (TDZ, BAP, 2iP, and Kin) were added to it, the number ranged from 2.3 to 3.4. Mitrofanova et al. [2] investigated the influence of BAP (0.5 to 2 mg·dm^{-3}) and TDZ (at concentrations of 0.66 to 1.98 mg·dm^{-3}) on the propagation of 13 *Clematis* varieties, including the number of nodes. A statistical analysis was not conducted between media within the same variety; however, the differences in the number of nodes between the lowest and highest cytokinin concentrations were at most twofold. In our recent studies, we obtained similar results: the number of nodes for plants propagated on MS medium was 2.2, while on the medium supplemented with BAP and IAA or GA$_3$, it was similar and ranged from 1.66 to 2.83.

Table 2. The influence of the combinations of BAP with IAA or GA$_3$ in the MS medium on the morphological traits of clematis 'Warszawska Nike'.

Morphological Traits	Control	BAP + IAA (mg·dm^{-3})				BAP + GA$_3$ (mg·dm^{-3})			
		0.5 + 1	0.5 + 2	1 + 1	1 + 2	0.5 + 1	0.5 + 2	1 + 1	1 + 2
Number of leaves	4.75 ab *	4.75 ab	6.91 ab	6.83 ab	10.16 a	4.37 ab	10.16 a	6.29 ab	7.20 ab
Shoot length (cm)	2.54 a	1.70 a	1.66 a	1.66 a	1.47 a	1.48 a	2.23 a	1.37 a	1.29 a
Number of nodals	2.00 a	2.16 a	2.08 a	1.66 a	2.25 a	2.16 a	2.83 a	2.12 a	2.08 a
Number of shoots	1.33 e	1.67 de	1.89 d	2.00 d	1.89 d	3.21 bc	3.67 a	3.46 ab	3.01 c
Fresh weight (g)	0.10 b–d	0.10 b–d	0.48 a	0.40 ab	0.22 b–d	0.38 abc	0.31 a–d	0.15 b–d	0.18 a–d

* Means followed by different letters in columns are significantly different at the 5% level according to Tukey's multiple ranges.

It was observed that the fresh weight of the plants decreased with increasing concentrations of IAA or GA$_3$ in the medium (Table 2). At IAA concentrations of 1 mg·dm^{-3} and low BAP concentrations (0.5 mg·dm^{-3}), the explants achieved nearly five times higher fresh weight compared to the control (0.1 g). A synergistic effect of cytokinin in combination with auxins was reported for *Clematis heynei* M.A. Rau [1], and *Stevia rebaudiana* Bertoni [16,17]. However, Gao et al. [18] achieved an effective adventitious bud proliferation of *Clematis* 'Julka' using MS medium supplemented with 1 mg·dm^{-3} BAP + 0.2 mg·dm^{-3} IBA and 0.2 mg·dm^{-3} GA$_3$. Raja Naika and Krishna [3] observed that the addition of auxin IBA (indole-3-butyric acid) in a higher concentration (above 0.8 mg·dm^{-3}) and FAP above 5 mg·dm^{-3} inhibited shoot organogenesis.

During the multiplication stage, the explants of *Clematis* 'Warszawska Nike' were unable to produce roots. Therefore, the multiplied shoots of clematis were transferred on rooting media supplemented with auxin IAA or IBA at different concentrations (Table 3, Figure 1e). Among the tested concentrations, 0.5 mg·dm^{-3} was the most effective for in vitro root induction (Figure 1f). After 6 weeks of culture, *Clematis* explants developed, at the base of the end of the shoots, an average of 1.75 roots per plant, with an average length of 3.10 cm and 4.41 cm, respectively. Upon comparison, the roots developed on the MS medium supplemented with IAA were whitish, brittle, and shorter compared to IBA. However, plants on the 0.5 mg·dm^{-3} IAA medium were taller compared to plants on the 0.5 mg·dm^{-3} IBA medium (2.48 cm). No statistically significant differences were observed in fresh mass when plantlets were rooted on MS medium supplemented with IAA or IBA, regardless of their concentrations. Despite this, the highest fresh weight was noted in plantlets grown on culture media with the lowest concentrations (0.5 mg·dm^{-3}) of auxins IAA and IBA (0.3 g and 0.25 g, respectively). According to Figiel-Kroczyńska et al. [19], IBA and IAA are often used for in vitro root initiation and to increase root number and length. The process of rooting stimulated by the presence of auxins in the medium influences the elongation of root hair cells by importing auxins into non-root-forming epidermal cells. The effectiveness of IAA in rooting has been reported by Raja Naika and Krishna [3] for *Clematis gouriana* Roxb. Explants rooted well on MS without plant growth regulators as well as on MS with the addition of 0.1–0.5 mg·dm^{-3} IAA. Chavan et al. [1] noticed that the MS medium with the addition of 1 mg·dm^{-3} IBA was able to produce about 8-9 roots per *Clematis heynei* M. A. Rau. plant.

After completing the rooting stage, *Clematis* seedlings rooted on media with different IAA or IBA content were transferred directly to pots filled with an appropriate substrate. The acclimatization rate was measured after two months. The percentage of plants acclimatized to the greenhouse conditions varied from 0 (for plantlets from MS + 2 mg·dm^{-3} IBA) to 75% (for plantlets from MS + 0.5 mg·dm^{-3} IAA). Moreover, a positive correlation was observed between the in vitro rooting rate and the acclimatization rate. In the case of IBA concentrations used, plant survivability was lower with a mean of 18.75% (Table 4). The highest percentage of acclimatized plants in this group (50%) was observed when MS with the addition of 0.5 mg·dm^{-3} IBA on the rooting stage was used (Figure 2). In contrast, on MS medium supplemented with IAA, the average percentage of acclimatized plants was 60.42%. The obtained results were also influenced by the conditions in the greenhouse during adaptation where the main problem was maintaining adequate humidity for the acclimatized plants. This issue was noted by Kreen et al. [7], who found that microcuttings require higher relative humidity (100%) than seedlings (80–90%) during acclimatization. This elevated humidity for stem cuttings increases their susceptibility to grey mold.

Table 3. In vitro rooting of clematis 'Warszawska Nike'.

Morphological Traits	Control	AUXIN (mg·dm^{-3})							
		IAA				IBA			
		0.5	1	2	4	0.5	1	2	4
Shoot length (cm)	2.54 a*	2.95 a	1.57 a	1.39 a	1.93 a	2.48 a	1.70 a	1.66 a	1.71 a
Fresh weight (g)	0.10 a	0.3 a	0.29 a	0.06 a	0.13 a	0.25 a	0.05 a	0.08 a	0.11 a
Number of roots	0.00 b	1.75 a	0.41 ab	0.16 b	0.75 ab	1.75 a	0.83 ab	0.33 ab	0.91 ab
Root length (cm)	0.00 b	3.10 ab	2.00 ab	0.18 b	3.19 ab	4.41 a	0.16 b	0.16 b	0.58 ab

* Means followed by different letters in columns are significantly different at the 5% level according to Tukey's multiple ranges.

Table 4. Percentage of plants [%] adapted to greenhouse conditions depending on the auxin content in the rooting medium.

Auxin	Concentrations (mg·dm^{-3})				Mean
	0.5	1	2	4	
IAA	75.0	58.3	50.0	58.3	60.42
IBA	50.0	16.7	0.0	8.3	18.75
Mean	62.5	37.5	25.0	33.3	39.60

Figure 2. Clematis 'Warszawska Nike' plants rooted on MS medium with the addition of 0.5 mg·dm^{-3} IAA after 4 months of growth in greenhouse conditions.

4. Conclusions

This paper presents a complete micropropagation protocol for *Clematis* 'Warszawska Nike'. The MS medium supplemented with 1 mg·dm^{-3} BAP showed good results in the case of shoot initiation (80%). The MS medium supplemented with 1 mg·dm^{-3} BAP and the MS medium with the addition from 0.5 mg·dm^{-3} BAP and 2 mg·dm^{-3} GA$_3$ proved to be very effective for the rapid proliferation/multiplication and growth of clematis explants. Well-rooted plants on MS medium with the addition of 0.5 mg·dm^{-3} IAA were adapted to field conditions and showed a 75% survival rate with normal morphology and growth characteristics. The developed in vitro regeneration protocol for *Clematis* 'Warszawska Nike' can help optimize the micropropagation of other plants belonging to the species under study and may be useful, for example, for the commercial propagation and ex situ conservation of this medicinal plant.

Author Contributions: Conceptualization, D.K.; methodology, D.K.; validation, D.K. and M.K.-M.; formal analysis, D.K. and M.K.-M.; investigation, D.K. and M.K.-M.; writing—original draft preparation, D.K. and M.K.-M.; writing—review and editing, D.K. and M.K.-M.; visualization, D.K. and M.K.-M.; supervision, D.K. All authors have read and agreed to the published version of the manuscript.

Funding: This research was supported by the West Pomeranian University of Technology, Grant No. 503-07-081-09/4.

Data Availability Statement: The data presented in this study are available on request from the corresponding author.

Acknowledgments: The authors are grateful to Wojciech Strugarek for providing laboratory facilities.

Conflicts of Interest: The authors declare no conflict of interest.

References

1. Chavan, J.J.; Kshirsagar, P.R.; Gaikwad, N.B. Rapid In Vitro Propagation Of *Clematis Heynei* Ma Rau: An Important Medicinal Plant. *Emir. J. Food Agric.* **2012**, *24*, 79–84. [CrossRef]
2. Mitrofanova, I.; Ivanova, N.; Kuzmina, T.; Mitrofanova, O.; Zubkova, N. In Vitro Regeneration Of *Clematis* Plants In The Nikita Botanical Garden Via Somatic Embryogenesis And Organogenesis. *Front. Plant Sci.* **2021**, *12*, 311. [CrossRef] [PubMed]
3. Raja Naika, H.; Krishna, V. In Vitro Micropropagation Of *Clematis gouriana* Roxb. From Nodal Stem Explants. *Asian Australas. J. Plant Sci. Biotechnol.* **2007**, *1*, 58–60.
4. Rabiza-Świder, J.; Skutnik, E.; Jędrzejuk, A. The Effect Of Preservatives On Water Balance In Cut Clematis Flowers. *J. Hortic. Sci. Biotechnol.* **2017**, *92*, 270–278. [CrossRef]
5. Ignacio Garcia S.J., J. The Contributions of European Jesuits to Environmental Sciences. *J. Jesuit. Stud.* **2016**, *3*, 562–576.
6. Leeds, E.; Toomey, M. *An Illustrated Encyclopedia of Clematis*; Timber Press: London, UK, 2001.
7. Kreen, S.; Svensson, M.; Rumpunen, K. Rooting of *Clematis* Microshoots And Stem Cuttings In Different Substrates. *Sci. Hortic.* **2002**, *96*, 351–357. [CrossRef]
8. Howells, J. The Genetic Background of Wilting Clematis. *Clematis* **1994**, 62–67.
9. Kruczek, A.; Krupa-Małkiewicz, M.; Ochmian, I. Micropropagation, Rooting, And Acclimatization Of Two Cultivars Of Goji (*Lycium chinense*). *Not. Bot. Horti Agrobot.* **2021**, *49*, 12271. [CrossRef]
10. Parzymies, M.; Dąbski, M. The Effect Of Cytokinin Types And Their Concentration On In Vitro Multiplication Of *Clematis viticella* (L.) And *Clematis integrifolia* 'Petit Faucon'. *Acta Sci. Pol. Hortorum Cultus* **2012**, *11*, 81–91.
11. Izadi Sadeghabadi, A.; Khalighi, A.; Ghasemi Pirbalouti, A.; Taqipour-Dehdkordi, M. Micropropagation of *Clematis orientalis* L. culture in vitro. *J. Med. Herbs* **2013**, *4*, 43–48.
12. Murashige, T.; Skoog, F. A Revised Medium For Rapid Growth And Bioassays For Tobacco Tissue Cultures. *Physiol. Plant.* **1962**, *15*, 473–497. [CrossRef]
13. Duta, M.; Posedaru, A. In Vitro Propagation *Clematis x Jackmanii*. *Bull. Univ. Agric. Sci. Vet. Med. Cluj-Napoca. Hortic.* **2008**, *65*, 462.
14. Evgenievna, K.O.; Evgenievna, M.A.; Sergeevna, K.I.; Vitalevna, T.S.; Ivanovich, T.I. Technologies of Reproduction Terry Varieties of *Clematis*. *J. Agric. Sci.* **2020**, *12*, 214. [CrossRef]
15. Gabryszewska, E. Effect Of Various Levels Of Sucrose, Nitrogen Salts And Temperature On The Growth And Development Of *Syringa vulgaris* L. Shoots In Vitro. *J. Fruit Ornam. Plant Res.* **2011**, *19*, 133–148.
16. Anbazhagan, M.; Kalpana, M.; Rajendran, R.; Natarajan, V.; Dhanavel, D. In Vitro Production of *Stevia rebaudiana* Bertoni. *Emir. J. Food Agric.* **2010**, *22*, 216–222. [CrossRef]
17. Rokosa, M.; Kulpa, D. Micropropagation Of *Stevia rebaudiana* Plants. *Cienc. Rural.* **2020**, *50*, 1–9. [CrossRef]

18. Gao, Y.; Mo, J.; Fu, Y.; Feng, S. Tissue Culture And Plant Regeneration of *Clematis* 'Julka'. *J. Nanjing For. Univ.* **2021**, *45*, 109.
19. Figiel-Kroczyńska, M.; Krupa-Małkiewicz, M.; Ochmian, I. Efficient Micropropagation Protocol Of Three Cultivars Of Highbush Blueberry (*Vaccinium corymbosum* L.). *Not. Bot. Horti Agrobot.* **2022**, *50*, 12856. [CrossRef]

Disclaimer/Publisher's Note: The statements, opinions and data contained in all publications are solely those of the individual author(s) and contributor(s) and not of MDPI and/or the editor(s). MDPI and/or the editor(s) disclaim responsibility for any injury to people or property resulting from any ideas, methods, instructions or products referred to in the content.

Article

Chemical Profile of Cell Cultures of *Kalanchoë gastonis-bonnieri* Transformed by *Agrobacterium rhizogenes*

María Guadalupe Barrera Núñez [1], Mónica Bueno [2], Miguel Ángel Molina-Montiel [3], Lorena Reyes-Vaquero [4], Elena Ibáñez [2] and Alma Angélica Del Villar-Martínez [1,*]

1. Centro de Desarrollo de Productos Bióticos, Instituto Politécnico Nacional, Yautepec de Zaragoza 62731, Morelos, Mexico; mgbarrera.nunez@gmail.com
2. Laboratory of Foodomics, Institute of Food Science Research, CIAL, CSIC, 28049 Madrid, Spain; monibuenofdez@gmail.com (M.B.); elena.ibanez@csic.es (E.I.)
3. Servicio Nacional de Sanidad, Inocuidad y Calidad Agroalimentaria, Jiutepec 62550, Morelos, Mexico
4. Conahcyt—Centro de Investigación y Asistencia en Tecnología y Diseño del Estado de Jalisco, Subsede Sureste, Mérida 97302, Yucatán, Mexico; lrvsaid@gmail.com
* Correspondence: adelvillarm@ipn.mx or almangel8166@gmail.com

Citation: Barrera Núñez, M.G.; Bueno, M.; Molina-Montiel, M.Á.; Reyes-Vaquero, L.; Ibáñez, E.; Del Villar-Martínez, A.A. Chemical Profile of Cell Cultures of *Kalanchoë gastonis-bonnieri* Transformed by *Agrobacterium rhizogenes*. *Agronomy* **2024**, *14*, 189. https://doi.org/10.3390/agronomy14010189

Academic Editors: Justyna Lema-Rumińska, Danuta Kulpa and Alina Trejgell

Received: 20 December 2023
Revised: 8 January 2024
Accepted: 11 January 2024
Published: 15 January 2024

Copyright: © 2024 by the authors. Licensee MDPI, Basel, Switzerland. This article is an open access article distributed under the terms and conditions of the Creative Commons Attribution (CC BY) license (https:// creativecommons.org/licenses/by/ 4.0/).

Abstract: *Kalanchoë gastonis-bonnieri* Raym.-Hamet & Perrier is a plant used for medicinal purposes in the treatment of several ailments. The aim of this study was to analyze the chemical profile of extracts from *K. gastonis-bonnieri* embryogenic calli, generated from genetically transformed roots by *Agrobacterium rhizogenes*. Putative transformants were verified by PCR. Hydroalcoholic extracts were obtained and the chemical profile was analyzed by LC-ESI-MS/MS. Root formation was obtained from 80% of infected seedlings. Fifteen root lines were isolated, and two lines showed prominent longitudinal growth and profuse branching in the B5 semi-solid medium. In all lines, the formation of nodules and later embryogenic callus was observed. Putative transgenic root lines were cultivated in free-plant growth regulators B5 medium. In the two selected lines, the PCR amplification of *rol*A, *rol*B, *rol*C, *rol*D, and *aux*1 genes was detected. The extract of embryogenic calli showed 60 chemical compounds tentatively identified, such as ferulic acid, quinic acid, neobaisoflavone, and malic acid, among others, and the chemical profile was different in comparison to wild-type extracts. This is the first study reporting the analysis of the chemical profile of hairy root extracts derived from *Kalanchoë gastonis-bonnieri*. This work displays the great potential for obtaining chemical compounds of pharmacological importance from hairy roots and facilitates the identification of new useful drugs against human chronic-degenerative diseases.

Keywords: embryogenic calli; genetic transformation; plant tissue culture; secondary metabolism

1. Introduction

The main drawback of obtaining bioactive compounds from plants is the variation in the accumulation of secondary metabolites due to development, plant growth cycles, and diversity of environmental conditions [1]. The biosynthesis and accumulation of chemical compounds of interest in highly specialized tissues occurs at specific stages of development [2]. The hairy roots induction through *Agrobacterium rhizogenes* infection is a biotechnological alternative to obtain specific secondary metabolites in vitro cultures free of plant growth regulators [3]. *A. rhizogenes* is a Gram-negative bacterium of the Rhizobiaceae family that induces hairy roots disease by infecting higher plants and inserting root *loci* (*rol* genes) from the root-inducing plasmid (Ri) into the plant genome [4]. From hairy roots, induction of embryogenic calli and regeneration of whole plants has been observed in species such as *Hypericum perforatum* [5], *Tylophora indica* [6], *Gentiana utriculosa* [7], and *Pentalinon andrieuxii* [8].

It has been reported that hairy roots synthesize chemical compounds that are not detected in wild plants [9]. Furthermore, some authors have reported that both shoots and transgenic plants from hairy roots accumulate higher levels of metabolites in comparison to wild plants. Tusevski et al. (2014) [5], reported the accumulation of naphthodiatrons and specific phenolic compounds in transgenic shoots of *Hypericum perforatum* from hairy roots. Vinterhalter et al. (2019) [10], reported a higher accumulation of xanthones in transgenic plants of *Gentiana utriculosa* regenerated by somatic embryogenesis derived from hairy roots. Hiebert-Giesbrecht et al. (2021) [8], reported the accumulation of terpenoids in leaf of transgenic plants, obtained from hairy roots, compared to wild *Pentalinon andrieuxii*.

The species of the genus *Kalanchoë* (Crassulaceae) are succulent plants, used in traditional medicine to treat several health conditions such as gastric ulcers, asthma, infections, tumors, and blood glucose regulation [11]. Additionally, it is an important ornamental plant. Various studies have been developed around the *Kalanchoe* genus, which provide knowledge on human health [12]. The importance of these plants is found in the diversity of chemical compounds that accumulate which represent the interest in the medicine industry, and as an ornamental plant due to the diversity of the colors and leaf shapes, these characteristics contribute to the economic importance of the *Kalanchoe* genus [13]. In any case, agronomic management for plant production is extremely important because chemical compounds accumulate at different parts of the plant and in response to environmental factors. The production of plants in controlled environments represents an option for maintaining germplasm and keeping plant diversity with all its benefits [14,15]. The application of biotechnological tools to develop technologies that can provide interesting metabolites and generate new plant materials with specific characteristics represents the option to make the most of natural resources. Thirukkumaran et al. [16], reported the application of technologies allowing the production of transgenic plants without selectable marker genes and described that marker-free transgenic *K. blossfeldiana* could be produced using *ipt*-type MAT vector system carrying the chimeric *ipt*. The transformation of *Kalanchoe pinnata* by *Agrobacterium tumefaciens* with ZsGreen1 by Cho et al. [17] selected optimum succulent species for future genetic transformation efforts and the development of an efficient transformation method using a novel fluorescent gene, was accomplished. This method achieves new cultivars of succulents with eye-catching colors or patterns in the leaves and flowers. The potential to develop new cultivars with predictable traits in a reduced period is a great advantage of the genetic transformation approach.

Kalanchoë gastonis-bonnieri Raym.-Hamet & H. Perrier is used in traditional Latin American medicine as a contraceptive and in the treatment of genital and urinary infections, diabetes, kidney infections, gastric ulcers, leishmaniasis, and cancer [18–21]. Few studies have reported the chemical profile of *K. gastonis-bonnieri* [19,21,22]. The aim of this study was to analyze the chemical profile of extracts from the embryogenic calli of *Kalanchoë gastonis-bonnieri* generated from hairy roots.

2. Materials and Methods

2.1. Plant Material

Kalanchoë gastonis-bonnieri was collected in Centro de Desarrollo de Productos Bióticos-IPN, Yautepec, Morelos, México (18°49'53" N, 99°05'37.40" W at 1064 m.a.s.l.). The in vitro plants were obtained from vegetative shoots (size: 2–3 cm) that grew from meristematic tissue, at the tip of acuminated adult plant leaves from wild-growing plants that were gently washed with tap water, then with a Tween 20 (Sigma Company, St. Louis, MO, USA) solution (1%) for 1 min, ethanol (70%) for 2 min and NaOCl (0.5%) during 17 min, and rinsed 3 times with sterile distilled water between each solution [23]. Vegetative shoots were cultivated on semi-solid MS (Sigma Company, St. Louis, MO, USA) medium [24], to which a sterile medium was added with 30 g/L sucrose (Sigma Company, St. Louis, MO, USA) and 3 g/L Phytagel (Sigma Company, St. Louis, MO, USA) and the pH was adjusted to 5.8 before autoclaving at 125 °C for 15 min. In glass Gerber-type containers with 20 mL of medium, cultures were incubated in a growth chamber at 25 ± 2 °C with a photoperiod of 16 h light/8 h dark, at 30 µmol/m^{2}s provided by cool white fluorescent tubes, for 35 d.

2.2. Induction of Transformed Roots

The *A. rhizogenes* strain A4 was used and cultured in YMB medium (2.8 g/L) with 3 g/L of phytagel (Sigma-Aldrich®, St. Louis, MO, USA) and incubated at 29 °C [25] for 3 days. Subsequently, it was kept at 4 °C and reseeded every 30 d. In vitro seedlings of *K. gastonis-bonnieri* were inoculated in the internodal zone by a longitudinal scalpel wound with the *A. rhizogenes* strain A4 and incubated in semi-solid MS medium [24] free of growth regulators added with 30 g/L of sucrose (Sigma Company, St. Louis, MO, USA), and 2.6 g/L of Phytagel (Phytotech, St. Lenexa, KS, USA) [26]. The transformation frequency was determined [27]. The bacterium was eliminated from the plant culture with cefotaxime (Phytotech, St. Lenexa, KS, USA) according to Tavassoli and Safipour-Afshar [28], and the cultures were maintained in semisolid MS medium [24] with cefotaxime for 30 days with subcultures every 7 days to eliminate *A. rhizogenes* residues. The bacteria-free cultures were transferred to a B5 liquid medium for subsequent analyses. Root segments were individualized and transferred to semi-solid B5 medium [29] phytohormones free, supplemented with 30 g/L sucrose, 2 g/L polyvinylpyrrolidone, and 2.6 g/L phytagel (Sigma Company, St. Louis, MO, USA).

Finally, after 55 days, the selected lines were transferred to liquid B5 medium, and the cultures were maintained at the above-mentioned conditions at 100 rpm and sub-cultured every 30 days.

2.3. Morphological Description of In Vitro Cultures

The analysis of the culture development was carried out as previously described [30]. The specific growth rate (μ) and the doubling time (T2) were determined as follows:

$$\mu = ln\,(X - X0/t - t0) \times 100 \qquad T2 = ln2/(\mu)$$

were, X: Final dry weight, X0: Initial dry weight, *t*: Final time, *t*0: Initial time. The plant material morphology was observed in a stereoscopic microscope (Nikon, SMZ-1500, Tokyo, Japan), coupled to a PC (Data image, DS33, Tokio, Japan) with a video camera, controller, and integrated interface. The plant material was disaggregated, and the samples were kept hydrated with B5 liquid culture medium. The samples were displayed in triplicate [31].

2.4. DNA Extraction and PCR Analysis

Genomic DNA from plant material was extracted using the method of Doyle and Doyle [32]. Plasmid DNA of *A. rhizogenes* was extracted with Wizard® Plus SV Minipreps DNA Purification System kit (Promega Corporation, A1460 Madison, WI, USA), following the manufacturer's protocol. DNA from the *A. rhizogenes* A4 strain was used as a positive control, DNA from wild-type *K. gastonis-bonnieri* plants was used as a negative control, and sterile distilled H_2O was used as a negative control, to develop the polymerase chain reaction (PCR). The amplification was carried out in a thermal cycler (Applied Biosystems, Gene Amp PCR Systems 9700, Waltham, MA, USA), using specific primers according to reports in each case. *rolA*: 5'-CGTTGTCGGAATGGCCCAGACC-3' and 3'-CGTAGGTCTGAATATTCCGGTCC-5' to amplify a 248 bp fragment, *rolB*: 5'-ACTATAGCAAACCCCTCCTGC-3' and 3'-TTCAGGTTTACTGCAGCAGGC-5', to amplify a 652 bp fragment [33], *rolC*: 5'-TGTGACAAGCAGCGATGAGC-3' and 3'-GATTGCAAACTTGCACTCGC-5' to amplify a 487 bp fragment, *rolD*: 5'-CCTTACGAATTCTCTTAGCGGCACC-3' and 3'-GAGGTACACTGGACTGAATCTGCAC-5' to amplify a 477 bp fragment [34] and *aux*1: 5'-CCAAGCTTGTCAGAAAACTTCAGGG-3' and 3'-CCGGATCCAATACCCAGCGCTTT-5' to amplify a 815 bp fragment [35]. DNA electrophoresis was performed in a 1.5% agarose gel at 95V for 60 min. The visualization of the amplified fragments was mixed with a SYBR®Green (Lonza, Hayward, CA, USA) solution. Electrophoresis was analyzed in a photo-documenter (ChemiDoc™ MP Imaging System BIO-RAD, 170-01402, Hercules, CA, USA).

2.5. Ultrasound Assisted Extraction (UAE)

Metabolites were extracted in an ultrasound bath Branson (Branson Ultrasonics™, 2510R-MTH, Brookfield, CT, USA) with automatic control of time and temperature and ultrasound frequency of 40 kHz. A total of 50 mg of dry and ground biomass were mixed in 2 mL of ethanol (80%, v/v), and were sonicated at 40 ± 5 °C for 30 min. Samples were centrifuged at 3500 rpm for 5 min. The supernatants were recovered and filtered through cellulose membranes (0.22 µm) (MILLEX® GS). Samples were dryness at 25 ± 2 °C and the dried extracts were stored at 4 °C until analyzed [36].

2.6. Chemical Characterization of Kalanchoë Gastonis-Bonnieri Extracts

The chemical profile of extracts was obtained by LC-ESI-MS/MS. Samples were solubilized in 500 µL of MeOH, HPLC grade (Sigma-Aldrich®) and filtered through nylon membrane, (0.45 µm, Agilent Technologies, Santa Clara, CA, USA). The mobile phase for gradient elution consists of two solvents: solvent A (0.1% formic acid (FA) Sigma-Aldrich® in H_2O) and solvent B (0.1% FA in CH_3CN/MeOH (1:1; v/v) Sigma-Aldrich®. The linear gradient profile was as follows: 95% A (5 min), 95–90% A (10 min), 90–50% A (55 min), 50–95% A (65 min), and 95% A (70 min). The injection volume was 10 µL. The flow rate (0.6 mL/min) was split 1:1 before the MS interface. Electrospray ionization analysis (ESI) was performed using a micrOTOF-Q II mass spectrometer (Bruker Daltonics, Bremen, Germany). The mass spectrometer was operated in negative ion mode with a capillary potential of 2.5 kV, gas temperature of 180 °C, drying gas flow of 6 L/min, and nebulizer gas pressure of 1.0 Bar. Detection was performed at 50–3000 m/z.

The tentative identification of compounds was based on the comparison of the MS fragmentation profile obtained by the analytical equipment, with the mass spectra of MassBank of North America (https://mona.fiehnlab.ucdavis.edu accessed on 15 July 2021) and Competitive Fragmentation Modeling for Metabolite Identification (http://cfmid3.wishartlab.com accessed on 30 July 2021).

2.7. Statistical Analysis

Relative abundance data of tentatively identified metabolites in *K. gastonis-bonnieri* extracts were analyzed by clustered color mapping, using a Pearson distance measurement mean bond clustering method, using the Heatmapper software (http://www.heatmapper.ca/, accessed on 15 September 2021).

3. Results

3.1. Morphology of the Kgb1 and Kgb2 Cultures and Molecular Analysis

K. gastonis-bonnieri seedlings were obtained from in vitro cultures. The infection with *A. rhizogenes A4* strain was accomplished and the root formation at the infection site was observed after 15 d with a transformation efficiency of 80%. Fifteen root lines were individualized in a semi-solid B5 medium, most of the lines were characterized by slow growth and poor branching, and the formation of cell aggregates was observed 30 d after isolation from the initial explant. In vitro cultures *of K. gastonis-bonnieri* were successfully initiated from aseptically vegetative shoots isolated from wild-growing plants. The in vitro shoots were subjected to *Agrobacterium*-mediated transformation (Figure 1). Figure 1a shows an uninfected explant (negative control), demonstrating that mechanical injury did not result in root formation. In Figure 1b–d, the response of different infected seedlings and the development of hairy roots that emerge from the infection site with plagiotropic growth is shown; Figure 1b shows abundant proliferation of hairy roots, while in Figure 1c, d explants with few roots were obtained; Figure 1d shows callus formation and few roots scarcely hairy developed from the infection site.

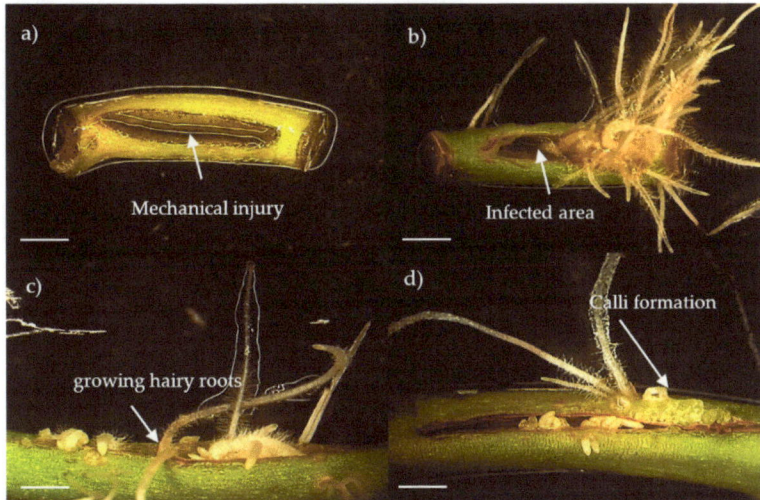

Figure 1. Hairy roots induction and appearance of embryogenic calli of *Kalanchoë gastonis-bonnieri*. in the internodal segment 15 days after infection. (**a**) Control: explant cut out, (**b–d**) Infected internodal segments. Reference bar = 1 mm.

A total of 15 lines were individualized and 2 lines, *Kgb*1 and *Kgb*2, were selected due to accelerated growth and abundant secondary roots. Figure 2 shows the follow-up of the development of *Kgb*1 and *Kgb*2 cell cultures at 9 d (Figure 2a), 18 d (Figure 2b), and 25 days (Figure 2c) of the subculture. The images were captured 90 days after remaining in liquid B5 medium. The asynchronous growth in different stages of somatic embryogenesis was observed. The globular (GB), torpedo (T), heart-shaped (H), and embryo (SE), structures were identified. Embryogenic calli cultures were morphologically heterogeneous, with asynchronous development, meaning that the growth of dedifferentiated cells, the cell differentiation, and the development of embryos occur at uncoordinated times, hence the cellular aggregates in *Kgb*-1 as *Kgb*-2 show different stages of embryogenesis through the 25-days of subculture. Figure 2d shows the initial embryogenic aggregates, which were observed since the roots were isolated from the infected explant and remained in a liquid medium. These structures were observed in all stages of culture. The lines currently remain stable with the same characteristics.

Figure 3 shows the amplified fragments of the *rol*A, *rol*B, *rol*C, *rol*D, and *aux*1 genes from the DNA of *Kgb*1 and *Kgb*2 lines; none of the analyzed genes was amplified from *K. gastonis-bonnieri* wild-type plants. These results suggest that the *Kgb*1 and *Kgb*2 lines were induced in the infection mediated by *A. rhizogenes* A4 strain. In this work, it is suggested that both TL-DNA and TR-DNA of *A. rhizogenes* A4 strain were inserted into *Kgb*1 and *Kgb*2 genomes. It has been reported that the response of a plant species to genetic transformation by *A. rhizogenes* is in the function of the integration and combined expression of *rol*A, *rol*B, *rol*C, and *rol*D genes [31,32].

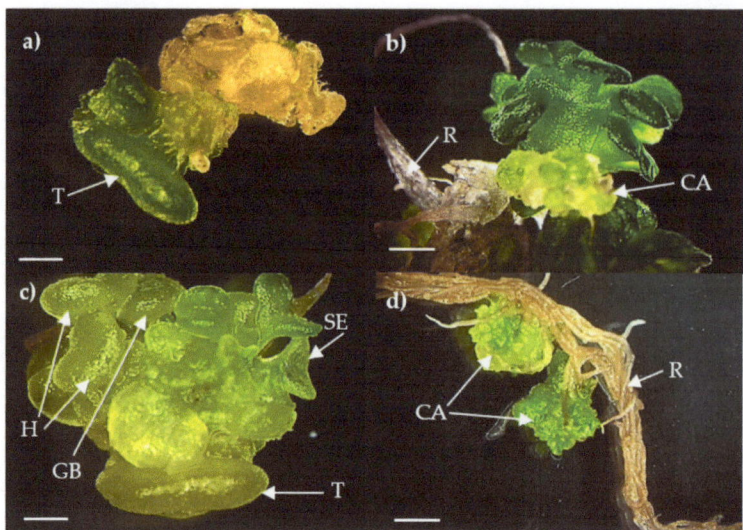

Figure 2. Transgenic embryogenic calli of *Kalanchoë gastonis-bonnieri*, *Kgb*1 and *Kgb*2. (**a**) 9, (**b**) 18, (**c**) 25 days of a subculture in a liquid B5 medium, (**d**) embryogenic aggregates, (CA) calli, (H) heart-shaped, (T) torpedo, (SE) somatic embryo, (R) transgenic root, and (GB) globular-shape. Reference bar = 1 mm.

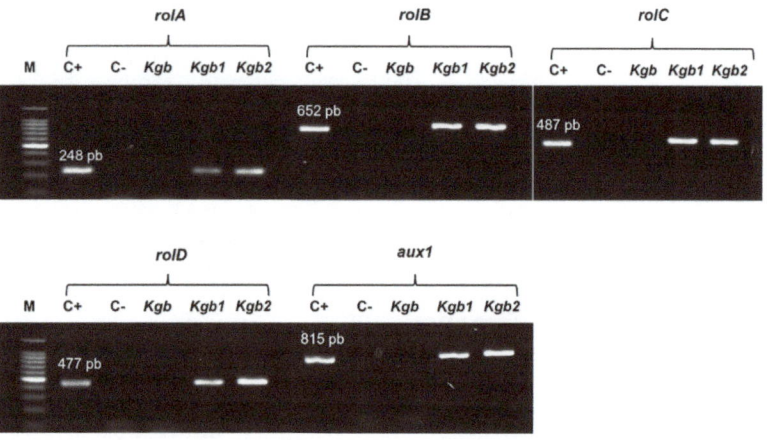

Figure 3. Amplification of *rol*A, *rol*B, *rol*C, *rol*D, and *aux*1 genes of *A. rhizogenes* from genomic DNA of embryogenic calli (*Kgb*1 and *Kgb*2 lines). (M) molecular weight marker 1Kb; (C+) DNA from *A. rhizogenes* A4 strain (positive control); (C−) water (negative control); (*Kgb*) DNA from wild-type *K. gastonis-bonnieri* (negative control); (*Kgb*1 and *Kgb*2) embryogenic callus lines.

3.2. Analysis of the Chemical Profile of Kgb1 and Kgb2 Extracts

Figure 4 shows changes during the cell growth of *Kgb*1 and *Kgb*2 lines. The stages were defined as follows: stage (I) 1–10 days of culture, as an adaptation period, and changes in cell growth were observed, stage (II) 11–21 d, increase accelerated biomass with doubling time for *Kgb*1 (T2) = 3.31 d and cell growth speed (μ) = 0.20 d^{-1}; while for *Kgb*2, T2 = 4.15 d and μ = 0.16 d^{-1}, stage (III) 21–25 d, a decrease in biomass growth was observed, for *Kgb*1 the T2 = 20.3 d and μ = 0.03 d^{-1}; finally, for *Kgb*2 T2 = 48.5 d, and μ = 0.01 d^{-1}; stage (IV) 25–35 d, considerable decrease in cell growth was observed in both lines. The chemical profile of *Kgb*1 and *Kgb*2 was analyzed at 9, 18 and 25 days of culture, which were selected taking quantity biomass as a selection criteria.

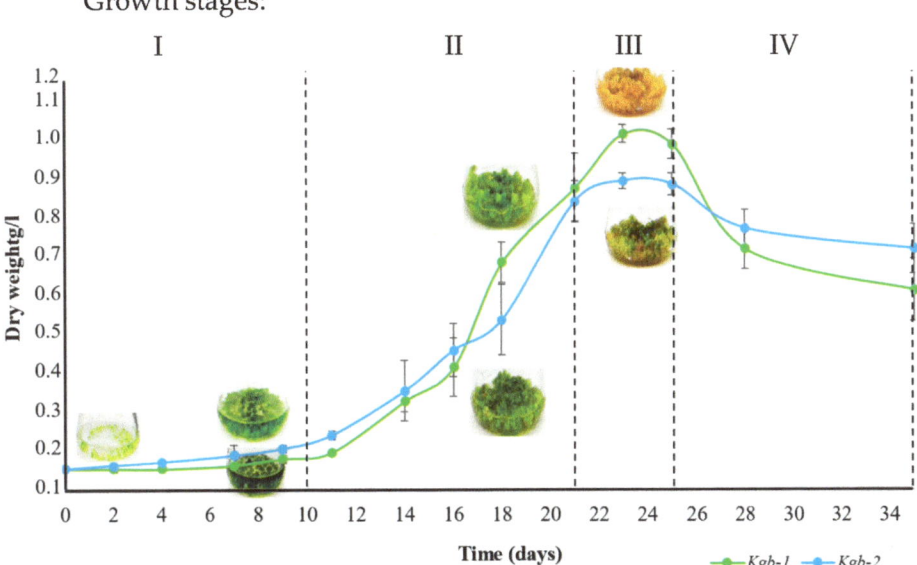

Figure 4. Changes during the cell growth of *Kgb*1 and *Kgb*2 lines in liquid B5 medium. The dotted line indicates the transition between growth stages.

Table 1 shows 60 tentatively identified metabolites in the extracts of *Kgb*1, *Kgb*2 lines at 9, 18, 25 days of culture, and the wild plant of *K. gastonis-bonnieri*; among them, 18 flavonoids, 11 fatty acids, 5 coumarins, 4 phenolic acids, 3 phenolic compounds, 3 terpenes, 4 carboxylic acids, 1 alkaloid, 2 amino acids, 1 carbohydrate and 8 compounds grouped as others, based on chemical structure. The highest number of flavonoids was identified in the wild type, while fatty acids and carbohydrate were mainly identified in *Kgb*1 and *Kgb*2 lines.

The metabolites were identified according to LC-ESI-MS/MS parameters: retention time, match factor values database, molecular formula, and monoisotopic mass. The tentative identification of compounds was based on the comparison of the MS fragmentation profile obtained by the analytical equipment, with the mass spectra of MassBank of North America and Competitive Fragmentation Modeling for Metabolite Identification.

The *Kgb*1 and *Kgb*2 lines showed a differential chemical profile compared to the wild plant. Particularly, in *Kgb*1 and *Kgb*2 extracts the following chemical compounds were detected: ribose-1-arsenate (0.7), 3-(4-hydroxy-3,5-dimethoxyphenyl)-1-(2,4,6-trihydroxy-3-methoxyphenyl)propane-1,2-dione (RT = 0.9), 2-(1,3-dihydroxy4-oxocyclohex-2-en-1-yl)-5-hydroxy-3,6,7-trimethoxy-4H-chromen-4-one (RT = 1.0), sarmentosin epoxide (RT = 1.1), 4-hydroxycoumarin (RT = 1.4), sarmentosin epoxide (RT = 1.1) D-glutamine (RT = 1.8), 1, 6-dihydroxy-3,7-dimethoxy-2-(3-methyl-2-butenyl)-8-(3-hydroxy-3-methyl-1E-butenyl)-xanthone (RT = 5.1), verbasoside (RT = 5.4), 4-coumaric acid (RT = 5.9), ferulic acid (6.0), hallactone B (RT = 6.3), 6,7,3′,4′-tetrahydroxyflavanone (RT = 6.6), epiafzelechin (2R,3R)(-) (RT = 6.9), naringenin (RT = 7.0), 3-hydroxytetradecanedioic acid (RT = 7.6), (9S,10E,12S,13S)-9,12,13-trihydroxy-10-octadecenoic acid (RT = 8.4), 13-HOTrE (RT = 11.3), ipecoside (RT = 12.3), altamisic acid (12.4), a-linolenic acid (RT = 12.8), linoleic acid (RT = 13.3), and heneicosanoic acid (RT = 14.9), p-coumaraldehyde (RT = 15.2). Although, neobavaisoflavone, malic acid, heneicosanoic acid, heliannuol A, 3-(4-hydroxy-3,5-dimethoxyphenyl)-1-(2,4,6-trihydroxy-3-methoxyphenyl)propane-1,2-dione,sarmentosin epoxide, were major compounds in the *Kgb*1 and *Kgb*2 cell lines, in comparison to wild plant.

Table 1. Tentatively identified compounds in *Kgb1* and *Kgb2* cell lines extracts of *Kalanchoe gastonis-bonnieri* (- means no detected, while +, ++ and +++ refer to different peak intensity).

RT (min)	Tentative Identification	Match Factor	Monoisotopic Mass	Main Fragments (*m/z*)	Parent Ion [M-H]-	Kgb1			Kgb2			Wild Type Plants	
						\multicolumn{6}{c	}{Days of Culture}	Leaves	Roots
						9	18	25	9	18	25		
	Flavonoids												
5.9	3,7-Dihydroxy-3',4'-dimethoxyflavone	79	314.0863	637.1386 638.1423 653.1331	313.0684	-	-	-	-	-	-	+	-
6.1	Syringetin-3-O-glucoside	78	508.1272	477.1005 508.1177 535.2139	507.1111	-	-	-	-	-	-	+	-
6.2	Quercetin-3-O-pentosyl (1-2) acetilpentosida	80	608.1295	327.0844 623.1247 623.2683	607.1259	-	-	-	-	-	-	+	-
6.2	3,4-dimethoxy-myricetin-3-O-dideoxyhexosyl(1-2)-dideoxyhexoside	80	638.1445	521.2012 623.1227 638.1445	637.1383	-	-	-	-	-	-	+	-
6.3	Guaijaverin	79	434.0762	434.0762 491.0744 519.2203	433.0762	-	-	-	-	-	-	+	-
6.5	Apigenin-6-C-glucoside-7-O-glucoside	90	594.1500	461.1064 506.0978 594.1500	593.1466	-	-	-	-	-	-	+	-
6.6	6,7,3',4'-Tetrahydroxyflavanone	86	288.0149	146.9666 288.1511 309.1295	287.1465	-	-	+	-	-	-	-	-
6.7	Kaempferol-3-O-arabinoside	95	418.0852	418.0852 491.1174 607.2728	417.0791	-	-	-	-	-	-	+	+
6.9	Epiafzelechin (2R,3R)(-)	80	274.1671	187.0955 289.1635 607.2757	273.1671	-	-	+	-	-	-	-	-

Table 1. Cont.

RT (min)	Tentative Identification	Match Factor	Monoisotopic Mass	Main Fragments (m/z)	Parent Ion [M-H]-	Kgb1			Kgb2			Wild Type Plants	
						\multicolumn{6}{c}{Days of Culture}	Leaves	Roots
						9	18	25	9	18	25		
7.0	Naringenin	80	272.1627	112.9843 289.1627 607.2727	271.1528	-	-	+	-	-	-	-	-
7.1	Kaempferide 3-glucuronide	95	476.0914	597.2469 607.2738 608.2776	475.0873	-	-	-	-	-	-	-	+
7.2	Diosmine	89	608.1664	475.0857 608.1683 643.1410	607.1641	-	-	-	-	-	-	+	-
7.7	Quercetin	90	302.0368	146.9683 157.0095 302.0331	301.0317	-	-	-	-	-	-	+	+
8.2	2′,7-Dihydroxy-4′-methoxy-8-prenylflavan 2′,7-diglucoside	83	664.2705	653.2342 680.2472 707.2874	663.2626	-	-	-	-	-	-	+	-
8.6	Fisetin	80	286.0392	112.9833 286.0392 315.0477	285.0360	-	-	-	-	-	-	+	+
10.5	Kaempferol-7-neohesperidoside	93	594.2728	197.9606 201.0350 594.2728	593.2716	+	-	-	+	-	-	+	-
11.1	Quercetin-3-O-vicianoside	94	596.2922	197.9612 596.2922 723.3794	595.2873	+	+	-	+	-	-	-	-
11.9	Neobavaisoflavone	90	322.1733	322.1762 406.1516 421.0987	321.1724	+	+	-	-	+++	-	-	+
	Fatty acids												
6.3	Deacetoxy (7)-7-Oxokhivorinic acid	78	520.2216	509.1912 520.2216 536.2101	519.2210	-	-	-	-	-	-	+	-

Table 1. Cont.

RT (min)	Tentative Identification	Match Factor	Monoisotopic Mass	Main Fragments (m/z)	Parent Ion [M-H]-	Kgb1 9	Kgb1 18	Kgb1 25	Kgb2 9	Kgb2 18	Kgb2 25	Wild Type Plants Leaves	Wild Type Plants Roots
7.6	3-Hydroxytetradecanedioic acid	83	274.171	146.9682 173.9999 274.1710	273.1676	-	-	+	-	-	-	-	-
8.4	(9S,10E,12S,13S)-9,12,13-Trihydroxy-10-octadecenoic acid	81	330.2366	174.0007 201.0377 330.2366	329.2320	-	+	-	-	-	-	-	-
8.7	Corchorifatty acid F	83	328.2277	201.0351 263.1309 328.2158	327.2141	+	+	+	+	+	+	+	-
9.2	(Z)-9,12,13-trihydroxyoctadec-15-enoic acid	80	330.9998	157.0100 330.2365 397.2198	329.2300	+	++	+	+	+	+	+	-
11.3	13-HOTrE	81	294.2131	275.2000 294.2131 613.9943	293.2095	+	+	+	+	-	+	-	-
11.7	Vernolic acid	83	296.2289	116.9269 296.2289 363.2136	295.2255	+	+	+	+	+	+	+	-
12.4	Altamisic acid	80		112.9851 135.9701 309.1725	279.1628	-	-	-	-	-	+	-	-
12.8	α-Linolenic acid	80	278.7812	135.9699 278.2192 400.2108	277.2160	+	-	-	-	-	+	-	-
13.3	Linoleic acid	80	280.2351	278.7270 280.2351 325.1854	279.2322	+	-	-	-	-	-	-	-
14.9	Heneicosanoic acid	90	326.1831	326.1873 339.2004 340.2043	325.1829	+++	-	-	-	+++	++	-	-

Table 1. Cont.

RT (min)	Tentative Identification	Match Factor	Monoisotopic Mass	Main Fragments (m/z)	Parent Ion [M-H]-	Kgb1			Kgb2			Wild Type Plants	
						\multicolumn{6}{c}{Days of Culture}	Leaves	Roots
						9	18	25	9	18	25		
	Coumarins												
1.4	4-Hydroxycoumarin	83	162.0495	117.0193 128.0353 292.1378	161.0451	+	-	-	-	-	-	-	-
4.9	3,4,5-Trihydroxy-6-[[2-oxo-6-(3-oxobutyl)-2H-chromen-7-yl]oxy]oxane-2-carboxylate	85	408.0981	341.0851 408.0981 443.0710	407.0945	-	-	-	+	-	-	+	-
6.3	Hallactone B	80	440.1090	112.9848 174.0002 440.1090	439.1031	-	-	-	+	-	-	-	-
11.4	Rugosal A	87	266.1482	134.8628 135.9684 817.1497	265.1446	-	-	+	+	-	+	++	+++
11.8	Corylifol A	85	390.2340	321.1705 390.2340 411.2125	389.2295	-	-	-	-	-	-	+	-
	Phenolic acids												
5.4	Caffeic acid	93	180.0365	135.0425 180.0365 755.1999	179.0325	-	-	-	-	-	-	+	+
5.7	1-Caffeoyl-4-deoxyquinic acid	90	338.0910	191.0536 338.0910 359.0706	337.0887	-	-	-	-	-	-	+	-
5.9	4-Coumaric acid	81	164.0467	164.0467 279.0519 475.1830	163.0378	+	-	+	+	-	+	-	-
6.0	Ferulic acid	86	194.0537	194.0537 309.0612 311.1114	193.0486	+	-	++	+	-	++	-	-
	Phenolic compounds												

Table 1. Cont.

RT (min)	Tentative Identification	Match Factor	Monoisotopic Mass	Main Fragments (m/z)	Parent Ion [M-H]-	Kgb1			Kgb2			Wild Type Plants	
						\multicolumn{6}{c	}{Days of Culture}	Leaves	Roots
						9	18	25	9	18	25		
5.2	Syringate	81	198.0453	198.0453 313.0536 431.1863	197.0428	-	-	-	-	-	-	+	-
5.5	Verbasoside	83	462.1709	415.1603 451.1389 462.1709	461.1640	+	-	+	+	-	+	-	-
15.2	p-Coumaraldehyde	81	147.8759	178.8796 220.9484 231.9439	146.9643	+	-	-	-	+	+	-	-
	Terpenes												
6.4	Kanokoside D	89	624.2753	577.2622 613.2428 624.2753	623.2694	-	-	-	+	-	-	-	+
10.3	Heliannuol A	94	250.1536	248.9578 250.1536 251.1476	249.1497	+	++	++	+	+++	+++	++	++
12.3	Ipecoside	80	565.3214	116.9282 554.2898 581.3077	564.3191	+	-	-	-	-	-	-	-
	Carboxylic acids												
1.0	Malic acid	80	134.0215	128.0336 341.1061 377.0831	133.0119	+++	+++	+++	-	-	+++	++	+
1.0	DL-Pyroglutamic acid	80	129.0345	290.0858 310.0687 403.1352	128.0336	-	+	-	+	-	-	-	-
1.2	Citrate	90	192.0301	111.0072 128.0334 173.0081	191.0175	-	-	-	-	-	-	+	-
1.3	D-(-)-Quinic acid	91	192.0579	157.0341 377.0813 379.0805	191.0520	-	-	-	-	++	-	+++	-

Table 1. Cont.

RT (min)	Tentative Identification	Match Factor	Monoisotopic Mass	Main Fragments (m/z)	Parent Ion [M-H]-	Kgb1			Kgb2			Wild Type Plants	
						\multicolumn{6}{c	}{Days of Culture}	Leaves	Roots
						9	18	25	9	18	25		
	Alkaloids												
6.1	Voacristine	80	384.0108	384.0108 481.1331 563.2139	383.0066	-	-	-	-	-	-	-	+
	Amino acids												
0.8	D-Glutamine	87	146.0324	179.0556 215.0324 307.1131	145.0598	+	-	-	+	-	-	-	-
4.3	Tryptophan	82	204.0841	204.0841 261.0370 271.0673	203.0800	-	-	-	-	-	-	+	-
	Carbohydrate												
1.1	Sarmentosin epoxide	90	291.0908	133.0134 200.0561 632.2048	290.0859	+++	++	++	+++	++	++	-	-
	Others												
0.7	Ribose-1-arsenate	89	273.9598	158.9785 273.9598 274.9575	272.9565	++	-	-	+	-	-	-	-
0.9	3-(4-hydroxy-3,5-dimethoxyphenyl)-1-(2,4,6-trihydroxy-3-methoxyphenyl)propane-1,2-dione	90	378.0889	341.1077 379.0830 404.1046	377.0854	-	+++	-	-	-	-	-	-
1.0	2-(1,3-dihydroxy-4-oxocyclohex-2-en-1-yl)-5-hydroxy-3,6,7-trimethoxy-4H-chromen-4-one	79	378.0822	341.1083 404.1043 470.1521	377.0846	+	-	-	-	-	-	-	-
4.4	Cusparine	82	307.1179	112.9835 296.0879 350.1422	306.1163	-	-	-	-	-	-	+	-

Table 1. Cont.

RT (min)	Tentative Identification	Match Factor	Monoisotopic Mass	Main Fragments (m/z)	Parent Ion [M-H]-	Kgb1			Kgb2			Wild Type Plants	
						\multicolumn{6}{c	}{Days of Culture}	Leaves	Roots
						9	18	25	9	18	25		
5.1	1,6-Dihydroxy-3,7-dimethoxy-2-(3-methyl-2-butenyl)-8-(3-hydroxy-3-methyl-1E-butenyl)-xanthone	86	440.1819	393.1746 429.1495 440.1819	439.1785	-	-	+	+	-	-	-	-
5.8	1-(2H-1,3-benzodioxol-5-yl)-2-[2,6-dimethoxy-4-(prop-2-en-1-yl)phenoxy]propyl benzoate	83	476.1831	429.1758 476.1831 521.1996	475.1800	-	-	-	-	-	-	+	-
8.5	3-(1,2-dihydroxypropyl)-1,6,8-trihydroxyanthracene-9,10-dione	83	330.2334	285.0364 286.0402 330.2334	329.2298	+	-	+	+	-	+	+	-
8.7	Isocyclocalamin	85	502.2151	491.1836 493.1828 518.2009	501.2109	-	-	-	-	-	-	-	+

Figure 5 shows the comparison of the relative abundance of the identified metabolites; the lowest concentration in red light, while the highest in green light. According to the row dendrogram, it is observed that the relative abundance of the compounds is different between extracts of the transformed lines *Kgb*1, *Kgb*2, leaf, and wild type root extracts. Column dendrogram allows for visualizing the formation of two main groups: on the left the extracts obtained from *Kgb*1 of 9, 18, and 25 days of culture and *Kgb*2 of 9 and 25 days, and on the right side the leaf and wild root extracts. Subgroups are formed between the *Kgb*1 and *Kgb*2 lines, showing the chemical compound diversity between the different culture days analyzed. Regarding the relative abundance of 4-coumaric acid, ferulic acid, verbasoside, and 3-(1,2-dihydroxypropyl)-1,6,8-trihydroxyanthracene-9,10-dione. They were similar at 9 and 25 days between *Kgb*1 and *Kgb*2, while the wild type of leaf and root extracts were similar in relation to the abundance of rugosal A, fisetin, quercetin, kaempferol-3-O-arabonoside, and y caffeic acid. Dendrogram also shows the identification of specific compounds in *Kgb*1 and *Kgb*2 extracts in the different culture periods. In the *Kgb*1-d9 extract, particular compounds were identified as ipecoside, linoleic acid, 2-(1,3-dihydroxy-4-oxocyclohex-2-en-1-yl)-5-hydroxy-3,6,7-trimethoxy-4H-chromen-4-one, and linolenic acid. In the *Kgb*1-d18 extract, specific compounds were identified as 3-(4-hydroxy-3,5-dimethoxyphenyl)-1-(2,4,6-trihydroxy-3-methoxyphenyl) propane-1,2-dione, 9,12,13-trihydroxy-10-octadecenoic acid, pyroglutamic acid, quercetin-3-O-vicianoside, and y 13-HOTrE, while in the *Kgb*1-d25 extract, the specific identified compounds were epiafzelechin (2R,3R), 6,7,3',4'-tetrahydroxyflavanone, and naringenin y 3-hydroxytetradecanedioic acid.

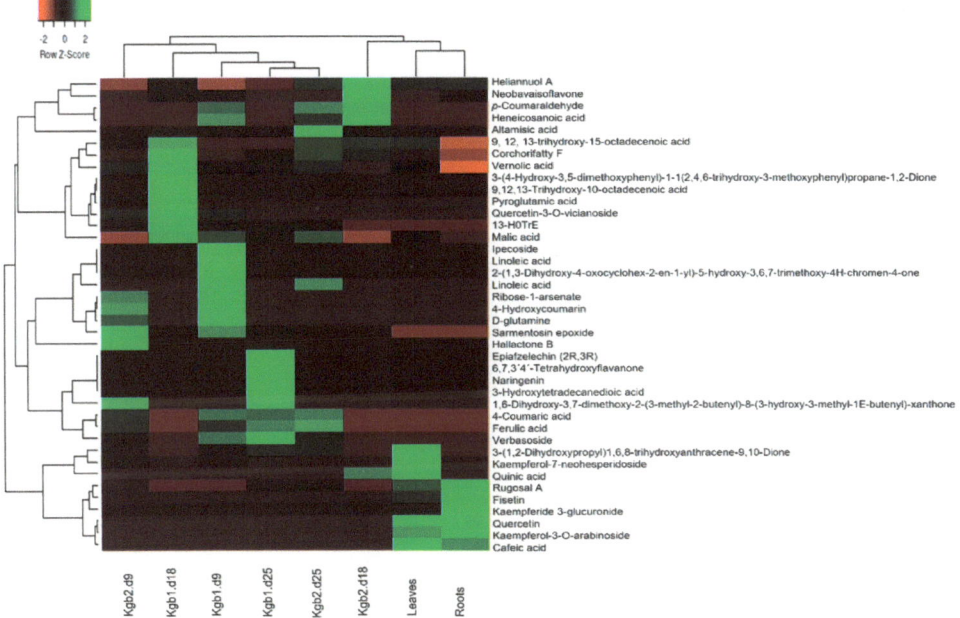

Figure 5. Heat map showing the relative abundance of the compounds tentatively identified in extracts of *Kgb*1, *Kgb*2, wild leaves and roots of *Kalanchoë gastonis-bonnieri*. Color code: light green (highest relative abundance); light red (lower relative abundance).

Likewise, specific compounds were observed in the leaf and/or root extracts: 3-(1,2-dihydroxypropyl)-1,6,8-trihydroxyanthracene-9,10-dione, kaempferol-7-neohesperidoside, rugosal A, fisetin kaempferide 3-glucuronide, quercetin, kaempferol-3-O-arabinoside, and caffeic acid. Similar relative compound abundance was observed for *Kgb*1 and *Kgb*2 extracts at 9 days of culture: ribose-1-arsenate, 4-hydroxycoumarin, D-glutamine, sarmentosin epoxide, 4-coumaric acid, and ferulic acid; at 25 days of culture: 4-coumaric acid, ferulic acid, and verbasoside.

4. Discussion

Some data on transformation efficiency have been reported: 72% in leaf explants of *Solanum erianthum* D. Don. [36]; 65% in shoot explants and 60 in leaf explants of *Althaea officinalis* [28]; 56% in leaf explants and 29% in seedlings' inter nodal explants of *Salvia bulleyana* [37]. According to this information, the transformation efficiency of *A. rhizogenes* strain A4 is diverse; in this work, A4 strain was capable of infecting *K. gastonis-bonnieri* tissue which led to genetically modified embryogenic calli. Also, Chaudhuri et al. [38] and Tavassoli and Safipour-Afshar [28] reported that transformation efficiency was obtained as a function of culture conditions, *Agrobacterium*-host interaction, age, and explant type. It has also been associated with actively dividing cells showing higher transformation rates.

In this work, 15 lines were individualized, and 2 lines, *Kgb*1 and *Kgb*2, were selected, due to accelerated growth and abundant secondary roots. *Kgb*1 and *Kgb*2 showed spontaneous callus formation. Vinterhalter et al. [10] reported calli and somatic embryo formation at 35 d of culture, from roots induced with the *A. rhizogenes* A4M70GUS strain, in *Gentiana utriculosa*. Hiebert-Giesbrecht et al. (2021) [8] reported in *Pentalinon andrieuxii* the formation of embryogenic callus 6 months after obtaining hairy roots with the *A. rhizogenes* ATCC15834 strain in leaves and hypocotyls. Results obtained in this work represent the possibility of whole plant regeneration; moreover, is an interesting tool to carry out advanced studies on the secondary metabolism of *K. gastonis-bonnieri* transgenic culture. The response of *K. gastonis-bonnieri* agrees with previous reports on the spontaneous formation of embryogenic calli from hairy roots obtained by *Agrobacterium* infection, on different plant species such as *Gentiana utriculosa* with the *A. rhizogenes* A4M70GUS strain [1,10], *Pentalinon andrieuxii* with the *A. rhizogenes* ATCC15834 strain [8]; all these strains carrying the same agropine-like *Ri* plasmid [39]. It has been suggested that abnormal morphological features of hairy roots can be a result of the combined participation of *rol* genes in plant cells since each gene might be associated with specific phenotypic alterations in *Kalanchoë* species; in addition, the response of each plant species against infection by *A. rhizogenes* is diverse [38]. In this work, it is suggested that both TL-DNA and TR-DNA of *A. rhizogenes* A4 strain were inserted into *Kgb*1 and *Kgb*2 genomes. It has been reported that the response of a plant species to genetic transformation by *A. rhizogenes* is in function of the integration and combined expression of *rol*A, *rol*B, *rol*C, and *rol*D genes [40,41].

The observed difference in the chemical profile of the transformed embryogenic calli extracts is possibly due to asynchronous growth in the different culture periods [42]. There is wide variability in the compounds reported in *K. gastonis-bonnieri*, which could be attributed to the analyzed plant material that includes the plant development stage, the collection season, and extraction conditions, even among the transformed cell lines analyzed in this work.

*Kgb*1 and *Kgb*2, at 9 days of culture, showed a greater number of detected compounds, in contrast to culture analyzed at 18 and 25 days, which could be related to the adaptation of subculture towards the beginning of a new cycle [43]; a lower number of compounds were detected at 18 days of culture which could be attributed to accelerated cell growth [43,44], therefore the cell metabolism is redirected toward multiplication and cell growth. Finally, at 25 days of culture, the increase of some compounds was observed, which could be attributed to the accumulation of compounds related to the embryogenic process. The accumulation and changes in metabolites detected are related to different stages of development and growth in the in vitro culture, which is also observed at different developmental stages of wild plants [45–47] as a part of plant development or in response to epigenetic factors.

The extracts of *Kgb*1 and *Kgb*2 lines and leaves showed corchorifatty F acid, a compound identified in rice and other cereals, with antibacterial activity [48]. In *Kgb*1 and *Kgb*2 extracts, sarmentosine epoxide was detected, which has shown antihepatotoxic activity [49]. In this work, malic acid, caffeic acid, rugosal A, campferol 3-O-arabinoside, quercetin, fisetin, and heliannuol were detected in leaves and roots of *K. gastonis bonnieri* which have not been previously reported in wild plants.

Interestingly, hairy root cultures synthesize compounds that have not been detected in wild plants [9]. Furthermore, some authors have reported that shoots and transgenic plants obtained from hairy roots accumulate specific metabolites compared to the wild plant. Tusevski et al. [5], reported an accumulation of naphthodiatrons and specific phenolic compounds in transgenic shoots obtained of hairy roots from *Hypericum perforatum*. Vinterhalter et al. [7] reported xanthones in transgenic *Gentiana utriculosa* plants regenerated by somatic embryogenesis from hairy roots. Hiebert-Giesbrecht et al. [8] reported terpenoids accumulation in leaf extracts of transgenic plants, which were obtained from hairy roots of *Pentalinon andrieuxii*.

The phenotypic changes showed in transformed cultures, could be due to (a) position and (b) number of copies of the T-DNA inserted in the genome of the host cell, (c) the regulation of gene expression, and (d) protein biosynthesis encoded by *rol* genes in the plant cell, among others [7,8].

The biological activity of some of compounds identified in this work has been reported, quinic and malic acid inhibit the growth of *S. aureus* and *P. aeruginosa* [50]; the neobavaisoflavone is a compound that has antioxidant, anti-inflammatory, and anticancer properties [51]; ferulic acid has a wide variety of effects, especially on oxidative stress, inflammation, vascular endothelial injury, fibrosis, apoptosis, and platelet aggregation [52]. The compounds detected in *Kgb*1 and *Kgb*2 suggest the effect of genetic transformation through the differential biosynthesis of chemical compounds. It has been reported that highly specialized plant organs such as roots and leaves are needed for the biosynthesis of phytochemicals [26]; therefore, *Kgb*1 and *Kgb*2 do not develop highly specialized organs, which could explain the limited accumulation of flavonoids.

5. Conclusions

This is the first report of the chemical profile of transformed cell cultures of *K. gastonis-bonnieri*. Specific compounds that have not been reported in *K. gastonis-bonnieri* wild-type plants were identified. In vitro, culture of genetically transformed embryogenic calli could be an alternative to produce metabolites of commercial interest in the pharmacological industry as treatment against various diseases, such as cancer. In addition, obtaining transgenic *K. gastonis-bonnieri* plants from embryogenic callus could be an opportunity in the ornamental industry since genetic modification could produce plants with different phenotype.

Author Contributions: M.G.B.N.: Conceptualization, Investigation, Methodology, Formal analysis, Writing (Revision—original and final draft. M.B.: Formal analysis, Writing (Revision—original and final draft). M.Á.M.-M.: Methodology, Formal analysis, Revision final draft. L.R.-V.: data curation. E.I.: Conceptualization, Methodology, Investigation, Methodology, Revision—final draft. A.A.D.V.-M.: Conceptualization, Investigation, Methodology, Writing (Revision—original and final draft), Formal analysis, Project administration, Supervision, Funding acquisition. All authors have read and agreed to the published version of the manuscript.

Funding: This work was supported by Instituto Politécnico Nacional, México (IPN/SIP 20220810, 20231185).

Data Availability Statement: The original contributions presented in the study are included in the article, further inquiries can be directed to the corresponding author.

Acknowledgments: MGBN (744161) acknowledge study grant from CONACYT.

Conflicts of Interest: The authors declare that they have no conflict of interest.

References

1. Chung, I.-M.; Rekha, K.; Rajakumar, G.; Thiruvengadam, M. Production of glucosinolates, phenolic compounds and associated gene expression profiles of hairy root cultures in turnip (*Brassica rapa* ssp. *rapa*). *3 Biotech* **2016**, *6*, 175. [CrossRef] [PubMed]
2. Li, Y.; Kong, D.; Fu, Y.; Sussman, M.R.; Wu, H. The effect of developmental and environmental factors on secondary metabolites in medicinal plants. *Plant Physiol. Biochem.* **2020**, *148*, 80–89. [CrossRef]
3. Abhyankar, G.; Suprasanna, P.; Pandey, B.N.; Mishra, K.P.; Rao, K.V.; Reddy, V.D. Hairy root extract of *Phyllanthus amarus* induces apoptotic cell death in human breast cancer cells. *Innov. Food Sci. Emerg. Technol.* **2010**, *11*, 526–532. [CrossRef]
4. Rekha, K.; Thiruvengadam, M. Secondary metabolite production in transgenic hairy root cultures of cucurbits. In *Transgenesis and Secondary Metabolism*; Reference Series in, Phytochemistry; Jha, S., Ed.; Springer: Cham, Switzerland, 2017. [CrossRef]
5. Tusevski, O.; Stanoeva, J.P.; Stefova, M.; Pavokovic, D.; Simic, S.G. Identification and quantification of phenolic compounds in *Hypericum perforatum* L. transgenic shoots. *Acta Physiol. Plant* **2014**, *36*, 2555–2569. [CrossRef]
6. Roychowdhury, D.; Halder, M.; Jha, S. Agrobacterium rhizogenes-mediated transformation in medicinal plants: Genetic stability in long-term culture. In *Transgenesis and Secondary Metabolism*; Reference Series in, Phytochemistry; Jha, S., Ed.; Springer: Cham, Switzerland, 2017. [CrossRef]
7. Vinterhalter, B.; Savić, J.; Zdravković-Korać, S.; Banjac, N.; Vinterhalter, D.; Krstić-Milošević, D. *Agrobacterium rhizogenes*-mediated transformation of *Gentiana utriculosa* L. and xanthones decussatin-1-O-primeveroside and decussatin accumulation in hairy roots and somatic embryo-derived transgenic plants. *Ind. Crops Prod.* **2019**, *130*, 216–229. [CrossRef]
8. Hiebert-Giesbrecht, M.R.; Avilés-Berzunza, E.; Godoy-Hernández, G.; Peña-Rodriguez, L.M. Genetic transformation of the tropical vine *Pentalinon andrieuxii* (Apocynaceae) via *Agrobacterium rhizogenes* produces plants with an increased capacity of terpenoid production. *In Vitro Cell Dev. Biol-Plant* **2021**, *57*, 21–29. [CrossRef]
9. Huang, S.H.; Vishwakarma, R.K.; Lee, T.T. Establishment of hairy root lines and analysis of iridoids and secoiridoids in the medicinal plant *Gentiana scabra*. *Bot. Stud.* **2014**, *55*, 17. [CrossRef] [PubMed]
10. Vinterhalter, B.; Krstić-Milošević, D.; Janković, T.; Pljevljakusic, D.; Ninković, S.; Smigocki, A.; Vinterhalter, D. *Gentiana dinarica* Beck. hairy root cultures and evaluation of factors affecting growth and xanthone production. *Plant Cell Tiss. Organ. Cult.* **2015**, *121*, 667–679. [CrossRef]
11. Stefanowicz-Hajduk, J.; Asztemborska, M.; Krauze-Baranowska, M.; Godlewska, S.; Gucwa, M.; Moniuszko-Szajwaj, B.; Stochmal, A.; Ochocka, J.R. Identification of flavonoids and bufadienolides and cytotoxic effects of *Kalanchoë daigremontiana* extracts on human cancer cell lines. *Planta Med.* **2020**, *86*, 239–246. [CrossRef]
12. Hernández-Caballero, M.E.; Sierra-Ramírez, J.A.; Villalobos Valencia, R.; Seseña-Méndez, E. Potential of *Kalanchoe pinnata* as a Cancer Treatment Adjuvant and an Epigenetic Regulator. *Molecules* **2022**, *27*, 6425. [CrossRef]
13. Christensen, B.; Sriskandarajah, S.; Serek MMüller, R. Transformation of *Kalanchoe blossfeldiana* with *rol*-genes is useful in molecular breeding towards compact growth. *Plant Cell Rep.* **2008**, *27*, 1485–1495. [CrossRef]
14. Fkiara, A.; Barba-Espín, G.; Bahij, R.; Müller, R.; Christensen, L.P.; Lütken, H. Elicitation of Flavonoids in *Kalanchoe pinnata* by *Agrobacterium rhizogenes*-mediated Transformation and UV-B Radiation. In *Medicinal Plants: Biodiversity, Sustainable Utilization and Conservation*; Khasim, S., Long, C., Thammasiri, K., Lütken, H., Eds.; Springer: Cham, Switzerland, 2020; pp. 395–403. [CrossRef]
15. Ramasamy, M.; Dominguez, M.M.; Irigoyen, S.; Padilla, C.S.; Mandadi, K.K. *Rhizobium rhizogenes*-mediated hairy root induction and plant regeneration for bioengineering citrus. *Plant Biotechnol. J.* **2023**, *21*, 1728–1730. [CrossRef]
16. Thirukkumaran, G.; Khan, R.S.; Chin, D.P.; Nakamura, I.; Mii, M. Isopentenyl transferase gene expression offers the positive selection of marker-free transgenic plant of *Kalanchoe blossfeldiana*. *Plant Cell Tiss. Organ. Cult.* **2009**, *97*, 237–242. [CrossRef]
17. Cho, K.H.; Vieira, A.E.; Kim, J.; Clark, D.G.; Colquhoun, T.A. Transformation of Kalanchoe pinnata by Agrobacterium tumefaciens with ZsGreen1. *Plant Cell Tissue Organ. Cult.* **2021**, *146*, 401–407. [CrossRef]
18. Yukes, J.; Balick, M. *Dominican Medicinal Plants: A Guide for Health Care Providers*, 2nd ed.; New York Botanical Garden: Bronx, NY, USA, 2010; pp. 7–14.
19. Costa, S.S.; Corrêa, M.F.P.; Casanova, L.M. A new triglycosyl flavonoid isolated from leaf juice of *Kalanchoë gastonis-bonnieri* (Crassulaceae). *Nat. Prod. Commun.* **2015**, *10*, 433–436. [CrossRef] [PubMed]
20. Abdalla, S.L.; Costa, S.S.; Gioso, M.A.; Casanova, L.M.; Coutinho, M.A.S.; Silva, M.F.A.; Botelho, M.C.D.S.N.; Díaz, R.S.G. Efficacy of a *Kalanchoë gastonis-bonnieri* extract to control bacterial biofilms and dental calculus in dogs. *Pesqui. Vet. Bras.* **2017**, *37*, 859–865. [CrossRef]
21. Palumbo, A.; Casanova, L.M.; Corrêa, M.F.P.; Da Costa, N.M.; Nasciutti, L.E.; Costa, S.S. Potential therapeutic effects of underground parts of *Kalanchoë gastonis-bonnieri* on benign prostatic hyperplasia. *Evid.-Based Complement. Altern. Med.* **2019**, *2019*, 6340757. [CrossRef]
22. Siems, K.; Jas, G.; Arriaga-Giner, F.J.; Wollenweber, E.; Dörr, M. On the chemical nature of epicuticular waxes in some succulent *Kalanchoë* and *Senecio* species. *Z. Naturforsch.* **1995**, *C 50*, 451–454. [CrossRef]
23. Vanegas, P.E.; Cruz-Hernández, A.; Valverde, M.E.; Paredes-López, O. Plant regeneration via organogenesis in marigold. *PCTOC* **2002**, *69*, 279–283. [CrossRef]
24. Murashige, T.; Skoog, F. A revised medium for rapid growth and bioassays with tobacco tissue cultures. *Physiol. Plant* **1962**, *15*, 472–497. [CrossRef]
25. Hoaykas, P.J.J.; Klapwijk, P.M.; Nuti, M.P.; Schilperoort, R.A.; Rörsch, A. Transfer of the *Agrobacterium tumefaciens* TI plasmid to avirulent Agrobacteria and to Rhizobium *ex planta*. *J. Gen. Microbiol.* **1977**, *98*, 477–484. [CrossRef]

26. Sharifi, S.; Sattari, T.N.; Zebarjadi, A.; Majd, A.; Ghasempour, H. The influence of Agrobacterium rhizogenes on induction of hairy roots and ß-carboline alkaloids production in *Tribulus terrestris* L. *Physiol. Mol. Biol. Plants* **2014**, *20*, 69–80. [CrossRef]
27. Torres-García, B.E.; Morales-Domínguez, J.F.; Fraire-Velázquez, S.; Pérez-Molphe-Balch, E. Generación de cultivos de raíces transformadas de la planta medicinal *Bidens odorata* Cav (Compositae) y análisis fitoquímico preliminar. *Polibotánica* **2018**, *46*, 241–257. [CrossRef]
28. Tavassoli, P.; Safipour-Afshar, A. Influence of different *Agrobacterium rhizogenes* strains on hairy root induction and analysis of phenolic and flavonoid compounds in marshmallow (*Althaea officinalis* L.). *3 Biotech* **2018**, *8*, 351. [CrossRef] [PubMed]
29. Gamborg, O.L.; Miller, R.A.; Ojima, K. Nutrient requirements of suspension cultures of soybean root cells. *Exp. Cell Res.* **1968**, *50*, 151–158. [CrossRef]
30. Kieran, P.; MacLoughlin, P.; Malone, D. Plant cell suspension cultures: Someengineering considerations. *J. Biotechnol.* **1997**, *59*, 39–52. [CrossRef]
31. Urquiza-López, A.; Álvarez-Rivera, G.; Ballesteros-Vivas, D.; Cifuentes, A.; Del Villar-Martínez, A.A. Metabolite profiling of rosemary cell lines with antiproliferative potential against human HT-29 colon cancer cells. *Plant Foods Hum. Nutr.* **2021**, *76*, 319–325. [CrossRef] [PubMed]
32. Doyle, J.J.; Doyle, J.L. A rapid DNA isolation procedure for small quantities of fresh leaf tissue. *Phytochem. Bull.* **1987**, *19*, 11–15.
33. Petrova, M.; Zayova, E.; Vlahova, M. Induction of hairy roots in *Arnica montana* L. by Agrobacterium rhizogenes. *Cent. Eur. J. Biol.* **2013**, *8*, 470–479. [CrossRef]
34. Savić, J.; Nikolić, R.; Banjac, N.; Zdravković-Korać, S.; Stupar, S.; Cingel, A.; Ćosić, T.; Raspor, M.; Smigocki, A.; Ninković, S. Beneficial implications of sugar beet proteinase inhibitor BvSTI on plant architecture and salt stress tolerance in *Lotus corniculatus* L. *J. Plant Physiol.* **2019**, *243*, 153055. [CrossRef]
35. Rana, M.M.; Han, Z.-X.; Song, D.-P.; Liu, G.-F.; Li, D.-X.; Wan, X.-C.; Karthikeyan, A.; Wei, S. Effect of medium supplements on *Agrobacterium rhizogenes* mediated hairy root induction from the callus tissues of *Camellia sinensis* var. *sinensis*. *Int. J. Mol. Sci.* **2016**, *17*, 1132. [CrossRef] [PubMed]
36. Sarkar, J.; Misra, A.; Banerjee, N. Genetic transfection, hairy root induction and solasodine accumulation in elicited hairy root clone of *Solanum erianthum* D. Don. *J. Biotechnol.* **2020**, *323*, 238–245. [CrossRef]
37. Wojciechowska, M.; Owczarek, A.; Kiss, A.K.; Grąbkowska, R.; Olszewska, M.A.; Grzegorczyk-Karolak, I. Establishment of hairy root cultures of *Salvia bulleyana* Diels for production of polyphenolic compounds. *J. Biotechnol.* **2020**, *318*, 10–19. [CrossRef]
38. Chaudhuri, K.N.; Ghosh, B.; Tepfer, D.; Jha, S. Spontaneous plant regeneration in transformed roots and calli from *Tylophora indica*: Changes in morphological phenotype and tylophorine accumulation associated with transformation by Agrobacterium Rhizogenes. *Plant Cell Rep.* **2006**, *25*, 1059–1066. [CrossRef]
39. Bahramnejad, B.; Naji, M.; Bose, R.; Jha, S. A critical review on use of *Agrobacterium rhizogenes* and their associated binary vectors for plant transformation. *Biotechnol. Adv.* **2019**, *37*, 107405. [CrossRef] [PubMed]
40. Ozyigit, I.I.; Dogan, I.; Artam Tarhan, E. Agrobacterium rhizogenes-mediated transformation and its biotechnological applications in crops. In *Crop improvement*; Springer: Boston, MA, USA, 2013; pp. 1–48.
41. Sarkar, S.; Ghosh, I.; Roychowdhury, D.; Jha, S. The effects of *rol* genes of *Agrobacterium rhizogenes* on morphogenesis and secondary metabolite accumulation in medicinal plants. In *Biotechnological Approaches for Medicinal and Aromatic Plants*; Kumar, N., Ed.; Springer: Singapore, 2018. [CrossRef]
42. Rattan, S.; Kumar, D.; Warghat, A.R. Growth kinetics, metabolite yield, and expression analysis of biosynthetic pathway genes in friable callus cell lines of *Rhodiola imbricata* (Edgew). *Plant Cell Tiss. Organ Cult.* **2021**, *146*, 149–160. [CrossRef]
43. Chiavegatto, R.B.; Castro, A.H.F.; Marçal, M.G.; Padua, M.S.; Alves, E.; Techio, V.H. Cell Viability, Mitotic Indexand Callus Morphology of *Byrsonima verbascifolia* (Malpighiaceae). *TropicalPlant Biol.* **2015**, *8*, 87–97. [CrossRef]
44. Sathish, S.; Venkatesh, R.; Safia, N.; Sathishkumar, R. Studies on growth dynamics ofembryogenic cell suspension cultures of commercially important Indica rice cultivars ASD16 and *Pusa basmati*. *3 Biotech* **2018**, *8*, 1–9. [CrossRef]
45. Pan, Y.; Li, L.; Xiao, S.; Chen, Z.; Sarsaiya, S.; Zhang, S.; ShangGuan, Y.; Liu, H.; Xu, D. Callus growth kinetics and accumulation of secondary metabolites of *Bletilla striata* Rchb.f. using a callus suspension culture. *PLoS ONE* **2020**, *15*, e0220084. [CrossRef] [PubMed]
46. Partap, M.; Kumar, P.; Ashrita; Kumar, P.; Kumar, D.; Warghat, A.R. Growth kinetics, metabolites production and expression profiling of picrosides biosynthetic pathway genes in friable callus culture of *Picrorhiza kurroa* Royle ex Benth. *Appl. Biochem. Biotechnol.* **2020**, *192*, 1298–1317. [CrossRef] [PubMed]
47. Wahyuni, D.K.; Rahayu, S.; Zaidan, A.H.; Ekasari, W.; Prasongsuk, S.; Purnobasuki, H. Growth, secondary metabolite production, and in vitro antiplasmodial activity of *Sonchus arvensis* L. callus under dolomite [$CaMg(CO_3)_2$] treatment. *PLoS ONE* **2021**, *16*, e0254804. [CrossRef] [PubMed]
48. Ang, H.; Mak, K.; Lum, M. Liquid chromatography mass spectrometry based high-throughput, unbiased profiling of upland and lowland rice varieties cultivated in Sabah. *Trans. Sci. Technol.* **2020**, *7*, 137–146.
49. Nahrstedt, A.; Walther, A.; Wray, V. Sarmentosin epoxide, a new cyanogenic compound from *Sedum cepaea*. *Phytochemistry* **1982**, *21*, 107–110. [CrossRef]
50. Adamczak, A.; Ożarowski, M.; Karpiński, T.M. Antibacterial activity of some flavonoids and organic acids widely distributed in plants. *J. Clin. Med.* **2020**, *9*, 109. [CrossRef] [PubMed]

51. Maszczyk, M.; Rzepka, Z.; Rok, J.; Beberok, A.; Września, D. Neobavaisoflavone may modulate the activity of topoisomerase inhibitors towards U-87 MG cells: An in vitro study. *Molecules* **2021**, *26*, 4516. [CrossRef]
52. Li, D.; Rui, Y.-X.; Guo, S.-D.; Luan, F.; Liu, R.; Zeng, N. Ferulic acid: A review of its pharmacology, pharmacokinetics and derivatives. *Life Sci.* **2021**, *284*, 119921. [CrossRef] [PubMed]

Disclaimer/Publisher's Note: The statements, opinions and data contained in all publications are solely those of the individual author(s) and contributor(s) and not of MDPI and/or the editor(s). MDPI and/or the editor(s) disclaim responsibility for any injury to people or property resulting from any ideas, methods, instructions or products referred to in the content.

Article

Effect of X-rays on Seedling Pigment, Biochemical Profile, and Molecular Variability in *Astrophytum* spp.

Piotr Licznerski [1], Justyna Lema-Rumińska [2,*], Emilia Michałowska [3], Alicja Tymoszuk [1] and Janusz Winiecki [4]

1. Laboratory of Ornamental Plants and Vegetable Crops, Faculty of Agriculture and Biotechnology, Bydgoszcz University of Science and Technology, 6 Bernardyńska St., 85-029 Bydgoszcz, Poland
2. Department of Environmental Biology, Faculty of Biological Science, Kazimierz Wielki University, 12 Ossolińskich Av., 85-093 Bydgoszcz, Poland
3. Institute of Forensic Genetics, Adam Mickiewicz Avenue 3/5, 85-071 Bydgoszcz, Poland
4. Medical Physics Department, Prof. Franciszek Lukaszczyk Memorial Oncology Center Bydgoszcz, 2 Romanowska St., 85-796 Bydgoszcz, Poland
* Correspondence: lem-rum@ukw.edu.pl

Citation: Licznerski, P.; Lema-Rumińska, J.; Michałowska, E.; Tymoszuk, A.; Winiecki, J. Effect of X-rays on Seedling Pigment, Biochemical Profile, and Molecular Variability in *Astrophytum* spp. *Agronomy* 2023, 13, 2732. https://doi.org/10.3390/agronomy13112732

Academic Editor: Junhua Peng

Received: 16 September 2023
Revised: 12 October 2023
Accepted: 27 October 2023
Published: 30 October 2023

Copyright: © 2023 by the authors. Licensee MDPI, Basel, Switzerland. This article is an open access article distributed under the terms and conditions of the Creative Commons Attribution (CC BY) license (https://creativecommons.org/licenses/by/4.0/).

Abstract: Cacti are important in agricultural economies and one of the most popular horticultural plant groups. The genus *Astrophytum* is one of the most valuable and desirable cacti for growers and collectors around the world. By selecting the appropriate breeding methods to induce variations in combination with modern biotechnology tools for rapid change detection, it is possible to meet the challenges of the modern world in creating new variability in plants. However, there exists a lack of research concerning the impact of ionizing radiation on cacti. The aim of the study was to assess the effects of X-rays at different doses (0 Gy—control, 15, 20, 25, and 50 Gy) on the dynamics of seed germination in vitro, changes in the color of seedlings, biochemical changes in the content of metabolites and changes at the molecular level in *Astrophytum* spp. 'Purple'. A significant effect of X-rays on the induction of genetic variation was observed. Remarkably high polymorphism rates were observed, ranging from 59.09% for primer S12 to a full 100.0% for S3 and S8, as determined by the SCoT (Start-Codon-Targeted) marker. In addition, a large variation in the content of plant pigments (anthocyanins, carotenoids, chlorophyll a, and chlorophyll b) was noted. Additionally, discernible alterations in the color of the tested cactus seedlings, assessed by the RHSCC catalog, were attributed to the impact of ionizing radiation. These findings hold promise for the application of radiomutation breeding in acquiring new cactus cultivars.

Keywords: anthocyanins; carotenoids; chlorophylls; ionizing radiation; molecular marker; SCoT; seeds

1. Introduction

The Cactaceae family is relatively large, consisting of 100 genera and approximately 2000 cactus species [1]. Most representatives of this family are found in the Americas, and the densest concentrations inhabit desert and arid regions, particularly in Mexico, the southwestern United States, and the southwestern part of the Andean region [2]. Around the world, cacti have been arousing great interest for years as ornamental plants that are particularly valued and desired by breeders and collectors. Modern breeding of ornamental plants is focused on the use of new, constantly improved, and advanced methods, such as the genetic modification of plants, and biotechnological tools, such as in vitro cultures and molecular markers [3]. By carefully choosing suitable breeding methods in conjunction with biotechnological tools, it is possible to fulfill the capacity to meet the market requirements posed not only by a wide range of consumers but also by a sublime group of collectors who pay special attention to new and competitive quality features in relation to previously known cultivars [4–6]. Although the most modern breeding methods are currently based on gene editing technology like CRISPR/Cas, mutagenic factors such as ionizing radiation (including X-rays), microwave radiation, high-energy photons, electrons, as well as gold or silver nanoparticles, and ethyl methanesulfonate (EMS) remain valuable tools for effecting genetic changes, especially in ornamental plants boasting large unknown genomes and in polyploid plants [7–11]. However, the literature lacks studies on the effect of ionizing radiation on cacti, which hold significant economic importance within the horticultural market [12].

The genus *Astrophytum* Lem. is one of the most valuable and relatively rare representatives of the Cactaceae family. This genus comprises six distinct species: *Astrophytum asterias*, *Astrophytum capricorne*, *Astrophytum caput-medusae*, *Astrophytum coahuilense*, *Astrophytum myriostigma*, and *Astrophytum ornatum*. This genus has a wide range of colors from green to brown. During periods of high humidity, *Astrophytum* specimens are green in color; however, during periods of drought, the cactus turns brown, merging with the ground. Flowers of this type are yellow (also with orange centers), oval, and fleshy (with lengths from 1.25 to 15 cm). The seeds are dark brown and shiny. Cacti of this type have spots covered with fine flocs [13].

In nature, species belonging to the genus *Astrophytum* can be found at low elevations, among shrubs and grasslands. This ecological niche aligns with the warm and subtropical steppe climate prevalent in regions such as Mexico and Texas [13].

A. asterias Lem. is one of the most desired species by producers and collectors around the world. It is also the most endangered cactus species in its native environment [14]. This species was first collected in 1843 by Baron von Karwinsky and described as *Achinas echinocactus* in 1845 by József Zuccarini [13]. Charles Lemaire in 1868 classified the *A. asterias* species in the genus *Astrophytum*. This species name for the plant is widely accepted by the scientific community up to this day [13].

The progressive degradation of the natural habitats of *A. asterias* due to competitive grasses has led to the full protection of this species [14]. Consequently, to protect the genetic resources of endangered species, including *Astrophytum asterias*, there arises a necessity for the development of new and effective methods of reproduction, including in vitro techniques [15]. Presently, numerous cultivars of this species have been meticulously bred for commercial purposes.

The use of modern breeding methods (such as the genetic modification of plants) and biotechnological tools (such as in vitro cultures and molecular markers) can be used to assess the quality of plant materials, particularly during the initial stages of breeding. This approach leads to the shortening of the breeding cycle and conscious and monitored gene transfer, increasing the efficiency of selection and, thus, significantly reducing the costs of obtaining new cultivars [16].

Morphological features, such as plant color, have proven effective both in exploring variability and in guiding plant breeding endeavors. Morphological markers, while cheap and easy due to their operation based on observation and visual analysis, have many limitations. The disadvantages of this marker system are associated with factors such as its dependence on environmental conditions, the limited number of markers, the late appearance of traits enabling assessment, or the dominant–recessive mode of inheritance of traits [17–19].

In contrast, molecular markers reveal differences at the DNA level, rendering them among the most reliable marker systems that have revolutionized biological sciences and fields related to forensic science [20–22].

One of the most innovative techniques is the SCoT (Start-Codon-Targeted) marker system, developed by the team of Collard and Mackill [23] and others [24,25]. This system is based on relatively short and conserved regions that surround the translation start codon—ATG (adenine–thymine–guanine)—in the plant genome. This marker system has extended to studies centered on genetic diversity in various plant species, including mango (*Mangifera indica* L.) [26], tomato (*Lycopersicum esculentum* Mill.) [27], peanut (*Arachis hypogaea* L.) [28], pistachio (*Pistacia vera* L.) [29], China grass (*Boehmeria nivea* L. Gaudich.) [30], quinoa (*Chenopodium quinoa* Willd) [31], maize (*Zea mays* L.) [32], and gerberas (*Gerbera jamesonii*) [33]. To the best of our knowledge, this is the first instance of research being conducted on the effects of ionizing radiation on the cactus species *Astrophytum* spp. 'Purple'.

The research aims to assess the impact of different doses of applied X-rays on several aspects of the cactus species *Astrophytum* spp. 'Purple'. These aspects include the dynamics of seed germination in vitro, color changes, biochemical changes (including secondary metabolites such as carotenoids and anthocyanins), as well as the assessment of the content of chlorophyll a and chlorophyll b. Additionally, molecular changes will be investigated utilizing the SCoT marker in *Astrophytum* spp. 'Purple'.

2. Materials and Methods

2.1. X-ray Treatment and the Scheme of Experiments

The research material comprised seeds of *Astrophytum* spp. 'Purple', obtained from our breeding efforts. These seeds underwent irradiation, utilizing X-rays with a nominal potential of 6 MV, administered through the Clinac 2300 CD accelerator situated at the Radiotherapy Department of the Oncology Center Prof. Franciszek Łukaszczyk in Bydgoszcz, Poland. The determination of the required exposure time was carried out in the Department of Medical Physics, using the Eclipse planning system by Varian.

The research encompassed radiation doses of 0 Gy (control), 15, 20, 25, and 50 Gy. During irradiation, 500 seeds for each combination (250 Petri dishes for each dose combination with 2 seeds) were contained within a 13.5 × 8 cm string bag (total 2500 seeds). These bags were situated atop solid water RW3 plates to enhance the secondary radiation interaction. Additionally, a bolus layer of known thickness, factored into the irradiation time calculations, was positioned atop the seeds.

After irradiation, the cactus seeds were subjected to presterilization procedures. This entailed a thorough rinsing under running water for 30 min, followed by rinsing with distilled water containing a drop of detergent for 5–10 min. Subsequently, the seeds were immersed in a 0.2% fungicide solution (62.5 WG Switch, Syngenta, Basel, Switzerland) to which detergent was added for around 17 h (on a shaker). A final soaking in a 0.5% fungicide solution, Amistar 250 S.C. (Syngenta, Basel, Switzerland), for 15 min was executed. After this regimen, the seeds were rinsed in distilled water for 5–10 min.

Continuing with the process, meticulous surface sterilization was performed under sterile conditions. The seeds underwent sequential treatments, including a 5–10 s exposure to 70% ethanol and a 30 min immersion in a 1.6% sodium hypochlorite solution; finally, they were rinsed three times for 10 min in sterile distilled water.

The subsequent stage entailed placing the prepared material, two seeds per specimen, onto MS medium [34]. This medium was solidified with agar (8 g dm^{-3}, BioMaxima S.A., Lublin, Poland), and the pH was adjusted to 5.8 before autoclaving. The seeds were positioned within 5.5 cm diameter Petri dishes. The in vitro cultures, sealed with parafilm, were maintained within a growth room. The conditions in this environment were maintained at a temperature of 24 ± 2 °C under a 16 h photoperiod utilizing Philips TLD 54/34 W lighting and an average quantum irradiance of 47.84 µmol m^{-2} s^{-1}.

Over the subsequent span of 8 weeks, daily observations were conducted to monitor the progression of the seed germination dynamics. The appearance of a germ root was considered a significant macroscopic symptom of seed germination. After 8 weeks, the seedlings obtained from the seeds were evaluated for color using the RHSCC color catalog [35]. In addition, measurements were taken to assess their size and weight, and the seedlings were set aside for further biochemical and molecular analyses. The experimental scheme is shown in Figure 1.

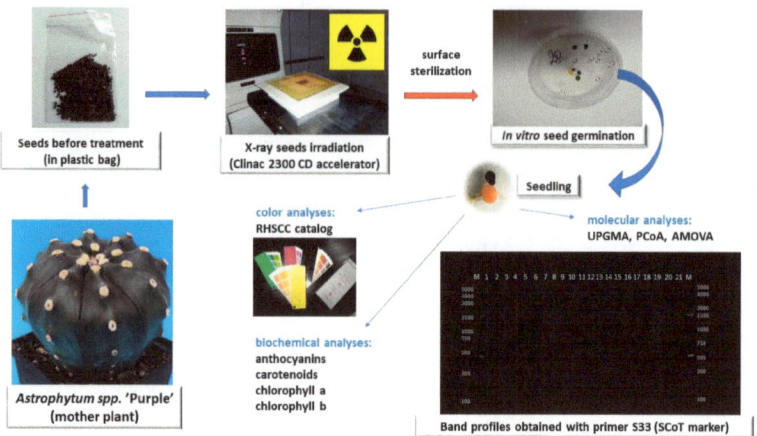

Figure 1. Scheme of the experiments.

2.2. Biochemical Analyses

Seedlings randomly selected from each combination of radiation dose and color as determined by the RHSCC color catalog [35] were used for the biochemical tests. The tests were performed in triplicate for each combination, except for the seedlings without chlorophyll in *Astrophytum* spp. 'Purple'. Due to the notably limited count of chlorophyll-free seedlings within *Astrophytum* spp. 'Purple', these were primarily reserved for genetic analyses.

In contrast, seedlings containing chlorophyll, categorized as green and brown seedlings, formed the primary focus of the biochemical testing. Additionally, creamy white seedlings (for doses of 0, 15, 20, and 25 Gy) or orange seedlings (for the 50 Gy dose) were included in the biochemical analysis.

2.3. Extraction of Plant Pigments from the Obtained Seedlings

The prepared plant material was carefully weighed using a precision analytical balance with an accuracy of 0.1 mg. Each seedling was subsequently crushed separately within a porcelain mortar, incorporating several dozen mg of quartz sand, which was ground alongside the seedling. For anthocyanin extraction, 3.5 mL of 1% HCl in methanol was added, while for the extraction of carotenoids, chlorophyll a, and chlorophyll b, 3.5 mL of 100% acetone was added.

The ensuing phase involved the quantitative filtration of the extracts, achieved through a funnel equipped with medium-grade qualitative filter paper, into 3.5 mL tubes. Absorption maxima were identified at wavelengths specific to each pigment (λ_{max}), and absorbance readings were taken at distinct wavelengths for the cumulative presence of the following: anthocyanins at 530 nm, carotenoids at 440 nm, and chlorophylls at 645 and 662 nm. The tests were carried out in triplicate for each combination of color and radiation dose, except in specified cases. The content of anthocyanins, carotenoids, and chlorophylls was assessed using the UV-VIS spectrophotometer (Shimadzu 1601PC, Kyoto, Japan) according to the modified procedure of Wettstein [36], Harborne [37], and Lichtenthaler and Buschmann [38], as referenced in Lema-Rumińska and Zalewska [39].

The algebraic method was used to quantify the concentration of total anthocyanins, using the following formula:

$$C_A = \frac{A_{530}}{h \cdot k} \left[g \cdot dm^{-3} \right]$$

where $k = 61.7$ (extinction coefficient for 3-cyanidine glycoside) and $h = 1$ cm (layer thickness).

The concentration of carotenoids was calculated according to the formula:

$$C_K = 4.695 \cdot A_{440} \ [\text{mg} \cdot \text{dm}^{-3}]$$

Chlorophyll a was calculated according to the formula:

$$C_a = 11.24 \cdot A_{662} - 2.04 \cdot A_{645} \ [\mu\text{g} \cdot \text{mL}]$$

Chlorophyll b was calculated according to the formula:

$$C_b = 20.13 \cdot A_{645} - 4.19 \cdot A_{662} \ [\mu\text{g} \cdot \text{mL}]$$

2.4. Molecular Analyses

DNA was isolated from the obtained 2-month-old seedlings using a ready-made DNA isolation kit—Genomic Mini AX Plant (Spin) from A&A Biotechnology (Gdańsk, Poland). For each combination of radiation dose and color as identified by the RHSCC color catalog, DNA isolation was performed according to the Genomic Mini AX Plant (Spin) protocol. Approximately 100 mg of fresh seedling weight was used for the extraction process. This plant material was placed within a 1.5 mL tube and homogenized with the FastPrep®-24 device (MP Biomedicals, Irvine, CA, USA). Subsequently, 900 µL of LS lysing suspension and 20 µL of Proteinase K solution were added. The whole sample was mixed and incubated at 50 °C for 10 min using a Biosan TS-100C thermoshaker (Biosan Medical-Biological Research and Technologies, Riga, Latvia) with continuous mixing at 1400 RPM. To ensure the thorough elimination of RNA, 2 µL of RNAse (10 mg/mL), sourced from A&A Biotechnology, was added. After incubation, the samples were vortexed for 2 min at 1000–1400 rpm. The samples were then subjected to centrifugation at 14,000× g for 5 min using a MPW-260R centrifuge (MPW Med. Instruments, Warsaw, Poland).

After centrifugation, 600 µL of the supernatant was collected and applied to a 2 mL Mini AX Spin column. These columns were spun for 30–60 s at 8000× g. After the used 2 mL tube was removed, the Mini AX Spin column was placed within a fresh 2 mL tube. Washing was initiated using the first wash buffer (W1) at a volume of 600 µL. The whole sample was subjected to centrifugation at 8000× g for 30–60 s. Similar steps were taken using the second wash buffer (W2) at a volume of 500 µL (A&A Biotechnology, Gdańsk, Poland).

The subsequent addition of an N-neutralizing buffer was omitted since the isolated DNA was not subjected to freezing. Following centrifugation, the utilized 2 mL tube was carefully removed, and the Mini AX Spin column was transferred to an elution tube. Subsequently, 150 µL of elution buffer E was loaded, and this mixture was allowed to incubate at room temperature for 5 min. After the incubation period, the system was centrifuged for 30–60 s at 8000× g. The column was then removed, and the tube containing the purified DNA was capped. DNA purity was measured using the NanoPhotometer® NP 80 (Implen GmbH, Germany). The isolated DNA was stored in a refrigerator at 6 °C.

For molecular analyses, seven SCoT primers were employed twice. The attributes of these primers are outlined in Table 1.

Table 1. Sequences of the SCoT primers used in the molecular analysis.

Primer	Sequence 5'-3'
S3	CAACAATGGCTACCACCG
S4	CAACAATGGCTACCACCT
S8	CAACAATGGCTACCACGT
S12	ACGACATGGCGACCAACG
S13	ACGACATGGCGACCATCG
S25	ACCATGGCTACCACCGGG
S33	CCATGGCTACCACCGCAG

PCR reactions were performed within a Thermal Cycler (BIO-RAD, model C1000 Touch™, Bio-Rad Laboratories, Hercules, CA, USA), employing a total volume of 25 µL. The final volume consisted of 1 µM of the single primer, 12.5 µL of 2×PCR MIX Plus kit (A&A Biotechnology, Gdańsk, Poland), which included 0.1 U/µL of Taq DNA polymerase, 4 mM $MgCl_2$, 0.5 mM of each dNTPs, 0.8 ng/µL of template DNA, and sterile water. The PCR reaction sequence was initiated with an initial denaturation step at 94 °C for 4 min. This was followed by 45 cycles: each cycle involved denaturation at 94 °C for 1 min, annealing at 50 °C for 1 min, and extension at 72 °C for 2 min. The last cycle was followed by the final extension step of 4 min at 72 °C. After the amplification process, 10 µL of the PCR reaction product was subjected to separation on a 1.5% agarose gel. The gel was stained with ethidium bromide in 1×TBE buffer.

Electrophoresis took place at 90 V for 20 min, followed by a subsequent run at 110 V for 90 min within the Standard Power Pack25 chamber (Biometra, Göttingen, Germany). The loading process involved the utilization of 6× TriTrack DNA Loading Dye (Thermo Fisher Scientific, Waltham, MA, USA) and the GenRuler Express DNA Ladder ready-to-use band size marker (Thermo Fisher Scientific).

After the procedure, the acquired products were archived utilizing the Gel Doc™XR+ archiving system (Bio-Rad Laboratories, Hercules, CA, USA). The labeling of the gel samples can be referenced in Table 2.

Table 2. Genotype designation based on molecular analysis.

No. of Samples	Dose of X-Radiation [Gy]	Color of Seedling
1	0	brown
2	0	green
3	0	creamy white
4	15	brown
5	15	green
6	15	creamy white
7	15	orange
8	15	red
9	20	brown
10	20	green
11	20	creamy white
12	20	orange
13	25	brown
14	25	green
15	25	creamy white
16	25	orange
17	50	brown
18	50	green
19	50	creamy white
20	50	orange
21	50	red

2.5. Statistical Analyses

The obtained results from both biochemical analyses and the morphological parameters were statistically analyzed using the analysis of variance at a significance level of

$p \leq 0.05$ and Fisher's F test using the Statistica 13.3 software (StatSoft Polska, Cracow, Poland). Molecular analyses were performed using GelAnalyzier 19.1 by Istvan Lazar Jr., and Istvan Lazar Sr. [40]. Within the molecular analyses, the SCoT marker loci for each genotype were counted using a binary system. In this system, the presence of a band was denoted as (1), while the absence of a band was represented as (0). The resultant matrix served as the basis for subsequent statistical calculations.

The dendrograms were created based on 0–1 binary matrices using agglomerative hierarchical clustering (AHC) with the unweighted pair–group average method (UP-GMA) (Statistica 13.3 software, StatSoft, Cracow, Poland). Population groups were distinguished based on the Huff et al. [41] genetic distance calculation, analysis of molecular variance (AMOVA) and principle cluster analysis (PCoA) estimates using GeneAlEx 6.5 software [42].

3. Results

3.1. In Vitro Seed Germination Dynamics following X-ray Exposure

The progression of seed germination in *Astrophytum* spp. 'Purple' subjected to X-ray treatments at doses of 0 Gy (control), 15, 20, 25, and 50 Gy was closely observed over the ensuing 8 weeks following the initial sterile sowing on MS medium [34]. During the first 3 days of the in vitro experiment, no observable macroscopic indicators of seed germination (germ roots) were evident in the *Astrophytum* spp. 'Purple' cacti that had been exposed to X-rays as well as in the control seeds (Figure 2).

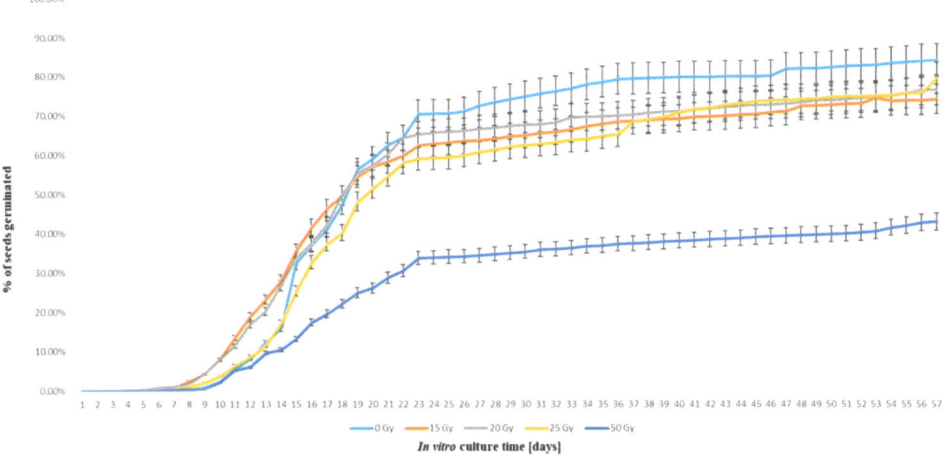

Figure 2. Percentage of seed germination ± standard deviation (mean ± SD) of *Astrophytum* spp. 'Purple' cactus seeds (control and X-ray-treated) during in vitro culture.

On the fourth day, the onset of germination was detected, marked by the emergence of a germ root in a single seedling from the pool that had previously been irradiated with a dose of 20 Gy. However, the commencement of germination for nonirradiated seeds and those exposed to higher doses (25 and 50 Gy) were notably delayed by approximately 3–4 days in comparison with the 15–20 Gy doses.

The subsequent days, spanning from the sixth to the nineteenth day, emerged as particularly fruitful in terms of seed germination for those subjected to X-rays at doses of 15 and 20 Gy. The number of germinated seeds for these two doses (15 and 20 Gy) was significantly higher ($p \leq 0.05$) than that of the nonirradiated seeds (0 Gy control) and the seeds treated with the 25 and 50 Gy doses. It was only on the 19th day of observation that the number of germinating nonirradiated seeds equaled that of the seeds treated with 20 Gy radiation. Subsequently, from the 22nd day onwards, the number of germinating control seeds began to exceed all other combinations, a trend that persisted throughout the remaining observation period. Between the 14th and 18th days of culture, as well as from the 36th day until the end of the observation period, the significantly ($p \leq 0.05$) least germinating seeds were identified among those irradiated at a dose of 50 Gy. In comparison with the control group, the difference in germinating seed count for those treated with X-rays at a 50 Gy dose was calculated to be 39.23%.

3.2. Evaluation of the Color of Seedlings and the Concentration of Plant Pigments after Seed Exposure to X-rays

Eight weeks after the irradiation of *Astrophytum* spp. 'Purple' cactus seeds with doses of 0 Gy (control), 15, 20, 25, and 50 Gy, and after initiating an in vitro culture, an evaluation of the seedlings' color was assessed (according to the RHSCC catalog). Additionally, the concentrations of plant pigments, including anthocyanins, carotenoids, chlorophyll a, and chlorophyll b, were determined. Conversely, the implementation of X-rays in *Astrophytum* spp. 'Purple' resulted in the manifestation of a newfound red and orange color in the seedlings (Figure 3).

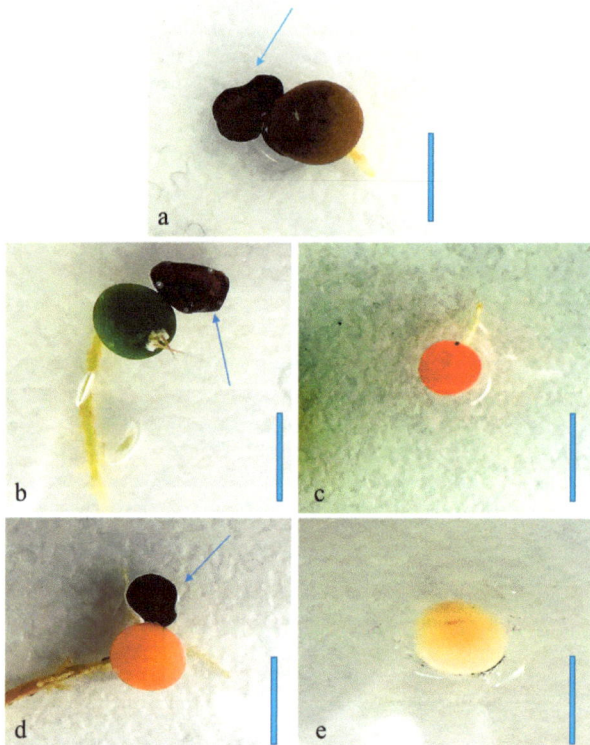

Figure 3. Colors of seedlings obtained as a result of exposure to X-rays (25 Gy) for *Astrophytum* spp. 'Purple' seeds: (**a**) brown color: Gray–Orange (176B, 176C); (**b**) green color: Yellow–Green (144A, 144C); (**c**) red color: Red Group (50C); (**d**) orange color: Orange Group (28C); (**e**) creamy yellow color: Yellow Group (158B); (scale bar = 1 cm; the color symbols in brackets are according to the RHSCC catalogue; the arrow indicates the seed coat).

The in vitro cultivation of *Astrophytum* spp. 'Purple' seeds resulted in the generation of 1869 seedlings, with 1425 originating from seeds exposed to varying X-rays doses (Figure 4).

Among these, the highest number of chlorophyll-free seedlings was obtained from seeds previously exposed to a radiation dose of 50 Gy (5.12%), whereas the lowest number was obtained from control seedlings not treated with the mutagenic factor (1.33%).

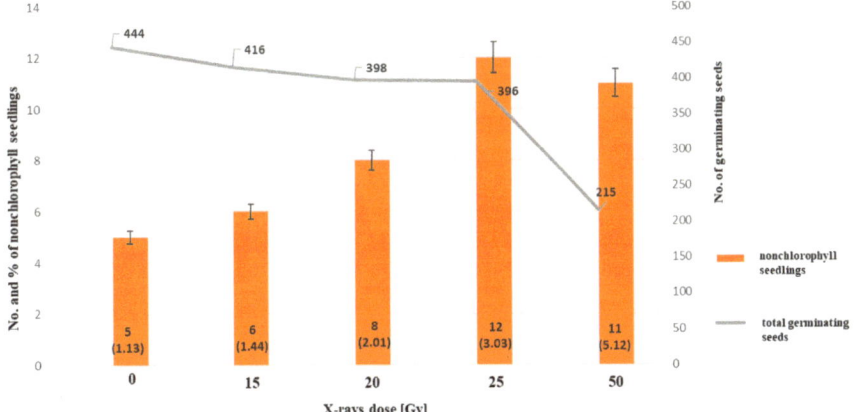

Figure 4. The number (and percentage) ± standard deviations (mean ± SD) of nonchlorophyll seedlings and total germinating seeds in *Astrophytum* spp. 'Purple' after X-rays application.

Given the limited number of nonchlorophyll seedlings (as tabulated in Figure 4), these were primarily designated for genetic analyses. Apart from these, among the seedlings containing chlorophyll (either green or brown), only creamy white seedlings were observed for doses of 0, 15, 20, and 25 Gy, while orange ones were noted for the 50 Gy dosage.

Concerning *Astrophytum* spp. 'Purple', all the creamy white seedlings, regardless of the radiation dose applied, exhibited an absence of anthocyanins, carotenoids, chlorophyll a, and chlorophyll b (Table 3). For the orange-colored, nonchlorophyll seedlings, the concentration of plant pigments was assessed due to their presence in these tested specimens.

Among the various seedlings, the highest concentration of anthocyanins and chlorophyll b was detected in the control seedlings (0 Gy dose), while the carotenoid and chlorophyll a concentrations were highest for seedlings resulting from prior irradiation of seeds with a dose of 15 Gy. On the contrary, the lowest levels of anthocyanins, carotenoids, and chlorophyll a were recorded from seedlings previously treated with a radiation dose of 50 Gy. The lowest concentration of chlorophyll b was observed in seedlings stemming from seeds subjected to 25 Gy radiation.

Across all the seedlings, irrespective of their color or the previous radiation dosage administered to the seeds, a reduced concentration of anthocyanins was evident in comparison with the control sample. The highest concentration of these pigments was present in a green seedling originating from a nonirradiated seed (30.47 mg dm^{-3}), while the lowest concentration (12.40 mg dm^{-3}) was identified in a green seedling derived from a seed previously exposed to X-rays at a dose of 50 Gy. A similar level of anthocyanin concentration to the control (brown seedling: 23.64 mg dm^{-3}; green seedling: 30.47 mg dm^{-3}) was apparent in material obtained from seeds treated with X-ray radiation of 25 Gy (brown seedling: 22.88 mg dm^{-3}; green seedling: 23.20 mg dm^{-3}).

Table 3. Seedling color (according to the RHSCC catalog) and the concentration of the sum of anthocyanins, carotenoids, and chlorophyll a and b per 1 g of fresh weight of the seedling in relation to the X-ray dose in *Astrophytum* spp. 'Purple'.

Dose of X-Radiation [Gy]	Color of Seedling (RHSCC)	Concentration [mg dm^{-3}]			
		Anthocyanins	Carotenoids	Chlorophyll a	Chlorophyll b
0	176 B, C	23.64 ± 0.00 c *	40.91 ± 0.03 d	41.53 ± 0.02 g	35.81 ± 0.19 a
	158 B	-	-	-	-
	144 A, C	30.47 ± 0.15 a	38.82 ± 0.00 f	45.58 ± 0.10 d	26.12 ± 0.01 e
15	176 B, C	17.42 ± 0.00 e	57.41 ± 0.05 a	75.21 ± 0.37 a	33.50 ± 0.19 b
	158 B	-	-	-	-
	144 A, C	13.56 ± 0.12 i	41.70 ± 0.01 d	51.01 ± 0.00 c	30.10 ± 0.02 c
20	176 B, C	15.45 ± 0.05 f	45.09 ± 0.00 c	43.06 ± 0.03 f	21.41 ± 0.20 f
	158 B	-	-	-	-
	144 A, C	24.37 ± 0.05 b	47.85 ± 0.02 b	66.50 ± 0.00 b	27.12 ± 0.12 d
25	176 B, C	22.88 ± 0.11 d	29.85 ± 0.02 f	35.76 ± 0.00 h	13.51 ± 0.05 h
	158 B	-	-	-	-
	144 A, C	23.20 ± 0.09 d	33.33 ± 0.00	44.57 ± 0.13 e	19.83 ± 0.14 g
50	176 B, C	14.18 ± 0.13 g,h	17.97 ± 0.02 g	24.74 ± 0.00 k	8.28 ± 0.62 i
	144 A, C	12.40 ± 0.04 j	8.33 ± 0.01 i	11.15 ± 0.00 i	4.06 ± 0.00 j
	28 C	14.68 ± 0.03 g	11.49 ± 0.11 h	7.88 ± 0.15 j	7.70 ± 1.96 i

* Means ± standard deviations within a column marked with the same letter do not differ significantly, with $p \leq 0.05$.

The most substantial carotenoid content was identified in the seedling stemming from a seed exposed to a radiation dose of 15 Gy, which displayed a brown color (57.41 mg dm^{-3}). In contrast, the lowest carotenoid levels were observed in the seedling originating from a seed treated with 50 Gy radiation, which exhibited an orange color (11.49 mg dm^{-3}). In comparison with the brown control sample (40.91 mg dm^{-3}), seedlings of the same brown color obtained from seeds treated with the mutagenic agent at 15 and 20 Gy doses exhibited higher concentrations of anthocyanins (57.41 and 45.09 mg dm^{-3}, respectively).

Considering the green control sample (38.82 mg dm^{-3}), seedlings sharing the same green color, obtained from seeds exposed to X-rays at 15 and 20 Gy doses, exhibited a higher concentration of carotenoids (41.70 and 47.85 mg dm^{-3}, respectively).

The highest chlorophyll a concentration (75.21 mg·dm^{-3}) was detected in samples extracted from the brown seedling whose seeds were subjected to X-ray irradiation at a dose of 15 Gy. Conversely, the lowest chlorophyll a concentration (7.88 mg dm^{-3}) was observed in the sample extracted from the orange seedling resulting from a seed treated with the mutagenic agent at 50 Gy.

Relative to the brown control seedling (41.53 mg dm^{-3}), samples derived from seeds exposed to radiation doses of 15 Gy (75.21 mg dm^{-3}) and 20 Gy (43.06 mg dm^{-3}) showcased increased chlorophyll a concentrations. Contrasting with the control sample that was characterized by a green color, extracts from seedlings of the same hue (resulting from seeds irradiated at 15 and 20 Gy doses) demonstrated amplified chlorophyll a levels (51.01 and 66.50 mg dm^{-3}, respectively).

Chlorophyll b displayed its highest concentration (35.81 mg dm^{-3}) in the extract from the brown control seedling, while the lowest concentration (13.51 mg dm^{-3}) was noted in the extract from the seedling originating from a seed subjected to a 25 Gy radiation dose.

The extract derived from the green seedling, stemming from a seed exposed to X-rays (15 Gy dose), displayed a higher anthocyanin concentration (30.10 mg dm^{-3}) compared with the control of the same color (26.12 mg dm^{-3}). A slightly lower chlorophyll b concentration was observed in the brown seedling resulting from a seed exposed to a radiation dose of 15 Gy (33.50 mg dm^{-3}) in comparison with the control seedling of the same color (35.18 mg dm^{-3}) that had not been exposed to the mutagenic agent.

3.3. Molecular Analysis of Cactus Seedlings Obtained In Vitro from Seeds Exposed to X-rays

The outcomes of the molecular investigation conducted using the SCoT marker are presented in Table 4. The cumulative count of the obtained products for *Astrophytum* spp. 'Purple' totaled 1544, averaging around 220.57 per primer. Notably, the highest number of products for *Astrophytum* spp. 'Purple' was observed with the S25 primer, while the lowest count was linked to S8. Band sizes exhibited a range spanning 353 to 4878 base pairs. In particular, a maximum of 31 loci were identified with primer S25. Remarkably high levels of polymorphism (ranging from 59.09% for primer S12 to 100.0% for S3 and S8) were ascertained following X-ray exposure.

Table 4. Number of products, band size range, number of loci, and polymorphisms obtained by molecular analysis using the SCoT marker in *Astrophytum* spp. 'Purple'.

Primer	No. of Products	Band Size Range [bp]	No. of loci			Total loci	Polymorphism [%]
			Monomorphic	Polymorphic	Specific		
S3	137	370–2830	0	21	1	22	100.00
S4	179	516–3613	1	22	1	24	95.83
S8	99	379–2338	0	22	3	25	100.00
S12	278	361–4878	9	11	2	22	59.09
S13	246	353–1900	3	12	4	19	84.21
S25	307	370–3040	3	27	1	31	90.32
S33	298	372–2776	5	18	3	26	80.77
Total	1544	-	21	133	15	169	-

The results of the UPGMA cluster analysis conducted on the examined genotypes within *Astrophytum* spp. 'Purple' is presented in Figure 5. Notably, the most significant genetic distance was observed with genotype 11 (creamy white seedling, resulting from a 20 Gy X-ray dose), which formed a distinct cluster apart from the rest. The correlation analysis between pigmentation and the molecular test results (UPGMA) showed that the creamy white color of seedlings was the most genetically distant from the other colors (genotypes 11, 3, 15, and 19), except for genotype 6, which was genetically close to the genotype 5 green seedling color (both obtained as a result of a radiation dose of 15 Gy). However, close genetic relatedness was noted for the orange and red genotypes (genotypes 7 and 8 and genotypes 20 and 21, respectively). Within the two subclusters, genotype 2—a green seedling that was not subjected to X-rays (control)—emerged as a separate cluster. Conversely, the smallest genetic distance was noted between genotypes 7 (orange) and 8 (red), both irradiated with a 15 Gy dose, as well as between genotypes 5 (green seedling) and 6 (creamy white seedling), both subjected to a 15 Gy dose. Similarly, a minimal genetic distance was recorded between genotype 20 (orange) and 21 (red), both exposed to a 50 Gy radiation dose.

A slightly different interpretation of the data was provided by the PCoA analysis of the *Astrophytum* spp. 'Purple' genotypes (1–21) (Figure 6). Among the tested plants, a distinct group was formed by genotypes 1, 2, and 4. High distinctiveness from other tested plants was revealed also for genotypes 9, 10, 13, and 14. The majority of the tested genotypes (3, 5, 6, 7, 8, 11, 12, 15, 16, 17, 18, 19, 20, 21) were arranged in the third uniform group. The smallest genetic distance was found for genotypes 7 and 8 and for genotypes 5 and 6.

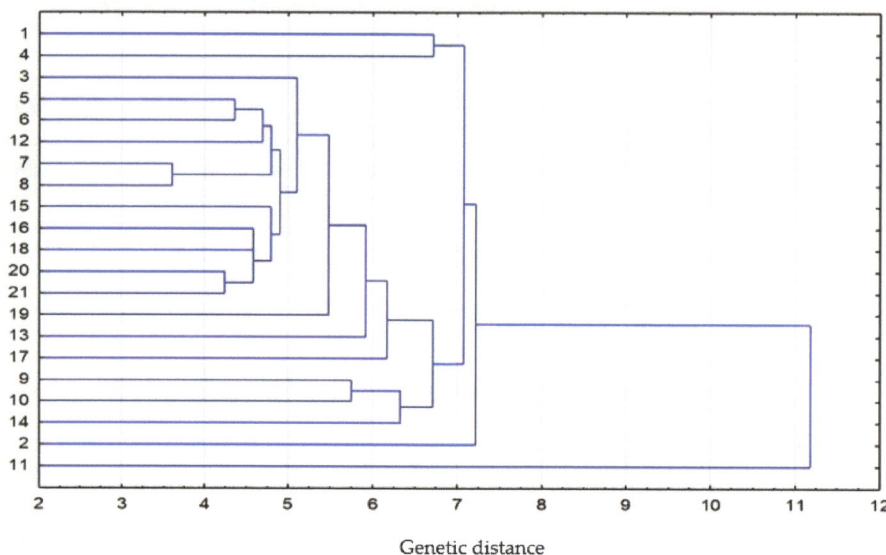

Figure 5. Dendrogram based on the estimation of the genetic distance coefficient and UPGMA clustering for *Astrophytum* spp. 'Purple' genotypes exposed to various X-ray doses (the scale shows the real genetic distance value; for genotype designation, see Table 2).

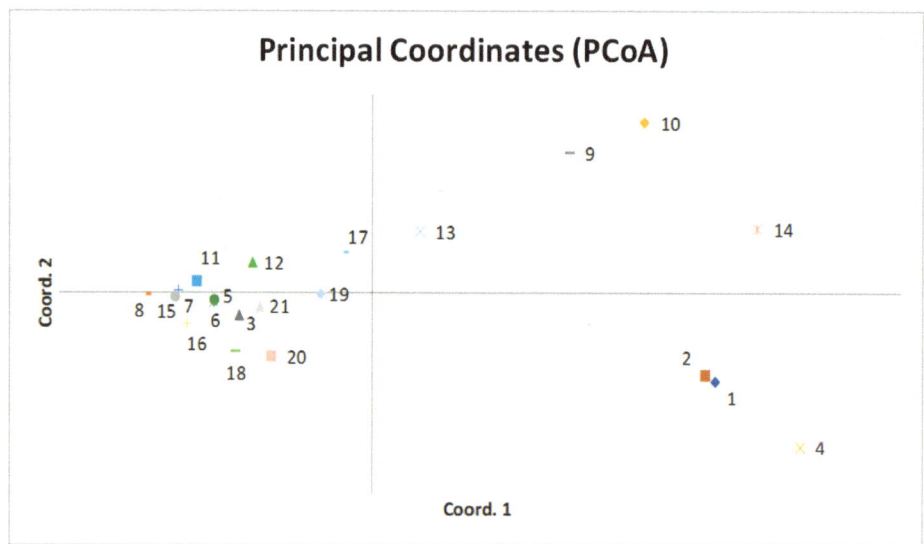

Figure 6. Graph of the principal coordinate analysis (PCoA) of *Astrophytum* spp. 'Purple' genotypes exposed to various X-ray doses based on SCoT analysis (for genotype designation, see Table 2).

The AMOVA analysis confirmed the occurrence of interspecific genetic variation. Molecular variance amounted to 100% among the tested *Astrophytum* spp. 'Purple' genotypes, according to the SCoT analysis (Table 5).

Table 5. Analysis of molecular variance (AMOVA) in the studied *Astrophytum* spp. 'Purple' genotypes (1–21) based on SCoT analysis.

Summary of AMOVA					
Source of variation	df	SS	MS	Est. Var.	%
Among populations	20	1381.43	69.07	23.02	100%
Within populations	42	0.000	0.00	0.00	0%
Total	62	1381.43		23.02	100%

4. Discussion

Inducing mutations into genes with ionizing radiation is a well-known method of increasing genetic diversity in crop breeding. X-rays have proven effective in inducing phenotypic changes in many ornamental plant species, including achimenes, alstroemeria, azalea, begonia, bougainvillea, chrysanthemum, dahlia, carnation, hibiscus, lily, calla, rose, and tulip [11,43–50]. However, the literature lacks studies on the effect of ionizing radiation on cacti.

Mutations can occur spontaneously but at a very low frequency, or they can be artificially induced through physical or chemical treatments [51]. Among the commonly used physical agents for inducing mutations, ionizing radiation holds a significant place. This radiation, due to its high energy, can penetrate deeply into tissues. A crucial consideration in researching the influence of radiation on mutation formation is the appropriate selection of plant tissue to be subjected to the mutagenic factor. It is important to note that irradiating nonmeristematic tissues increases the chance of obtaining genetically altered mutants that are homogeneous and genetically stable. This phenomenon occurs as these mutants come from single mutated cells, and the new color may cover the entire plant organ, such as the inflorescence [52]. Conversely, when dealing with meristematic tissues, genetically heterogeneous chimeras often emerge, as frequently observed in chrysanthemums [47].

Numerous research teams have delved into the effects of radiation on seeds. Generally, a high dose of X- or γ-irradiation is understood to impede plant development, thereby retarding growth [53–56]. Kumar et al. [57] indicated that radiation significantly inhibits the germination of seeds exposed to higher doses. However, scientists also indicate that a properly selected small dose can benefit plants [56–60]. Consequently, preliminary studies are imperative to ascertain the optimal radiation doses. In the context of cacti, these ideal doses remain unknown [51].

The only study concerning the effect of radiation on cactus seeds, specifically γ-radiation exposure, was reported by Boujghagh et al. [51]. Prickly pear seeds (*Opuntia ficus-indica*) originating from two regions, Aït Baamrane (southern Morocco) and SkhourRhamna (Marrakech), were exposed to gamma irradiation at varying doses: 50, 100, 150, and 200 Gy. Germinated seeds were counted every 2 weeks for a period of 4 months. Notably, the control group displayed a much higher germination rate (97%, equivalent to 0.75 seeds daily) than the seeds previously exposed to radiation. In contrast, the irradiated seeds showed a reduction in germination proportionate to the increase in radiation dose.

For seeds from the Aït Baamrane region, the germination rate was lower, with figures of 37% for a 200 Gy dose, 39% for 150 Gy, and 49% for 100 Gy. Likewise, for the SkhourRhamna region, the germination rates stood at 37% and 42% for a 200 Gy dose, 39% for 150 Gy, and 52% for 100 Gy. This irradiation resulted not only in a decrease in germination capacity but also extended the time frame for germination. This phenomenon could potentially be attributed to the damage incurred by the seeds, which, by nature, had an early germination tendency and relatively thin seed coats. In addition, the irradiated seeds exhibited reduced water content in comparison with those unaffected by mutagenic agents [58]. As the experiment concluded, the percentage of nongerminating seeds escalated as the applied radiation dose increased.

Based on those outcomes, it was determined that the optimal radiation dose for Opuntia seeds should fall within the range of 50–125 Gy. Within this range, germination capacities of 70% (for Aït Baamrane seeds) and 50% (for SkhourRhamna seeds) could be achieved. Unlike this previous study, no research has been conducted thus far on the effect of radiation on the seeds of the *Astrophytum* spp. 'Purple' cacti. In these studies, the varying radiation doses led to fluctuations in the number of seedlings observed, along with variations in the acceleration or deceleration of seed germination.

Similar to the study by Boujghagh et al. [51], our research also showed a significant effect of X-rays on seed germination in the tested *Astrophytum* spp. 'Purple'. Exposure of seeds to X-rays resulted in a lower percentage of germinating seeds compared with the control (0 Gy). The highest dose of 50 Gy resulted in the lowest percentage of germinating seeds. However, in contrast to the research on prickly pear, despite the adverse effect of X-rays on the number of seedlings obtained, the use of low doses of radiation had the beneficial effect of shortening the germination time. Doses of 15 and 20 Gy accelerated the formation of seedlings by 3–4 days compared with the control (0 Gy). Similarly, in line with the experiment by Boujghagh et al. [51], an increase in the radiation dose used (25–50 Gy) delayed germination by 1–2 days compared with the control.

Research by Boujghagh et al. [51] on the germination of *Opuntia ficus-indica* seeds irradiated with γ-radiation in vivo suggests that for cacti, the optimal dose should range from 50 to 125 Gy, yielding 70% and 50% germination, respectively. It is worth noting that such high doses of ionizing radiation are lethal for most plant species; for example, in the case of large-flowered chrysanthemums, a dose of 25 Gy is already sublethal [61].

In addition to the selection of the appropriate plant material for irradiation, the selection of the appropriate radiation dose also assumes significance in the context of X-ray applications. Suyitno et al. [62] obtained five groups of mutants resulting from the treatment of orchid seeds (*Spathoglottis plicata*) with X-rays at dosages of 0 rad, 6 rad, 12 rad, 18 rad, and 24 rad (1 rad = 0.01 Gy). Mutations encompassed variations in the leaf structure (thick, unopened leaves), root system (a proliferation of lateral roots), and shoot development (feeble, swollen shoots), pigmentation shifts (yellow leaf coloration, white leaf spots), and accelerated flowering. Their findings demonstrated that doses between 12 rad and 18 rad had the capacity to spur morphological diversity, particularly pertaining to roots, leaves, and shoots.

Our investigation into cacti showed a pronounced influence of X-rays in increasing the number of nonchlorophyll-bearing seedlings in *Astrophytum* spp. 'Purple'. In this species, the percentage of nonchlorophyll seedlings displaying orange, creamy white, and red colorations surged in tandem with escalating X-ray dosages (ranging from 1.44% at 15 Gy to 5.12% at 50 Gy). In the case of our research involving *Astrophytum* spp. 'Purple', the percentage of nonchlorophyll seedlings was amplified proportionally with the dosage of radiation administered, with the highest recorded at the 50 Gy threshold.

The study conducted by Miler et al. [9] equally underscored the significance of selecting an appropriate radiation dose. Miler et al. [9] studied the effect of irradiation at different levels (5, 10, and 15 Gy) on the ovules of the large-flowered chrysanthemum cultivars 'Professor Jerzy' and 'Karolina'.

The assessment encompassed growth and flowering parameters, wherein the most satisfactory results were achieved at a dosage of 10 Gy. This particular dose exhibited the most effectiveness in eliciting stable mutations in inflorescence color and form, devoid of undesirable side effects, such as a delay or extension of cultivation time. In the context of the 'Professor Jerzy' cultivar, the new phenotypes of inflorescences emerged as dark yellow and pinkish, while in the case of 'Karolina', the mutation led to orange–red inflorescences. Elevating the radiation dosage demonstrated a negative correlation with the ability to regenerate the explants.

Findings from the investigation involving *Astrophytum* spp. 'Purple' cacti showed that the percentage of nonchlorophyll seedlings was most pronounced at the 50 Gy dose. Our research aligns with this, affirming the relationship between escalated dosage and heightened phenotypic changes.

In ornamental plants like chrysanthemums, the color of their inflorescence hinges on the presence and quantity of various pigments, including anthocyanins, carotenoids, flavones, and flavonols, predominantly found in the L1 and L2 layers [47,63]. Anthocyanins are responsible for generating the blue, red, and violet colors, while flavonols and flavones contribute to the white, cream, and yellow colors. Carotenoids influence colors ranging from yellow to red and orange [39]. Chlorophyll assumes a crucial role in converting light energy into chemical energy, serving as an indicator of plant metabolism efficiency and health. Carotenoids, conversely, play a major role in photoprotection and defense against oxidative stress in plant cells [64–67]. The emergence of new inflorescence colors is often attributed to modifications in the biosynthetic pathways of plant pigments [68].

While anthocyanins are commonly found in higher plants, in cacti and many other species within the Caryophyllales order, they are replaced by betalains. The biosynthesis of anthocyanins can be blocked at a late stage during the transition from dihydroflavonols to anthocyanins. Sakuta et al.'s [69] research on dihydroflavonol 4-reductase (DFR) and anthocyanidin synthase (ANS) isolation and functional characterization revealed that the absence of anthocyanins is attributed to the suppression of these enzymes. However, in the transgenic cultivar *Astrophytum myriostigma*, the possibility of anthocyanin biosynthesis, instead of betalains, in flower petals was confirmed [69].

The content of plant pigments holds significant sway over the resulting coloration of plants. Our studies have demonstrated that the relationship between the radiation dose used and the pigment content differs in the case of *Astrophytum* spp. 'Purple', where an inverse correlation was observed between increasing X-ray doses and the concentration of plant pigments. A noteworthy exception was observed at the 15 Gy dose, which resulted in an increased concentration of carotenoids and chlorophyll a in brown seedlings. Conversely, the 50 Gy dose yielded the lowest concentrations of anthocyanins, chlorophyll b, and carotenoids in green seedlings, as well as chlorophyll a in orange seedlings.

Dhawi et al. [70] established that low radiation doses had a positive impact on photosynthetic pigments in date palms (*Phoenix dactylifera*), while higher doses exerted a negative effect. Their study also indicated that chlorophyll a and carotenoids were more susceptible to magnetic fields than chlorophyll b. Similarly, Pick Kiong Ling et al. [71] found lower chlorophyll levels in seedlings exposed to γ-radiation, as compared to nonirradiated sweet orange (*Citrus sinensis*) seedlings. Nonetheless, the chlorophyll content appeared to be largely unaffected by low doses of γ-radiation. Conversely, Abu et al. [72] showed an increase in chlorophyll a, chlorophyll b, and total chlorophyll levels in γ-irradiated *Vigna unguiculata* [70–72]. In the context of *Astrophytum* spp. 'Purple' our studies noted a decrease in plant pigment concentration with escalating X-ray doses, excluding the 15 Gy dosage.

The determination and differentiation of newly acquired cultivars from their mother plants often rely on multiple approaches. Morphological traits are analyzed to establish distinctness, which is often accompanied by an assessment of the qualitative and quantitative composition of the plant pigments [39]. Furthermore, biotechnological tools, such as molecular markers, are employed to enhance objectivity and reliability [73]. Genetic markers play a pivotal role in determining genetic distance in mutation breeding.

Our research validated a remarkably high genetic diversity using SCoT markers in X-irradiated *Astrophytum* spp. 'Purple' seedlings. In parallel studies, X-rays have emerged as the most effective mutagen against gold nanoparticles and microwaves in *L. spectabilis* 'Valentine' [11]. In particular, the 20 Gy radiation dose demonstrated remarkable efficacy in inducing mutations in the golden heart. Nonetheless, the number of plants exhibiting genetic changes remained relatively modest, with only 3.3% displaying alterations in DNA content post-X-ray irradiation. The number of polymorphisms ranged from 0% to 6.25%, depending on the SCoT primer, with an average of 2.4 polymorphic loci.

In our research, a notably extensive genetic distance was observed among cactus seedlings irradiated with a 50 Gy dose. Within the examined cacti, considerably higher levels of polymorphism in seedlings were discerned, ranging from 59.09% to 100% in the case of *Astrophytum* spp. 'Purple'. It is worth noting that a similar genetic distance was generated by the SCoT marker for separate species belonging to the genus *Astrophytum* and *Frailea* [74]. Similar levels of polymorphism (PPB%), ranging from 75% to 100%, were demonstrated by Nasri et al. [10] in chrysanthemum mutants generated through the action of the chemical mutagen EMS using the IRAP (Inter-Retrotransposon Amplified Polymorphism) marker. It also turned out that the largest genetic distances between seedlings exposed to X-rays in *Astrophytum* spp. 'Purple' occurred for creamy white genotypes, which may be the result of differences related to the pigment biosynthesis pathways.

5. Conclusions

X-rays had a noticeable impact on the germination dynamics of *Astrophytum* spp. 'Purple'; doses of 15 and 20 Gy accelerated the onset of seed germination by 3–4 days in comparison with nonirradiated (control) seeds. Conversely, seeds exposed to higher doses (25–50 Gy) germinated 1–2 days later than the control seeds. The maximum count of germinating seeds was observed in the nontreated seeds (control—0 Gy), while the 50 Gy radiation dose exhibited the most pronounced influence in diminishing the number of germinating seeds relative to the control (by 39.23%) and other radiation doses.

The application of X-rays influenced the percentage of nonchlorophyll seedlings in *Astrophytum* spp. 'Purple', with the percentage of nonchlorophyll seedlings increasing in correspondence with the radiation dose, peaking at the 50 Gy dose. The dose of 20 Gy is optimal for *Astrophytum* spp. 'Purple' due to the significantly higher percentage of nonchlorophyll seedlings with a simultaneous high percentage of germinating seedlings. The application of X-rays introduced a novel seedling color (red and orange), a phenomenon previously unobserved in *Astrophytum* spp. 'Purple'. In addition, the effect of X-rays on seedling color and the varied concentration of plant pigments in seedlings was visible. As the X-ray dose escalated, a decrease in plant pigment concentration was discerned, even within a specific color category. An exception was the 15 Gy irradiation, which amplified the carotenoid and chlorophyll a content in brown seedlings. Notably, the utilization of 50 Gy irradiation resulted in the lowest concentration of anthocyanins, chlorophyll b, and carotenoids in green seedlings, as well as chlorophyll a in orange seedlings. Creamy white seedlings exhibited no presence of anthocyanins or carotenoids and chlorophylls.

The significant influence of X-rays on the genetic diversity of seedlings was established through the application of the SCoT marker, with polymorphism levels proving remarkably high. In addition, the UPGMA analysis showcased substantial genetic divergence among the examined genotypes exposed to X-rays. Thus, X-rays can be a valuable tool for breeding new cultivars in *Astrophytum* spp.

Author Contributions: Conceptualization, J.L.-R. and P.L.; methodology, J.L.-R., P.L. and J.W.; validation, J.L.-R., P.L. and J.W.; formal analysis, J.L.-R., P.L., E.M. and A.T.; investigation, P.L. and J.L.-R.; resources, P.L.; data curation, P.L. and J.L.-R.; writing—original draft preparation, P.L. and J.L.-R.; writing—review and editing, P.L. and J.L.-R.; visualization, J.L.-R. and P.L.; supervision, J.L.-R. All authors have read and agreed to the published version of the manuscript.

Funding: This work was supported by the Polish Minister of Science and Higher Education under the program "Regional Initiative of Excellence" in 2019–2023 (Grant No. 008/RID/2018/19).

Data Availability Statement: The data presented in this study are available on request from the corresponding author.

Acknowledgments: The authors would like to thank Translmed Publishing Group (T|P|G) for the English correction of the manuscript.

Conflicts of Interest: The authors declare no conflict of interest.

References

1. Anderson, E.F. *The Cactus Family*; Timber Press: Portland, OR, USA, 2001.
2. Nyffeler, R.; Eggli, U. Comparative Stem Anatomy and Systematics of *Eriosyce sensu lato* (Cactaceae). *Ann. Bot.* **1997**, *80*, 767–786. [CrossRef]
3. Schum, A. Mutation breeding in ornamentals: An efficient breeding method? *Acta Hortic.* **2003**, *612*, 47–60. [CrossRef]

4. Czekalski, M. Ogólna uprawa roślin ozdobnych. *Wydaw. Uniw. Przyr. We Wrocławiu* **2010**, *3*, 191–192.
5. Kulpa, D. *Somatyczna Embriogeneza w Kulturach Chryzantemy Wielokwiatowej (Chrysanthemum × grandiflorum) (Ramat.) Kitam.*; Wydawnictwo Uczelniane Zachodniopomorskiego Uniwersytetu Technologicznego w Szczecinie: Szczecin, Poland, 2012; pp. 7–76.
6. Botelho, F.B.; Rodrigues, C.S.; Bruzi, A.T. Ornamental Plant Breeding. *Ornam. Hortic.* **2015**, *21*, 9–16. [CrossRef]
7. Holme, I.B.; Gregersen, P.L.; Brinch-Pedersen, H. Induced Genetic Variation in Crop Plants by Random or Targeted Mutagenesis: Convergence and Differences. *Front Plant Sci.* **2019**, *10*, 1468. [CrossRef] [PubMed]
8. Tymoszuk, A.; Kulus, D. Silver nanoparticles induce genetic, biochemical, and phenotype variation in chrysanthemum. *Plant Cell Tissue Organ Cult.* **2020**, *143*, 331–344. [CrossRef]
9. Miler, N.; Jędrzejczyk, I.; Jakubowski, S.; Winiecki, J. Ovaries of Chrysanthemum Irradiated with High-Energy Photons and High-Energy Electrons Can Regenerate Plants with Novel Traits. *Agronomy* **2021**, *11*, 1111. [CrossRef]
10. Nasri, F.; Zakizadeh, H.; Vafaee, Y.; Mozafari, A.A. In vitro mutagenesis of *Chrysanthemum morifolium* cultivars using ethylmethanesulphonate (EMS) and mutation assessment by ISSR and IRAP markers. *Plant Cell Tissue Organ Cult.* **2021**, *149*, 657–673. [CrossRef]
11. Kulus, D.; Tymoszuk, A.; Jędrzejczyk, I.; Winiecki, J. Gold nanoparticles and electromagnetic irradiation in tissue culture systems of bleeding heart: Biochemical, physiological, and (cyto) genetic effects. *Plant Cell Tissue Organ Cult.* **2022**, *149*, 715–734. [CrossRef]
12. Novoa, A.; Le Roux, J.J.; Richardson, D.M.; Wilson, J.R.U. Level of environmental threat posed by horticultural trade in Cactaceae. *Conserv. Biol.* **2017**, *31*, 1066–1075. [CrossRef]
13. Damude, N.; Poole, J. *Status Report on Echinocactus asterias (Astrophytum asterias)*; U.S. Fish and Wildlife Service: Albuquerque, NM, USA, 1990.
14. Martínez-Ávalos, J.G.; Golubov, J.; Mandujano, M.C.; Jurado, E. Causes of individual mortality in the endangered star cactus *Astrophytum asterias* (Cactaceae): The effect of herbivores and disease in Mexican populations. *J. Arid Environ.* **2007**, *71*, 250–258. [CrossRef]
15. Pérez-Molphe-Balch, E.; Santos-Díaz, M.; Ramírez-Malagón, R.; Ochoa-Alejo, N. Tissue culture of ornamental cacti. *Sci. Agricola.* **2015**, *72*, 540–561. [CrossRef]
16. Święcicki, W.K.; Surma, M.; Koziara, W.; Skrzypczak, G.; Szukała, J.; Bartkowiak-Broda, J.; Zimny, J.; Banaszak, Z.; Marciniak, K. Nowoczesne technologie w produkcji roślinnej—Przyjazne dla człowieka i środowiska. *Pol. J. Agron.* **2011**, *7*, 102–112.
17. Andersen, J.R.; Lubberstedt, T. Functional markers in plants. *Trends Plant Sci.* **2003**, *8*, 554–560. [CrossRef] [PubMed]
18. Weising, K.; Nybom, H.; Wolff, K.; Kah, G. *DNA Fingerprinting in Plants Principles, Methods, and Applications*; Taylor & Francis Group: Abingdon, UK, 2005; pp. 2–73. [CrossRef]
19. Nadeem, M.A.; Nawaz, M.A.; Shahid, M.Q.; Doğan, Y.; Comertpay, G.; Yıldız, M.; Hatipoğlu, R.; Ahmad, F.; Alsaleh, A.; Labhane, N.; et al. DNA molecular markers in plant breeding: Current status and recent advancements in genomic selection and genome editing. *Biotechnol. Biotechnol. Equip.* **2017**, *32*, 261–285. [CrossRef]
20. Kumar, V.; Rajvanshi, S.K.; Yadav, R.K. Potential Application of molecular markers in improvement of vegetable crops. *Int. J. Adv. Biotechnol. Res.* **2009**, *5*, 690–707. [CrossRef]
21. Kesawat, M.S.; Das, B.K. Molecular markers: It's application in crop improvement. *J. Crop Sci. Biotechnol.* **2009**, *12*, 168–178. [CrossRef]
22. Al-Samarai, F.R.; Al-Kazaz, A.A. Molecular markers: An introduction and applications. *Eur. J. Mol. Biotechnol.* **2015**, *9*, 118–130. [CrossRef]
23. Collard, B.C.Y.; Mackill, D.J. Start codon targeted (SCoT) polymorphism: A simple. novel DNA marker technique for generating gene-targeted markers in plants. *Plant Mol. Biol. Rep.* **2009**, *27*, 86–93. [CrossRef]
24. Vivodík, M.; Gálová, Z.; Balážová, Z.; Petrovičová, L. Start Codon Targeted (SCoT) polymorphism reveals genetic diversity in european old maize (*Zea mays* L.) Genotypes. *Potravinarstvo* **2016**, *10*, 563–569. [CrossRef]
25. Amom, T.; Nongdam, P. The use of molecular marker methods in plants: A review. *Int. J. Curr. Res. Rev.* **2017**, *9*, 1–7. [CrossRef]
26. Gajera, H.P.; Bambharolia, R.P.; Domadiya, R.K.; Patel, S.V.; Golakiya, B.A. Molecular characterization and genetic variability studies associated with fruit quality of indigenous mango (*Mangifera indica* L.) cultivars. *Plant Syst. Evol.* **2014**, *300*, 1011–1020. [CrossRef]
27. Shahlaei, A.; Torabi, S.; Khosroshahli, M. Efficacy of SCoT and ISSR marekers in assesment of tomato (*Lycopersicum esculentum* Mill.) genetic diversity. *Int. J. Biosci.* **2014**, *5*, 14–22.
28. Xiong, F.; Zhong, R.; Han, Z. Start Codon Targeted polymorphism for evaluation of functional genetic variation and relationships in cultivated peanut (*Arachis hypogaea* L.) genotypes. *Mol. Biol. Rep.* **2011**, *38*, 3487–3494. [CrossRef] [PubMed]
29. Baghizadeh, A.; Dehghan, E. Efficacy of SCoT and ISSR markers in assessment of genetic diversity in some Iranian pistachio (*Pistacia vera* L.) cultivars. *Pist. Health J.* **2018**, *1*, 37–43. [CrossRef]
30. Satya, P.; Karan, M.; Jana, S.; Mitra, S.; Sharma, A.; Karmakar, P.G.; Ray, D.P. Start codon targeted (SCoT) polymorphism reveals genetic diversity in wild and domesticated populations of ramie (*Boehmeria nivea* L. Gaudich.), a premium textile fiber producing species. *Meta Gene* **2015**, *3*, 62–70. [CrossRef] [PubMed]
31. Lema-Rumińska, J.; Miler, N.; Gęsiński, K. Identification of new polish lines of *Chenopodium quinoa* (Willd.) by spectral analysis of pigments and a confirmation of genetic stability with SCoT and RAPD markers. *Acta Sci. Pol. Hortorum Cultus* **2018**, *17*, 75–86. [CrossRef]

32. Sadek, M.S.E.; Shafik, D.I. Genetic relationships among maize inbred lines as revealed by start codon targeted (SCoT) analysis. *J. Innov. Pharm. Biol. Sci.* **2018**, *5*, 103–107.
33. Saidi, A.; Hajkazemian, M.; Emami, S.N. Evaluation of genetic diversity in gerbera genotypes revealed using SCoT and CDDP markers. *Pol. J. Nat. Sci.* **2020**, *35*, 21–34.
34. Murashige, T.; Skoog, F. A revised medium for rapid growth and bioassays with tobacco tissue cultures. *Physiol. Plant.* **1962**, *15*, 473–497. [CrossRef]
35. RHSCC; The Royal Horticultural Society: London, UK, 1966.
36. Wettstein, D. Chlorophyll-letale und der submikroskopische Formwechsel der Plastiden. *Exp. Cell Res.* **1957**, *12*, 427–506. [CrossRef] [PubMed]
37. Harborne, J.B. Comparative biochemistry of the flavonoids. *Phytochemistry* **1967**, *6*, 1569–1573. [CrossRef]
38. Lichtenthaler, H.K.; Buschmann, C. Chlorophylls and carotenoids: Measurement and characterizacion by UV-VIS Spectroscopy. In *Current Protocols in Food Analytical Chemistry*; Wiley: Hoboken, NJ, USA, 2001; pp. F4.3.1–F4.3.8.
39. Lema-Rumińska, J.; Zalewska, M. Studies on flower pigments of chrysanthemum mutants: Nero and Wonder groups. *Acta Sci. Pol. Hortorum Cultus* **2004**, *3*, 125–135.
40. Lazar, I., Jr.; Lazar, I., Sr. GelAnalyzer 19.1. Available online: www.gelanalyzer.com (accessed on 1 April 2023).
41. Huff, D.R.; Peakall, R.; Smouse, P.E. RAPD variation within and among natural populations of outcrossing buffalograss *Buchloe dactyloides* (Nutt) Engelm. *Theor. Appl. Genet.* **1993**, *86*, 927–934. [CrossRef] [PubMed]
42. Peakall, R.; Smouse, P.E. GenAlEx 6.5: Genetic analysis in Excel. Population genetic software for teaching and research-an update. *Bioinformatics* **2012**, *28*, 2537–2539. [CrossRef] [PubMed]
43. Datta, S.K.D. Improvement of ornamental plants through induced mutation. In *Recent Advances in Genetics and Cytogenetics*; Farook, S.A., Khan, I.A., Eds.; Premier Publishing House: Hyderabad, India, 1989.
44. Jerzy, M.; Lubomski, M. Adventitious shoot formation on ex vitro derived leaf explants of Gerbera jamesonii. *Sci. Horic.* **1991**, *47*, 115–124. [CrossRef]
45. Zalewska, M. In vitro adventitious bud techniques as a tool in creation of new chrysanthemum cultivars. In *Floriculture. Role of tissue Culture and Molecular Techniques*; Datta, S.K., Chakrabarty, D., Eds.; Pointer Publishers: Jaipur, India, 2010; Volume 196.
46. Ahloowalia, B.S.; Maluszyński, M. Induced mutations—A new paradigm in plant breeding. *Euphytica* **2001**, *118*, 167–173. [CrossRef]
47. Zalewska, M.; Tymoszuk, A.; Miler, N. New chrysanthemum cultivars as a result of in vitro mutagenesis with the application of different explant types. *Acta Sci. Pol. Hortorum Cultus* **2011**, *10*, 109–123.
48. Sedaghathoor, S.; Sharifi, F.; Eslami, A. Effect of chemical mutagens and X-rays on morphological and physiological traits of tulips. *Int. J. Exp. Bot.* **2017**, *86*, 252–257.
49. Reznik, N.; Subedi, B.S.; Weizman, S.; Friesem, G.; Carmi, N.; Yedidia, I.; Sharon-Cohen, M. Use of X-ray mutagenesis to increase genetic diversity of *Zantedeschia aethiopica* for Early Flowering, Improved Tolerance to Bacterial Soft Rot, and Higher Yield. *Agronomy* **2021**, *11*, 2537. [CrossRef]
50. Jankowicz-Cieślak, J.; Hofinger, B.J.; Jarc, L.; Junttila, S.; Galik, B.; Gyenesei, A.; Ingelbrecht, I.L.; Till, B.J. Spectrum and Density of Gamma and X-ray Induced Mutations in a Non-Model Rice Cultivar. *Plants* **2022**, *11*, 3232. [CrossRef] [PubMed]
51. Boujghagh, M.; Bouharroud, R.; Mouhib, M. Germination of cactus (*Opuntia ficus-indica*) seeds irradiated with various doses of radiations mutagenic treatment. *Acta Hortic.* **2015**, *1067*, 75–84. [CrossRef]
52. Broertjes, C.; van Harten, A. *Applied Mutation Breeding for Vegetatively Propagated Crops*; Elsevier: Amsterdam, The Netherlands, 1988; pp. 29–59.
53. Mensah, J.K.; Akomeah, P.A.; Ekpekurede, E.O. Gamma irradiation induced variation of yield parameters in Cowpea (*Vigna unguiculata* (L.) Walp. *Glob. J. Pure Appl. Sci.* **2005**, *11*, 327–330. [CrossRef]
54. Animasaun, D.A.; Morakinyo, J.A.; Mustapha, O.T. Assessment of the effects of gamma irradiation on the growth and yield of *Digitaria exilis*. *J. Appl. Biosci.* **2014**, *75*, 6164–6172. [CrossRef]
55. Deshmukh, S.B.; Bagade, A.B.; Choudhari, A.K. Induced Mutagenesis in Rabi Sorghum. *Int. J. Curr. Microbiol. Appl. Sci.* **2018**, *6*, 766–771.
56. Dada, K.E.; Animasaun, D.A.; Mustapha, O.T.; Bado, S.; Foster, B.P. Radiosensitivity and biological effects of gamma and X-rays on germination and seedling vigour of three *Coffea arabica* varieties. *J. Plant Growth Reg.* **2022**, *42*, 1582–1591. [CrossRef]
57. Kumar, A.; Mishra, M.N.; Kharkwal, M.C. Induced mutagenesis in black gram (*Vigna mungo* L. Hepper). *Indian J. Genet.* **2007**, *67*, 41–46.
58. Rojas-Aréchiga, M.; Vázquez-Yanes, C. Cactus seed germination: A review. *J. Arid. Environ.* **2000**, *44*, 85–104. [CrossRef]
59. Al-Enezi, N.A.; Al-Khayri, J.M. Alterations of DNA, ions and photosynthetic pigments content in date palm seedlings induced by X-irradiation. *Int. J. Agric. Biol.* **2012**, *14*, 329–336.
60. Al-Enezi, N.A.; Al-Khayri, J.M. Effect of X-irradiation on proline accumulation, growth, and water content of date palm (*Phoenix dactylifera* L.) seedlings. *J. Biol. Sci.* **2012**, *12*, 146–153. [CrossRef]
61. Jerzy, M.; Zalewska, M. Flower colour recurrence in chrysanthemum and gerbera mutants propagated in vitro from meristems and leaf explants. *Acta Hortic.* **1997**, *447*, 611–614. [CrossRef]
62. Suyitno, A.; Aziz, P.; Kumala, D.; Endang, S. Improvement of genetic variability in seedlings of *Spathoglottis plicata* orchids through X-ray irradiation. *Biodiversitas* **2017**, *18*, 20–27. [CrossRef]

63. Bush, S.R.; Earle, E.D.; Langhans, R.W. Plantlets from petal segments, petal epidermis, and shoot tips of the periclinal chimera *Chrysanthemum morifolium* 'Indianapolis'. *Am. J. Bot.* **1976**, *63*, 729–737. [CrossRef]
64. Bartley, G.E.; Scolnik, P.A. Plant carotenoids: Pigments for photoprotection, visual attractant and human health. *Plant Cell* **1995**, *7*, 1027–1038. [PubMed]
65. Demmig-Addams, B.; Gilmore, A.M.; Addams, W.W. Carotenoids 3: In vivo function of carotenoids in higher plants. *FASEB J.* **1996**, *10*, 403–412. [CrossRef] [PubMed]
66. Lefsrud, M.G.; Kopsell, D.A.; Kopsell, D.E.; Curran-Celentano, J. Air temperature affects biomass and carotenoid pigment accumulation in kale and spinach grown in a controlled environment. *HortScience* **2005**, *40*, 2026–2030. [CrossRef]
67. Kopsell, D.A.; Kopsell, D.E.; Curran-Celentano, J. Carotenoid pigments in kale are influenced by nitrogen concentration and form. *J. Sci. Food Agric.* **2007**, *87*, 900–907. [CrossRef]
68. Tanaka, Y.; Sasaki, N.; Ohmiya, A. Biosynthesis of plant pigments: Anthocyanins, betalains and carotenoids. *Plant J.* **2008**, *54*, 733–749. [CrossRef]
69. Sakuta, M.; Tanaka, A.; Iwase, K.; Miyasaka, M.; Ichiki, S.; Yoriko, M.H.; Yamagami, T.Y.; Nakano, T.; Yoshida, K.; Shimada, S. Anthocyanin synthesis potential in betalain-producing Caryophyllales plants. *J. Plant Res.* **2021**, *134*, 1335–1349. [CrossRef]
70. Dhawi, F.; Al-Khayri, J.; Hassan, E. Static Magnetic Field Influence on Elements Composition in Date Palm (*Phoenix dactylifera* L.). *Res. J. Agric. Biol. Sci.* **2009**, *5*, 161–166.
71. Pick Kiong Ling, A.; Chia, J.Y.; Hussein, S.; Harun, A.R. Physiological Responses of *Citrus sinensis* to Gamma Irradiation. *World Appl. Sci. J.* **2008**, *5*, 12–19.
72. Abu, J.D.; Duodu, G.; Minnaar, A. Effects of X-irradiation on some physicochemical and thermal properties of cowpea (*Vigna unguiculata* L. Walp) starch. *Food Chem.* **2006**, *95*, 386–393. [CrossRef]
73. Miler, N.; Zalewska, M. Somaclonal variation of chrysanthemum propagated in vitro from different explants types. *Acta Sci. Pol. Hortorum Cultus* **2014**, *13*, 69–82.
74. Lema-Rumińska, J.; Michałowska, E.; Licznerski, P.; Kulpa, D. Genetic diversity of important horticultural cacti species from the genus *Astrophytum* and *Frailea* established using ISSR and SCoT markers. *Acta Sci. Pol. Agric.* **2022**, *21*, 3–14.

Disclaimer/Publisher's Note: The statements, opinions and data contained in all publications are solely those of the individual author(s) and contributor(s) and not of MDPI and/or the editor(s). MDPI and/or the editor(s) disclaim responsibility for any injury to people or property resulting from any ideas, methods, instructions or products referred to in the content.

Article

Production of Black Cumin *via* Somatic Embryogenesis, Chemical Profile of Active Compounds in Callus Cultures and Somatic Embryos at Different Auxin Supplementations

Ahmed E. Higazy [1], Mohammed E. El-Mahrouk [1], Antar N. El-Banna [2], Mosaad K. Maamoun [3], Hassan El-Ramady [4,*], Neama Abdalla [5,6,*,†] and Judit Dobránszki [6,†]

[1] Department of Horticulture, Faculty of Agriculture, Kafrelsheikh University, Kafr El-Sheikh 33516, Egypt; ahmed.ebad@rocketmail.com (A.E.H.); threemelmahrouk@yahoo.com (M.E.E.-M.)
[2] Genetics Department, Faculty of Agriculture, Kafrelsheikh University, Kafr El-Sheikh 33516, Egypt; antarsalem@yahoo.com
[3] Department of Breeding and Genetics of Vegetables, Aromatic & Medicinal Plants, Agriculture Research Center, Horticultural Research Institute, Giza 12619, Egypt; mossadmaamoun@arc.sci.eg
[4] Soil and Water Department, Faculty of Agriculture, Kafrelsheikh University, Kafr El-Sheikh 33516, Egypt
[5] Plant Biotechnology Department, Biotechnology Research Institute, National Research Centre, 33 El Buhouth St., Dokki, Giza 12622, Egypt
[6] Centre for Agricultural Genomics and Biotechnology, FAFSEM, University of Debrecen, 4400 Nyíregyháza, Hungary; dobranszki@freemail.hu
* Correspondence: hassan.elramady@agr.kfs.edu.eg (H.E.-R.); neama_ncr@yahoo.com (N.A.)
† These authors have contributed equally to this work and share last authorship.

Citation: Higazy, A.E.; El-Mahrouk, M.E.; El-Banna, A.N.; Maamoun, M.K.; El-Ramady, H.; Abdalla, N.; Dobránszki, J. Production of Black Cumin *via* Somatic Embryogenesis, Chemical Profile of Active Compounds in Callus Cultures and Somatic Embryos at Different Auxin Supplementations. *Agronomy* **2023**, *13*, 2633. https://doi.org/10.3390/agronomy13102633

Academic Editors: Justyna Lema-Rumińska, Danuta Kulpa, Alina Trejgell and Moshe Reuveni

Received: 1 September 2023
Revised: 14 October 2023
Accepted: 16 October 2023
Published: 17 October 2023

Copyright: © 2023 by the authors. Licensee MDPI, Basel, Switzerland. This article is an open access article distributed under the terms and conditions of the Creative Commons Attribution (CC BY) license (https://creativecommons.org/licenses/by/4.0/).

Abstract: Black cumin or *Nigella sativa* L. is a medicinal plant of the Ranunculaceae family that has enormous importance. It has traditionally been used to cure a lot of diseases since ancient times. In the current study, the effects of different auxins on callus induction and subsequent somatic embryo formation of *N. sativa* L. cv. Black Diamond were examined. The best result of callus induction was observed when cotyledon explants were incubated in a Murashige and Skoog (MS) medium supplemented with 1.0 mg L^{-1} α-naphthaleneacetic acid (NAA). The formation of somatic embryos was achieved efficiently from cotyledon-derived calli cultured on a 2 mg L^{-1} Indole-3-butyric acid (IBA)-containing medium. Furthermore, histological analysis of embryogenic calli was used to detect the presence of different developmental stages of somatic embryos. In contrast to the calli and embryos of *N. sativa* 'Black Diamond', which initiated in the dark, light was necessary for the complete differentiation of callus and embryo cultures into shoots/developed plants. Hypocotyl-derived calli and embryos were successfully differentiated on IBA at 2.0, 1.0 mg L^{-1}, and NAA at 2.0 mg L^{-1}. To the best of our knowledge, this work can be considered the first report on the differentiation of *N. sativa* 'Black Diamond' somatic embryos into developed plants. Moreover, the metabolic profiles of secondary products of *N. sativa* 'Black Diamond' callus and embryo cultures originated from the best auxin treatments identified and were compared with that of intact seeds. Callus cultures of *N. sativa* 'Black Diamond' contained thymoquinone (TQ) in a significant percentage of the peak area (2.76%). Therefore, callus cultures could be used as a perfect alternative source of TQ for pharmaceutical and therapeutic purposes. In addition, fatty acids and/or their esters were recorded as the major components in callus and embryo cultures. These vital compounds could be isolated and used for numerous industrial applications.

Keywords: black cumin; Ranunculaceae; medicinal plants; in vitro cultures; auxins; phytochemicals; gas chromatography; thymoquinone; antibacterial; anticancer; antioxidant

1. Introduction

Nigella sativa L. is classified as one of the most important medicinal plants containing volatile and fixed oils in its seeds. It is an annual herbaceous plant belonging to the Ranunculaceae family [1]. It is widely grown for its black seeds in the countries bordering the Mediterranean Sea, Middle East, Southern areas of Europe, Pakistan, Iran, India, and Egypt due to its nutritional, medicinal, and industrial properties [2,3]. It originated from South and Southwest Asia, North Africa, and the Mediterranean region [4]. *N. sativa* seed is described as a medicinal herb. It has largely been used in folk medicine in Arabic and Asian regions for the remediation of numerous ailments, such as cough, fever, headache, toothache, gastrointestinal problems, diarrhea, rheumatism, influenza, diabetes, and hypertension [5]. Due to the highly valuable functional nutrients in black cumin seed, its extract can fortify yogurt [6], honey [7], can be used as a putative therapeutic agent [8], or a supplementary in the broiler industry [9].

Recently, *N. sativa* has drawn the attention of scientists to the therapeutic values and pharmacological effects of its seeds. The seeds have a wide range of biological active secondary metabolites, containing TQ, dithymoquinone (DTQ), thymohydroquinone (THQ), thymol, and carvacrol, which have pharmaceutical potential [10–12]. Analgesic, anti-inflammatory, anti-allergic, anticancer, anti-asthmatic, hypoglycemic, hypotensive, antioxidant activity, hepatoprotective effect, immunity stimulation, and antifungal potential have been reported for this important medicinal plant [13]. However, TQ, the essential component of *N. sativa* oil, is the most important one among the other isolated compounds. TQ exhibited significant antibacterial potential by inhibiting the bacterial biofilm formation against several human pathogenic bacteria, and it showed anticancer potential and hepatoprotective activity as well [14–16].

Medicinal plants are rich resources of naturally occurring bioactive compounds that are widely used as food additives, medicaments, agrochemicals, and perfumes [17]. However, secondary metabolites have various biological properties; their biosynthesis depends on genetics, geographical area, climate, and environmental conditions. In addition, their allocation is very restricted compared to primary metabolites, and many of these compounds occur in nature in very low quantities. Therefore, great efforts have been made via plant biotechnological approaches towards optimizing the culture conditions to maximize the secondary metabolite production needed to support industrial production [17]. Biotechnological investigations on this plant species have been carried out [10]. These studies focused on callus induction for secondary metabolite production [18–22]; callus differentiation into regenerated shoots [23]; phytochemical elicitation in callus cultures under salinity stress [4]; enhancing somatic embryogenesis (SE). However, the conversion of somatic embryos into shoots was not detected [24]. Callus and embryo cultures could be employed to produce valuable phytochemicals in a short period of time, under controlled and sterile conditions and even out of the growing season [18]. The extract obtained from callus cultures of *N. sativa* showed considerable antimicrobial activity against some bacterial strains. Moreover, thymol content in the extract of callus cultures was examined [10]. Somatic embryogenesis was proven to be an important technique that offers an alternative pathway for germplasm conservation, mass clonal propagation of elite plants in a short time, genetic transformation, and synthetic seed production [25].

The previous reports proved that the requirements of plant growth regulators (PGRs) needed for inducing somatic embryogenesis depend on certain cultivars or genotypes [26]. It is known that exogenously applied auxins could enhance somatic embryogenesis by affecting the endogenous content of auxins in the cultured explant or tissue such as IAA [26]. In addition, the mechanism of natural accrued auxin on SE is related to the type of exogenous auxin, which added to the medium [27]. The effect of synthetic auxins on SE could be observed as NAA in *Picea abies* and *P. omorika* [28], 2,4-D (2,4-dichlorophenoxyacetic acid) in Coffea [29], and IBA in *Digitalis lanata* [30].

Therefore, the present investigation aimed to study the influence of exogenous auxins (i.e., IBA, NAA, and 2,4-D), added separately in various concentrations to MS medium [31], on the induction of callus and somatic embryos in *N. sativa* 'Black Diamond' from hypocotyl and cotyledon explants. Moreover, we aim to evaluate the metabolomic profile of *N. sativa* 'Black Diamond' calli and somatic embryos compared to seeds, which are traditionally used in the pharmacy. This is in order to examine and enhance the accumulation of the secondary natural products in these types of tissues.

2. Materials and Methods

This study was conducted in 2021 at the Physiology and Breeding of Horticultural Crops Laboratory, Department of Horticulture, Faculty of Agriculture, Kafrelsheikh University, Egypt.

2.1. Plant Material and Seed Germination

Seeds of pure diploid (2 n = 12) line, of *Nigella sativa* L. cv. Black Diamond, originating from more than five years self-pollination in breeding program at the Faculty of Agriculture, Kafrelsheikh University, Egypt, were used as starting plant material in this research [32]. Black Diamond is improved cultivar originating from local cultivar under registration as the first commercial cultivar in Egypt. Seeds were kept (cold stratification) at 4 °C for three weeks to break dormancy and in order to enhance germination. The cold treatment was reported to be the most efficient for optimal seedling growth [33]. After then, seeds were washed thoroughly using tap water that contained few drops of polyoxyethylene-sorbitan monolaurate ((Tween-20), Loba Chemical Company, Mumbai, India). In the laminar air flow, under aseptic conditions, the seeds were surface sterilized by using 70% ethanol for 2 min then dipped for 15 min in 0.1% mercury chloride ($HgCl_2$) containing 2–3 drops of Tween-20. After rinsing 3 times with sterile-distilled water, seeds were cultured for germination in 350 mL glass jars containing 50 mL of half-strength MS basal medium containing 3% (w/v) sucrose, 0.7% (w/v) Duchefa agar-agar (Hofmanweg 71, 2031 BH Haarlem, The Netherlands). The pH of the medium was set at 5.8, and then the medium was autoclaved for 20 min at 121 °C. The cultures were incubated at 22 ± 2 °C for 10 days under dark conditions followed by 16/8 h photoperiod supplied by cool-white fluorescent lights at photosynthetic photon flux density (PPFD), 30 µmol m^{-2} s^{-1}, for 10 days.

2.2. Callus Induction and Differentiation, and Somatic Embryo Formation and Conversion to Plantlets

Hypocotyl and cotyledon explants were taken from 3-week-old in vitro seedlings of *N. sativa* 'Black Diamond', and were used for callus induction and somatic embryo formation. All explants were cut into 2–3 mm long segments and thereafter cultured on MS medium including sucrose (3%) w/v, 0.7% (w/v) agar, and supplemented with different concentrations of IBA, NAA, and 2,4-D at 0, 1, 2, and 3 mg L^{-1}, added separately to the different media. Plant growth regulator free (PGR-free) medium was considered as a control. The pH of the medium was adjusted to 5.8 before autoclaving at 121 °C for 15 min. Explants were inculcated in sterile glass Petri dishes (70 × 15 mm) containing 25 mL of medium. Cultures were maintained at 22 °C in the dark. Callus percentage (%), callus fresh weight (g) and callus diameter (cm), somatic embryo percentage (%), and number of somatic embryos per callus were recorded after six weeks of culture. After then, the cultures were kept at 25 °C and 30 µmol m^{-2} s^{-1} PPFD light intensity and 16 h/d lighting period. Callus differentiation percentage and embryo conversion to plantlet percentage were recorded after 6 weeks of culture. All cultures were sub-cultured after three weeks on the same medium in dark and light conditions, accordingly.

Equations used in the current study:

Callus percentage (%) = (No. of explant gave callus/total No. of explants) × 100

Somatic embryo percentage (%) = (No. of explant gave embryo/total No. of explants) × 100

Callus differentiation percentage (%) = (No. of callus gave shoot/total No. of callus) × 100

Embryo conversion to plantlets percentage (%) = (No. of embryo conversion to plantlets/Total No. of embryo) × 100

The tissue culture experiments were factorial from three factors (auxin type and concentration, and type of explant), and they were organized in a completely randomized design.

Three auxins, four concentrations, and two explant types were examined in the current study. There were 10 explants per Petri dish (replicate) and 4 replicates per treatment.

2.3. Histological Study of the Embryogenic Callus

The histological analysis was carried out according to Boissot et al. [34]. Embryogenic calli were isolated from 6-week-old callus culture of *N. sativa* 'Black Diamond'. Embryos were fixed for 24 h in a solution of absolute alcohol, glacial acetic acid, and formaldehyde (90: 5: 5, $v/v/v$). Then, the samples were desiccated in a graded sequence of ethanol (70, 95, and 100%) for 1 h each; after that, they were embedded in paraffin wax. Sections of thickness of 15–20 µm were obtained using a rotary microtome (American optical rotary microtome, model 820, New York, NY, USA) and fixed to the slides with albumin. Sections were stained in toluidine blue for 12 h. Then, they were cleared in xylol and mounted in Canada balsam to be ready for microscopic examination. Ten sections were made for each sample on one slide, and then the best section was photographed. Observations were made using a Leica Aristoplan light microscope (Neu-Isenburg, Germany) with Leica DC 300 F digital imaging.

2.4. Gas Chromatography Analysis of Extracts from Seeds, Calli, and Embryos

The extracts were prepared from 2 g of seeds, callus, and somatic embryos of *N. sativa* 'Black Diamond' at globular stage; the latter ones originated from the best two treatments for callus induction (1 and 3 mg L^{-1} NAA) and embryo (2 and 3 mg L^{-1} IBA) developed from cotyledon explant, examined via gas chromatography (GC-MS) analysis. The samples were taken from three replicates. All samples were dried and finely powdered using an electrical grinder (Moulinex—French-DP706G Zerkleinerer La Moulinette Deluxe, France) and soaked in chloroform–methanol (C/M 2:1 v/v) in a ratio of 1:5 (w/v) at ordinary room temperature. The mixture of solvent and samples was covered with aluminum foil. After this, it was shaken for 24 h. The mixture was filtered using a Bucher funnel, and the residue was pressed to obtain a maximum amount of the filtrate. The combined filtrates were mixed with 2.5 g anhydrous sodium sulfate (Na_2SO_4) to remove traces of water and kept in a refrigerator for 2 h. After that, the extract was filtered through a Whitman filter paper (No. 1) and evaporated until dryness on a rotary evaporator (Heidolph—Laborota 4000 eco, Darmstadt, Germany) at 40 °C to remove chloroform, transferred into glass dark bottles, and kept upon completion of the oil for subsequent analysis. Finally, each extract was stored in a refrigerator at 4 °C [35]. Analysis of the extracts were carried out using Gas Chromatography GC-HP (Hewlett Packard, Palo Alto, CA, USA) 6890, with FID detector (flame ionizing) and DB-23 Column (50%—cyanopropyl—methylpolysiloxane), 30 m × 0.32 mm, ID = 0.25 µm film thickness. The carrier gas was nitrogen (1 mL min^{-1} gas flow).

2.5. Statistical Analyses

The obtained results of tissue culture experiments (callus and embryo induction, differentiation, and development) were analyzed using multiple-way ANOVA. ANOVA analysis was conducted using CoStat (version 6.311) statistical CoHort software (Berkeley, CA, USA). The mean separations were conducted using LSD and Duncan's multiple range tests, and significance was measured at $p \leq 0.05$.

3. Results

3.1. Callus Induction and Somatic Embryo Formation

The obtained data in Table 1 showed highly significant differences for the triple interaction between the three factors under study on callus induction measurements. Callus was induced on all tested media from both hypocotyl and cotyledon explants of *N. sativa* 'Black Diamond'. Callus was also initiated on the PGR-free medium in 40–50% from both studied explants. The highest value of callus percentage (100%) was recorded for the cotyledon explant when cultured on an MS medium supplemented with 2.0 mg L^{-1} IBA or 1.0 mg L^{-1} NAA. Callus diameter was the highest (2.7 cm) for hypocotyl explant cultured on MS medium supplemented with 1.0 mg L^{-1} NAA. Explants produced the highest callus fresh weight (4.4 for hypocotyl and 4.5 for cotyledon, respectively) on the medium described above. The best treatment noticed for callus induction was the MS medium supplemented with 1.0 mg L^{-1} NAA followed by 3.0 mg L^{-1} NAA, and cotyledon was the superior explant (Figure 1A). Regarding embryo formation from calli derived from hypocotyl or cotyledon explants, the embryo percentage and number of embryos had high significant differences for the triple interaction between the three factors studied. Both the hypocotyl and cotyledon explants achieved a maximum value for the embryo percentage (98%) when were cultured on the MS medium supplemented with 2.0 mg L^{-1} IBA or NAA as well as 3.0 mg L^{-1} IBA. Calli derived from either hypocotyl or cotyledon explants failed to produce embryos on the MS medium supplemented with 1.0 or 2 mg L^{-1} 2,4-D. The number of embryos was significantly enhanced for the cotyledon explant cultured on IBA-supplemented media (2.9 or 2.8 embryos) and for hypocotyl on the 3 mg L^{-1} IBA-containing medium (2.8 embryos). The best results for embryo formation were obtained for cotyledon-derived calli on 2 mg L^{-1} IBA followed by 3 mg L^{-1} IBA-containing media (Figure 1D).

Table 1. Effect of some auxins added at different concentrations on callus induction and embryo formation from hypocotyl and cotyledon explants of N. sativa 'Black Diamond'.

Auxins (mg L^{-1})	Callus (%) Hyp.	Callus (%) Cot.	Callus Diameter (cm) Hyp.	Callus Diameter (cm) Cot.	Callus Fresh Weight (g) Hyp.	Callus Fresh Weight (g) Cot.	Embryo (%) Hyp.	Embryo (%) Cot.	Number of Embryos Hyp.	Number of Embryos Cot.
IBA										
0.0	50 c	40 c	0.3 i	0.4 i	1.3 i	0.65 i	6.6 f	3.3 f	0.33 c	0.33 c
1.0	99 a	99 a	2.2 de	1.5 g	3.1 g	2.3 hi	67 d	77 c	1.9 b	2.8 a
2.0	99 a	100 a	1.7 f	1.6 fg	3.2 fg	3.3 ef	98 a	98 a	1.9 b	2.9 a
3.0	97 a	99 a	1.7 f	2.2 de	3.2 fg	3.2 fg	98 a	98 a	2.8 a	2.8 a
NAA										
0.0	50 c	40 c	0.3 i	0.4 i	1.3 i	0.65 i	6.6 f	3.3 f	0.33 c	0.33 c
1.0	98 a	100 a	2.7 a	2.4 b	4.4 a	4.5 a	89 b	67 d	2.8 a	1.9 b
2.0	99 a	98 a	1.3 h	1.6 fg	2.4 h	3.5 cd	98 a	98 a	2.7 a	1.9 b
3.0	98 a	98 a	2.4 b	2.3 bcd	3.4 de	4.2 b	78 c	88 b	2.9 a	1.9 b
2,4-D										
0.0	50 c	40 c	0.3 i	0.4 i	1.3 i	0.65 i	6.6 f	3.3 f	0.33 c	0.33 c
1.0	97 a	89 b	2.3 bcd	1.3 h	4.2 b	3.6 c	0 g	0 g	1.9 b	1.9 b
2.0	98 a	89 b	2.4 bc	1.2 h	3.5 cd	2.2 i	0 g	0 g	1.9 b	1.9 b
3.0	97 a	99 a	2.1 e	2.1 e	3.4 de	2.2 i	57 e	68 d	1.8 b	1.9 b
Significance										
A	***		***		***		***		***	
E	**		***		***		***		**	
C	***		***		***		***		N.S	
A × E	***		***		***		***		***	
A × C	***		***		***		***		***	
C × E	***		***		***		***		***	
A × C × E	***		***		***		***		***	

, * significant at $p \leq 0.01$, and 0.001, respectively according to Duncan's multiple range tests followed by ANOVA. Values followed by the same letters in the same column under the two explants were not significantly different. A = auxins; C = concentrations; E = explants. A × C × E indicates the significance of the triple interaction between the three factors (auxins, concentrations and explants); Hyp = hypocotyl and Cot = Cotyledon.

3.2. Differentiation of Calli into Shoots/Plants and Embryo Conversion to Plantlets

Hypocotyl- and cotyledon-derived calli of N. sativa 'Black Diamond', initiated in the dark, were differentiated into shoots in light on all media under investigation, except 2,4-D-supplemented media (Table 2; Figure 1B,C). Moreover, the somatic embryos, produced in the dark on the callus originating from both hypocotyl and cotyledon explants, were differentiated into complete developed plants after they were put into light in all tested media, except 2,4-D at 1 or 2 mg L^{-1} (Figure 1D,F). Callus differentiation and embryo conversion to plantlets percentages showed high significant differences for the triple interaction between the three factors studied. Hypocotyl-derived calli were differentiated successfully with the highest significant value (97.7%) on the MS medium supplemented with 2.0 mg L^{-1} IBA. However, the embryos produced on hypocotyl-derived calli were converted to plantlets by percentage (37.7%) on the MS medium supplemented with either 1.0 mg L^{-1} IBA or 2.0 mg L^{-1} NAA, while embryos produced on cotyledon-derived calli were converted to plantlets at percentages of 35.0% and 37.7%, respectively, on the MS medium supplemented with 2.0 or 3.0 mg L^{-1} IBA. On the other hand, the embryos on embryogenic calli originating from each of the hypocotyl and cotyledon explants did not differentiate on the MS medium supplemented with 1 or 2 mg L^{-1} 2,4-D.

Figure 1. Callus and embryo formation of *N. sativa* 'Black Diamond'; (**A**) callus induction of cotyledon on 1 mg L^{-1} NAA; (**B**) callus differentiation; (**C**) callus differentiation in light conditions (2x * 10x); (**D**) somatic embryogenesis on 2 mg L^{-1} IBA (red arrows refer to embryos); (**E**) embryo conversion to plantlets (2x * 10x); (**F**) complete plantlet developed from somatic embryo.

Table 2. Effect of some auxins at different concentrations on callus differentiation percentage and somatic embryo conversion to plantlet percentage from each of the hypocotyl and cotyledon explants of *N. sativa* 'Black Diamond'.

Auxins (mg L^{-1})	Embryo Conversion to Plantlets (%)		Callus Differentiation (%)	
	Hyp.	Cot.	Hyp.	Cot.
IBA				
0.0	3.3 g	3.3 g	10.0 g	3.3 h
1.0	37.7 a	29.4 c	47.0 c	8.4 f
2.0	23.0 d	35.0 ab	97.7 a	77.7 b
3.0	23.0 de	37.7 a	79.7 b	77.7 b
NAA				
0.0	3.3 g	3.3 g	10.0 g	3.3 h
1.0	24.0 d	9.33 f	47.7 c	13.0 e
2.0	37.7 a	27.4 c	49.4 c	50.0 c
3.0	33.7 b	27.7 cd	77.7 b	27.7 d
2,4-D				
0.0	3.3 g	3.3 g	10.0 g	3.3 h
1.0	0 h	0 h	0 g	0 g
2.0	0 h	0 h	0 g	0 g
3.0	8.4 f	9.4 f	0 g	0 g
Significance				
A	***		***	
E	*		***	
C	***		***	
A × E	***		***	
A × C	***		***	
C × E	***		***	
A × C × E	***		***	

*, *** significant at $p \leq 0.05$, and 0.001, respectively according to Duncan's multiple range tests followed by ANOVA. Values followed by the same letters in the same column under the two explants were not significantly different. A = auxins; C = concentrations; E = explants. A × C × E indicates the significance of the triple interaction between the three factors (auxins, concentrations and explants).

3.3. Histological Analysis of the Embryogenic Calli

The histological micrograph of the embryonic calli of N. sativa 'Black Diamond' shows the ideal developmental stages of somatic embryos (i.e., globular, heart, torpedo shaped, and cotyledonary-stage embryos) (Figure 2).

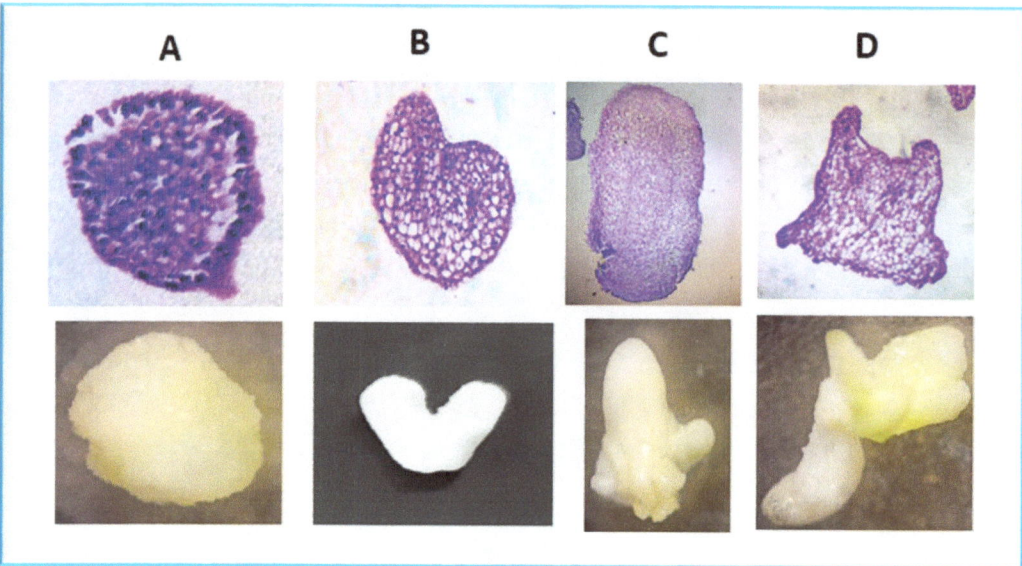

Figure 2. Histological micrograph of somatic embryos in N. sativa 'Black Diamond' at different stages of development: globular stage (**A**); heart stage (**B**); torpedo stage (**C**); cotyledon stage (**D**). The upper part of the figure shows a microscopic examination of embryonic tissues to confirm the different stages of growth and development of the embryos, and this was taken at 40× magnification. The lower part of the figure shows the different stages of living embryos separated from the explant studied under the binocular microscope at 15× magnification.

3.4. Gas Chromatography Analyses of Callus, Embryo, and Seed Extracts

We chose cotyledon-derived calli of N. sativa 'Black Diamond' embryos formed on them for phytochemical component analyses. Both types of samples originated from the best treatments of callus induction and embryo formation. Callus and embryo extracts were examined and compared to seed extract to determine their phytochemical components via GC–MS spectrophotometry (Tables 3–5). The analysis of the extract of cotyledon-derived calli, induced on MS medium with 1.0 and 3.0 mg L^{-1} NAA, showed the presence of a flavonoid, TQ from polyphenols, fatty acids (i.e., oleic, linoleic, and palmitic acids), their salts and esters (i.e., methyl palmitate, linoleol chloride, and oleic acid methyl ester), and amines and their oxides (benzyl amine, onamine 12, and myristamine oxide). The best results for all phytochemicals were mostly recorded for 3.0 mg L^{-1} NAA. The most important bioactive constituent in the callus culture extract was TQ. The percentage of this compound in the extract of calli produced on 3.0 mg L^{-1} NAA (2.76%) was nearly twice its value for calli initiated on 1.0 mg L^{-1} NAA (1.58%). Unsaturated fatty acids accompanied with their derivatives, salts, and esters were noticed to be the major component of the N. sativa 'Black Diamond' callus extract. They recorded a percentage of 83% from the total compounds from callus induced on 3.0 mg L^{-1} NAA. The other compounds were amines and their oxides; they represented the rest ratio of the callus extract (Table 3).

Table 3. Phytochemical compounds of *N. sativa* 'Black Diamond' calli derived from cotyledon explant cultured on 1 and 3 mg L^{-1} NAA.

No.	Compound	Molecular Weight	RT (min)	Area (%) 1 mg L^{-1} NAA	Area (%) 3 mg L^{-1} NAA
1	Thymoquinone	164	7.18	1.58	2.76
2	Onamine 12	213	11.96	8.80	8.14
3	Anastrozole	293	15.05	1.28	-
4	Myristamine oxide	241	15.76	2.98	-
5	Sulfobetaine 14	363	15.77	-	2.86
6	Methyl Palmitate	270	19.61	5.98	7.22
7	Hexadecanoic Acid (Palmitic acid)	256	20.30	4.00	4.32
8	2,3-Dehydro methyl linoleate	294	22.29	10.09	14.62
9	Oleic acid methyl ester	296	22.38	9.2	13.07
10	Elaidic acid methyl ester	296	22.48	2.11	2.75
11	Benzyl amine	234	22.56	2.18	1.87
12	Methyl stearate	298	22.80	1.72	2.41
13	Octadecadienoic acid (Linoleic acid)	280	22.98	5.49	6.85
14	Oleic Acid (cis-9-Octadecenoic acid)	282	23.05	8.21	-
15	Linoleol chloride	298	23.06	-	11.16
16	2,2'-methylenebis [4-methyl-6-tert-butylphenol]	340	27.02	1.17	1.33
17	Oleic acid	282	27.10	1.17	-
18	Diisooctyl phthalate	390	28.66	30.63	20.64
19	Oleic acid, 3-(octadecyloxy) propyl ester	529	34.29	1.57	-
20	4H-1-Benzopyran-4-One,2-(3,4 Dimethoxyphenyl)-3,5-Dihydroxy-7-Methoxy	344	35.25	1.83	-

Thymoquinone (C$_{10}$H$_{12}$O$_2$)

Onamine 12 (C$_{14}$H$_{31}$N)

Myristamine oxide (C$_{16}$H$_{35}$NO)

Oleic acid methyl ester (C$_{19}$H$_{36}$O$_2$)

Methyl Palmitate (C$_{17}$H$_{34}$O$_2$)

Sulfobetaine 14 (C$_{19}$H$_{41}$NO$_3$S)

Anastrozole (C$_{17}$H$_{19}$N$_5$)

Benzyl amine (C$_7$H$_9$N)

Hexadecanoic Acid (Palmitic acid) (C$_{16}$H$_{32}$O$_2$)

Oleic acid (C$_{18}$H$_{34}$O$_2$)

Octadecadienoic acid (Linoleic acid) (C$_{18}$H$_{32}$O$_2$)

Elaidic acid methyl ester (C$_{19}$H$_{36}$O$_2$)

Table 3. Cont.

No.	Compound	Molecular Weight	RT (min)	Area (%)	
				1 mg L^{-1} NAA	3 mg L^{-1} NAA

Diisooctyl phthalate (C$_{24}$H$_{38}$O$_4$)	Oleic Acid (cis-9-Octadecenoic acid) (C$_{18}$H$_{34}$O$_2$)	Methyl stearate (C$_{19}$H$_{38}$O$_2$)
Oleic acid, 3-(octadecyloxy) propyl ester (C$_{39}$H$_{76}$O$_3$)	2,2′-methylenebis [4-methyl-6-tert-butylphenol] (C$_{23}$H$_{32}$O$_2$)	Linoleoyl chloride (C$_{18}$H$_{31}$ClO)

Sources: https://webbook.nist.gov/ and https://pubchem.ncbi.nlm.nih.gov/ accessed on 10 October 2023.

Table 4. Phytochemical compounds of *N. sativa* 'Black Diamond' embryos produced on calli derived from cotyledon explant cultured on 2 and 3 mg L^{-1} IBA, respectively.

No.	Compound	Molecular Weight	RT (min)	Area (%)	
				2 mg L^{-1} IBA	3 mg L^{-1} IBA
1	Nizatidine	331	11.96	6.42	5.59
2	Bisabolol oxide II	238	14.89	5.66	-
3	Bisabolone oxide A	236	15.40	4.91	-
4	Myristamine oxide	241	15.76	2.12	1.89
5	alpha Bisabolol oxide A	238	16.56	38.10	1.79
6	Methyl Palmitate	270	19.61	4.58	7.15
7	Palmitic Acid	256	20.30	2.72	3.75
8	Octadecadienoic acid, methyl ester	294	22.29	8.08	14.94
9	Elaidic acid methyl ester	296	22.38	8.72	14.72
10	11-Octadecenoic acid, methyl ester	296	22.48	-	3.20
11	1-Morpholino-2-(Benzylamino) Propane	234	22.56	-	1.20
12	Methyl stearate	298	22.79	-	2.86
13	Linoelaidic acid	280	22.98	3.77	6.85
14	Oleic acid -1,2,3,7,8-	282	23.05	6.31	8.72
15	Trans-13-Octadecenoic acid	282	27.10	-	1.50
16	Diisooctyl phthalate	390	28.66	8.61	25.84

Alpha Bisabolol oxide A (C$_{15}$H$_{26}$O$_2$)	Bisabolol oxide II (C$_{15}$H$_{26}$O$_2$)	Bisabolone oxide A (C$_{15}$H$_{24}$O$_2$)

Table 4. Cont.

No.	Compound	Molecular Weight	RT (min)	Area (%) 2 mg L⁻¹ IBA	Area (%) 3 mg L⁻¹ IBA

Nizatidine ($C_{12}H_{21}N_5O_2S_2$)

Methyl Palmitate ($C_{17}H_{34}O_2$)

Octadecadienoic acid, methyl ester ($C_{19}H_{34}O_2$)

11-Octadecenoic acid, methyl ester ($C_{19}H_{36}O_2$)

Elaidic acid methyl ester ($C_{19}H_{36}O_2$)

1-Morpholino-2-(Benzylamino) Propane

Diisooctyl phthalate ($C_{24}H_{38}O_4$)

Trans-13-Octadecenoic acid ($C_{18}H_{34}O_2$)

Oleic acid-1,2,3,7,8-($C_{18}H_{34}O_2$)

Sources: https://webbook.nist.gov/ and https://pubchem.ncbi.nlm.nih.gov/ accessed on 10 October 2023.

Table 5. Phytochemical compounds of *N. sativa* 'Black Diamond' seed extract.

No.	Compound	Molecular Weight	RT (min)	Area (%)
1	Thymol	134	5.25	0.37
2	Thymoquinone	164	7.37	0.98
3	Methyl myristate	242	22.09	0.30
4	Palmitoleic acid	268	25.77	0.55
5	Methyl palmitate	270	26.27	16.00
6	Methyl Heptadecanoate	284	28.16	0.17
7	Methyl linoleate	294	29.42	38.38
8	Methyl oleate	296	29.60	25.03
9	Methyl stearate	298	30.04	8.01
10	8,11-Octadecadienoic acid, methyl ester	294	30.33	0.26
11	Methyl octadecadienoate	294	31.11	0.31
12	Methyl Eicosadienoic	322	32.92	7.09
13	Methyl gadoleate	324	33.02	1.45
14	Heneicosanoic acid	326	33.49	0.73
15	Pentacosylic acid	382	39.74	0.37

Thymol ($C_{10}H_{14}O$)

Thymoquinone ($C_{10}H_{12}O_2$)

Methyl myristate ($C_{15}H_{30}O_2$)

Table 5. Cont.

No.	Compound	Molecular Weight	RT (min)	Area (%)
	Methyl linoleate ($C_{19}H_{34}O_2$)	Methyl Heptadecanoate ($C_{18}H_{36}O_2$)		Palmitoleic acid ($C_{16}H_{30}O_2$)
	Methyl octadecadienoate ($C_{19}H_{34}O_2$)	Methyl oleate ($C_{19}H_{36}O_2$)		8,11-Octadecadienoic acid, methyl ester ($C_{19}H_{34}O_2$)
	Heneicosanoic acid ($C_{21}H_{42}O_2$)	Methyl gadoleate ($C_{21}H_{40}O_2$)		Methyl Eicosadienoic ($C_{21}H_{38}O_2$)

Sources: https://webbook.nist.gov/ and https://pubchem.ncbi.nlm.nih.gov/ accessed on 10 October 2023.

Embryos examined for phytochemical compounds were formed from cotyledon-derived calli on the MS medium supplemented with 2 and 3 mg L^{-1} IBA as the best two treatments for embryo formation. In contrast to the callus extract, the obtained data indicated the absence of TQ in the embryo extract (Table 4). However, some compounds were found in both the callus and embryo extracts, i.e., myristamine oxide, methyl palmitate, palmitic acid, elaidic acid methyl ester, methyl stearate, oleic acid, and diisooctyl phthalate. Moreover, fatty acids and their derivatives are considered to be the main ingredients of *N. sativa* 'Black Diamond' embryo extracts. They represent about 90% of the total compounds measured in embryos, which originated from the medium containing 3 mg L^{-1} IBA.

A rare occurrence of thymol was detected in the seed extract at 0.37% in all compounds (Table 5). Moreover, the seed extract contained 0.98% TQ. Fatty acids in the form of salts or esters represented the main component of the *N. sativa* 'Black Diamond' seed extract, as well as in callus and embryo extracts. The most important fatty acids that could be found in the seed extract were oleic acid and linoleic acid.

4. Discussion

Medicinal plants represent a spectacular store of bioactive compounds with various pharmacological properties. Folk medicine is a wealthy source of remedies. Traditionally, *N. sativa* seeds have been consumed to cure several health problems [36]. *N. sativa* seeds have been utilized for years as a food preservative and spice, flavoring in bakery products and cheese, in nutraceuticals and pharmaceutical products, and in functional foods [37].

In vitro plant cell culture techniques and biotechnological approaches constitute an invaluable, sustainable, and environmental substitute for the production of these bioactive compounds to diminish the use of compounds, which synthesize chemically, while decreasing the excessive usage of the available natural resources. Plant cell culture methods allow for the conservation of plant species, as well as the enhancement of metabolite biosynthesis, and the possibility of modifying the synthesis pathways [17]. Differences in the chemical composition of callus tissue and seeds of intact plants have been described for a number of plant species [22]. Nevertheless, very few studies have focused on metabolic comparisons between cell/tissue cultures (callus cultures and/or embryo cultures), and original plants have been reported, even though comparisons of the biological active metabolite content of cultured cells and tissues with that of the normal plants are of great importance.

Callus induction of *N. sativa* was reported for the first time by Banerjee and Gupta [38] but without metabolic identification and/or quantification of the phytochemical content. After years, callus cultures were proliferated from the stem, young leaf, petiole, and root of *N. sativa* plantlet on a solidified MS medium supplemented with 2.15 mg L^{-1} kinetin (Kin) and 1 mg L^{-1} 2,4-D in dark conditions [20], and the optimum growth rate of callus (115.4 ± 2.8 mg day^{-1}) was observed for the leaf explant. This result was in agreement with those obtained by Chand and Roy [39] and Al-Ani [19], who indicated that the leaf-derived calli had the highest growth rates. Moreover, the combination of 0.5 mg L^{-1} from both benzylaminopurine (BAP) and NAA resulted in effective callus induction from leaf explants of *N. sativa* [23]. In addition, callus cultures for *N. sativa* were successfully initiated on the MS medium supplemented with 1 mg L^{-1} NAA and 3 mg L^{-1} BAP, which recorded 80.41% for callus induction and 0.31 mm day^{-1} for the callus growth rate [4]. However, the hypocotyl explant was the superior explant for callus induction by percentage (81.78%), with a callus growth rate of 0.33 mm day^{-1}. Bibi et al. [18] reported an increase in the callus induction frequency up to 88%, when cotyledon explants were cultured on the MS medium containing 4.0 mg L^{-1} from both NAA and thidiazuron (TDZ). Chaudhry et al. [21] declared that the highest frequency of callus induction (82%) of *N. sativa* was observed for epicotyle explants on the MS medium containing 1.0 mg L^{-1} NAA and 2.0 mg L^{-1} Kin. In another study on a medicinal plant of *Nigella damascena* L., Klimek-Chodacka et al. [40] reported that 83% and 100% callus formation were achieved from hypocotyl and cotyledon explants, respectively, on the MS medium supplemented with 3 mg L^{-1} BAP and 0.5 mg L^{-1} NAA. These findings are in agreement with our recently obtained results, which proved that cotyledon-derived explants of *N. sativa* 'Black Diamond' were more efficient at inducing callus than hypocotyl-derived ones and on the NAA-containing medium. In the current investigation, the supplementation of NAA at 1.0 mg L^{-1} gave the best results for callus induction of Black Diamond.

Callus cultures can serve as a means for the production of bioactive compounds in vitro as they have antioxidant potential or activity, due to the presence of flavonoids and phenolic compounds, like carvacrol, thymoquinone, and thymol. Because of its bioactivities, the industrial production of such health-promoting natural products is a main target through callus culture systems [4]. The callus obtained from the young leaf explant of *N. sativa* was proven to have great potential for TQ production [20].

Furthermore, callus can also be used as an explant to establish somatic embryogenesis, induce rhizogenesis, or can be differentiated into shoots, depending on the type and concentration of PGRs and the culture media [18,41]. Klimek-Chodacka et al. [40] found that shoots could regenerate and develop from 95% of hypocotyl-derived calli of *N. damascena* after transferring them on hormone-free media, regardless of the callus induction medium used before. However, in the present investigation, calli derived both from hypocotyl and cotyledon explants of Black Diamond and embryos formed on them could be regenerated into shoots/plantlets on the MS basal medium but in low percentages of 3.3–10% compared with the other auxin treatments (8.4–97.7%).

Somatic embryos are powerful biotechnological tools that can be employed for various applications, such as for clonal micropropagation, plant improvement, and germplasm conservation. They also provide an excellent system to study the early development of plant morphogenesis and genetic transformation [40]. Few reports concerning the somatic embryogenesis of *N. sativa* have been published [24]. They examined some combinations of 2,4-D and NAA and they found that somatic embryos could be induced in *N. sativa*, but their conversion into plants has never been observed. In this work, somatic embryos could be initiated from cotyledon-derived calli of *N. sativa* 'Black Diamond' on the 2 mg L^{-1} IBA-supplemented medium.

Plant polyphenols are secondary metabolites with bioactivity, and they are produced in response to stress conditions in plants to mitigate the harmful effects of free radicals. The low-molecular-weight phenolic acids and flavonoids are important classes of polyphenols such as TQ. TQ has prominent antioxidant activity and is of pharmacological interest in the treatment of many human diseases [18,42]. So, plant in vitro technologies are widely applied to enhance the production of such high-value-added natural products as natural antioxidants [43].

Concerning the pharmaceutical importance of *N. sativa* seeds, the consumption of black cumin seed extract was confirmed to control many problems, such as cough, break up renal calculi, delay the carcinogenic process, and treat abdominal pain, diarrhea, and flatulence. It was also reposted to have anti-inflammatory and antioxidant effects [44]. This extract showed significant antioxidant and anti-inflammatory potential. Most of the pharmacological effects of *N. sativa* seeds are due to the quinine constituent, of which TQ is mainly abundant [45]. Several reports have confirmed TQ as one of the main ingredients of *N. sativa* seed extracts where it was found ranging from 8 to 27% [20]. Another study reported that TQ content in commercial black cumin seed oil (BSO) products varied from 0.07% to 1.88% wt/wt, where the TQ content in those products differed depending on both the oil source and extraction method [13]. In the current work, it was found that thymol could be found in the seed extract of Black Diamond but not detected in callus or embryo extracts. On the contrary, Al-Ani [19] confirmed the production of thymol in the leaf-derived calli of *N. sativa*. However, the major constituents in the *N. sativa* seed extract are esters of fatty acids, as reported by Mahmmoud and Christensen [36].

In the current study, GC-MS analysis was utilized to compare the secondary product metabolite profile of the extract from intact *N. sativa* 'Black Diamond' seeds with callus or embryo culture extracts. In this concern, HPLC analysis, using the standard TQ sample, indicated that the extract of the leaf callus of *N. sativa* 'Black Diamond' contained the highest amount of TQ (8.78 mg mL^{-1}). This content was 12-times higher than that measured in the seed extract (0.74 mg mL^{-1}) [20]. Therefore, callus cultures could be used as an alternative source of TQ, especially when seeds are not available. TQ is the major active compound in black cumin seed oil [13]. From the obtained results, TQ is one of the most active products that could be identified in callus and seed extracts of *N. sativa* 'Black Diamond', but it was not detected in the embryo extract.

Thirty-two fatty acids (99.9%) have been identified in the extracted fixed oil of *N. sativa*, while the major fatty acids were linoleic acid and oleic acid [46,47]. *N. sativa* seed oil was reported to contain a mixture of oleic and linoleic acids. They have a particularly significant role in lowering high blood pressure [36]. Oleic acid could be used in the industry as an emulsifying or solubilizing agent for the prevention of oxidation in oils [48].

5. Conclusions

Modern plant biotechnology techniques provide scientists with plant cells and tissue cultures, which allow for maximizing the production of active natural compounds from medicinal plants. There is increasing interest to study the biochemical properties of proliferated cell cultures under controlled artificial conditions and to compare the results with those of native plant species. An efficient protocol for enhancing the callus biomass of *N. sativa* 'Black Diamond' was developed. Callus cultures of *N. sativa* 'Black Diamond' were successfully induced from cotyledon explants cultured on the MS medium supplemented with 1.0 mg L^{-1} NAA. Callus cultures had potency for the further production of health-promoting natural products. In addition to the applications of somatic embryos mentioned before, they should be reinvestigated to be employed for the production of secondary metabolites of this remarkable medicinal plant. The formation of somatic embryos was achieved from cotyledon-derived calli of *N. sativa* 'Black Diamond' on the 2 mg L^{-1} IBA-supplemented MS medium. This is the first report on the successful conversion of somatic embryos into plants in *N. sativa* 'Black Diamond'. Furthermore, studying the metabolic profile of callus, embryo, and seed extracts of *N. sativa* 'Black Diamond' to identify phytochemicals that might be found in the extract is not enough. But, also, the quantification of secondary products is considered urgent to elucidate the exact amount of these metabolites in in vitro cultures compared with the intact plant, in order to produce them on a large scale, industrially. Therefore, optimizing the culture conditions via biotechnological techniques is needed in the future to support the industrial production of the most valuable secondary products of *N. sativa* 'Black Diamond' through calli and/or embryo cultures.

Author Contributions: Conceptualization, M.E.E.-M.; methodology, A.E.H. and M.E.E.-M.; software, A.E.H.; validation, M.E.E.-M., A.N.E.-B. and N.A.; formal analysis, M.E.E.-M. and A.N.E.-B.; investigation, A.E.H.; resources, M.E.E.-M.; data curation, N.A.; writing—original draft preparation, N.A. and J.D.; writing—review and editing, N.A. and J.D.; visualization, M.E.E.-M. and J.D.; supervision, M.E.E.-M., A.N.E.-B. and M.K.M.; project administration, H.E.-R.; funding acquisition, H.E.-R. All authors have read and agreed to the published version of the manuscript.

Funding: Project no. TKP2021-EGA-20 has been implemented with the support provided by the Ministry of Culture and Innovation of Hungary from the National Research, Development and Innovation Fund, financed under the TKP2021-EGA funding scheme.

Data Availability Statement: Not applicable.

Acknowledgments: The authors thank the staff members of the Physiology and Breeding of Horticultural Crops Laboratory, Department of Horticulture, Faculty of Agriculture, Kafrelsheikh University, Kafr El-Sheikh, Egypt, for supporting the completion of this work.

Conflicts of Interest: The authors declare no conflict of interest.

References

1. Telci, İ.; Özek, T.; Demirtaş, İ.; Özek, G.; Yur, S.; Ersoy, S.; Yasak, S.; Gül, F.; Karakurt, Y. Studies on black cumin genotypes of Turkiye: Agronomy, seed and thymoquinone yields. *J. Appl. Res. Med. Aromat. Plants* **2023**, *35*, 100494. [CrossRef]
2. Golkar, P.; Nourbakhsh, V. Analysis of genetic diversity and population structure in *Nigella sativa* L. using agronomic traits and molecular markers (SRAP and SCoT). *Ind. Crops Prod.* **2019**, *130*, 170–178. [CrossRef]
3. Dalli, M.; Azizi, S.; Kandsi, F.; Gseyra, N. Evaluation of the in vitro antioxidant activity of different extracts of *Nigella sativa* L. seeds, and the quantification of their bioactive compounds. *Mater. Today Proc.* **2021**, *45*, 7259–7263. [CrossRef]
4. Golkar, P.; Bakhshi, G.; Vahabi, M.R. Phytochemical, biochemical, and growth changes in response to salinity in callus cultures of *Nigella sativa* L. *Vitr. Cell. Dev. Biol. Plant* **2020**, *56*, 247–258. [CrossRef]
5. Hannan, A.; Rahman, A.; Sohag, A.A.M.; Uddin, J.; Dash, R.; Sikder, M.H.; Rahman, S.; Timalsina, B.; Munni, Y.A.; Sarker, P.P.; et al. Black Cumin (*Nigella sativa* L.): A Comprehensive Review on Phytochemistry, Health Benefits, Molecular Pharmacology, and Safety. *Nutrients* **2021**, *13*, 1784. [CrossRef] [PubMed]
6. Nazari, A.; Zarringhalami, S.; Asghari, B. Influence of germinated black cumin (*Nigella sativa* L.) seeds extract on the physicochemical, antioxidant, antidiabetic, and sensory properties of yogurt. *Food Biosci.* **2023**, *53*, 102437. [CrossRef]
7. Kemal, M.; Esertaş, Ü.Z.Ü.; Kanbur, E.D.; Kara, Y.; Özçelik, A.E.; Can, Z.; Kolaylı, S. Characterization of the black cumin (*Nigella sativa* L.) honey from Türkiye. *Food Biosci.* **2023**, *53*, 102760. [CrossRef]
8. Burdock, G.A. Assessment of black cumin (*Nigella sativa* L.) as a food ingredient and putative therapeutic agent. *Regul. Toxicol. Pharmacol.* **2022**, *128*, 105088. [CrossRef] [PubMed]
9. Fathi, M.; Hosayni, M.; Alizadeh, S.; Zandi, R.; Rahmati, S.; Rezaee, V. Effects of black cumin (*Nigella sativa*) seed meal on growth performance, blood and biochemical indices, meat quality and cecal microbial load in broiler chickens. *Livest. Sci.* **2023**, *274*, 105272. [CrossRef]
10. Kazmi, A.; Khan, M.A.; Ali, H. Biotechnological approaches for production of bioactive secondary metabolites in *Nigella sativa*: An up-to-date review. *Int. J. Sec. Metab.* **2019**, *6*, 172–195. [CrossRef]
11. Zaky, A.A.; Shim, J.H.; Abd El-Aty, A.M. A Review on Extraction, Characterization, and Applications of Bioactive Peptides from Pressed Black Cumin Seed Cake. *Front. Nutr.* **2021**, *8*, 743909. [CrossRef]

12. Sakdasri, W.; Sila-ngam, P.; Chummengyen, S.; Sukruay, A.; Ngamprasertsith, S.; Supang, W.; Sawangkeaw, R. Optimization of yield and thymoquinone content of screw press-extracted black cumin seed oil using response surface methodology. *Ind. Crops Prod.* **2023**, *191*, 115901. [CrossRef]
13. Alkhatib, H.; Mawazi, S.M.; Al-Mahmood, S.M.A.; Zaiter, A.; Doolaanea, A.A. Thymoquinone content in marketed black seed oil in Malaysia. *J. Pharm. Bioallied Sci.* **2020**, *12*, 284–288. [CrossRef] [PubMed]
14. El-Sayed, S.A.E.-S.; Rizk, M.A.; Yokoyama, N.; Igarashi, I. Evaluation of the in vitro and in vivo inhibitory effect of thymoquinone on piroplasm parasites. *Parasites Vectors* **2019**, *12*, 37. [CrossRef]
15. Khurshid, Y.; Syed, B.; Simjee, S.U.; Beg, O.; Ahmed, A. Antiproliferative and apoptotic effects of proteins from black seeds (*Nigella sativa*) on human breast MCF-7 cancer cell line. *BMC Complement. Med. Ther.* **2020**, *20*, 5. [CrossRef]
16. Abd-Rabou, A.A.; Edris, A.E. Cytotoxic, apoptotic, and genetic evaluations of *Nigella sativa* essential oil nanoemulsion against human hepatocellular carcinoma cell lines. *Cancer Nanotechnol.* **2021**, *12*, 28. [CrossRef]
17. Mohaddab, M.; El Goumi, Y.; Gallo, M.; Montesano, D.; Zengin, G.; Bouyahya, A.; Fakiri, M. Biotechnology and In Vitro Culture as an Alternative System for Secondary Metabolite Production. *Molecules* **2022**, *27*, 8093. [CrossRef] [PubMed]
18. Bibi, A.; Khan, M.A.; Adil, M.; Mashwani, Z.U.R. Production of callus biomass and antioxidant secondary metabolites in black cumin. *J. Anim. Plant Sci.* **2018**, *28*, 1321–1328.
19. Al-Ani, N.K. Thymol Production from Callus Culture of *Nigella sativa* L. *Plant Tissue Cult. Biotech.* **2009**, *18*, 181–185. [CrossRef]
20. Alemi, M.; Sabouni, F.; Sanjarian, F.; Haghbeen, K.; Ansari, S. Anti-inflammatory effect of seeds and callus of *Nigella sativa* L. extracts on mix glial cells with regard to their thymoquinone content. *AAPS PharmSciTech.* **2013**, *14*, 160–167. [CrossRef]
21. Chaudhry, H.; Fatima, N.; Ahmad, I.Z. Establishment of callus and cell suspension cultures of *Nigella sativa* L. for thymol production. *Int. J. Pharm. Pharm. Sci.* **2014**, *6*, 788–794.
22. Landa, P.; Maršík, P.; Vaněk, T.; Rada, V.; Kokoška, L. In vitro antimicrobial activity of extracts from the callus cultures of some *Nigella* species. *Biologia* **2006**, *61*, 285–288. [CrossRef]
23. Hoseinpanahi, S.; Majdi, M.; Mirzaghaderi, G. Effects of growth regulators on in vitro callogenesis and regeneration of black cumin (*Nigella sativa* L.). *Iran. J. Rangel. Forest Plant Breed.* **2016**, *24*, 232–242.
24. Elhag, H.; El-Olemy, M.M.; Al-Said, M.S. Enhancement of Somatic Embryogenesis and Production of Developmentally Arrested Embryos in *Nigella sativa* L. *Hort. Sci.* **2004**, *39*, 321–323. [CrossRef]
25. Kim, D.H.; Kang, K.W.; Sivanesan, I. Influence of auxins on somatic embryogenesis in *Haworthia retusa* Duval. *Biologia* **2019**, *74*, 25–33. [CrossRef]
26. Vondráková, Z.; Krajňáková, J.; Fischerová, L.; Vágner, M.; Eliášová, K. Physiology and role of plant growth regulators in somatic embryogenesis. In *Vegetative Propagation of Forest Trees*; Park, Y.S., Bonga, J.M., Moon, H.K., Eds.; National Institute of Forest Science: Seoul, Republic of Korea, 2016; pp. 123–169.
27. Simon, S.; Petrášek, J. Why plants need more than one type of auxin. *Plant Sci.* **2011**, *180*, 454–460. [CrossRef]
28. Hazubska-Przybył, T.; Ratajczak, E.; Obarska, A.; Pers-Kamczyc, E. Different Roles of Auxins in Somatic Embryogenesis Efficiency in Two Picea Species. *Int. J. Mol. Sci.* **2020**, *21*, 3394. [CrossRef] [PubMed]
29. de Morais Oliveira, J.P.; da Silva, A.J.; Catrinck, M.N.; Clarindo, W.R. Embryonic abnormalities and genotoxicity induced by 2,4-dichlorophenoxyacetic acid during indirect somatic embryogenesis in Coffea. *Sci. Rep.* **2023**, *13*, 9689. [CrossRef]
30. Bhusare, B.P.; John, C.K.; Bhatt, V.P.; Nikam, T.D. Induction of somatic embryogenesis in leaf and root explants of *Digitalis lanata* Ehrh.: Direct and indirect method. *S. Afr. J. Bot.* **2020**, *130*, 356–365. [CrossRef]
31. Murashige, T.; Skoog, F. A revised medium for rapid growth and bio assays with tobacco tissue cultures. *Physiol. Plant.* **1962**, *15*, 473–497. [CrossRef]
32. El-Mahrouk, M.E.; Maamoun, M.K.; Dewir, Y.H.; Omran, S.A.; EL-Banna, A.N. Morphological and molecular characterization of induced mutants in *Nigella sativa* L. using irradiation and chemical mutagens. In *The 9th Plant Breeding International Conference*; Faculty of Agriculture, Benha University: Benha, Egypt, 2015.
33. El-Mahrouk, M.E.; Maamoun, M.K.; Dewir, Y.H.; El-Banna, A.N.; Rihan, H.Z.; Salamh, A.; Al-Aizari, A.A.; Fuller, M.P. Synchronized Seed Germination and Seedling Growth of Black Cumin. *HortTechnology* **2022**, *32*, 182–190. [CrossRef]
34. Boissot, N.; Valdez, M.; Guiderdoni, E. Plant regeneration from leaf and seed-derived calli and suspension cultures of the African perennial wild rice, *Oryza longistaminata*. *Plant Cell Rep.* **1990**, *9*, 447–450. [CrossRef] [PubMed]
35. Folch, J.; Lees, M.; Sloane Stanley, G.H. A simple method for the isolation and purification of total lipides from animal tissues. *J. Biol. Chem.* **1957**, *226*, 497–509. [CrossRef]
36. Mahmmoud, Y.A.; Christensen, S.B. Oleic and linoleic acids are active principles in *Nigella sativa* and stabilize an E(2)P conformation of the Na,K-ATPase. Fatty acids differentially regulate cardiac glycoside interaction with the pump. *Biochim. Biophys. Acta* **2011**, *1808*, 2413–2420. [CrossRef]
37. Hassanien, M.F.; Assiri, A.M.; Alzohairy, A.M.; Oraby, H.F. Health-promoting value and food applications of black cumin essential oil: An overview. *J. Food Sci. Technol.* **2015**, *52*, 6136–6142. [CrossRef]
38. Banerjee, S.; Gupta, S. Morphogenesis in tissue cultures of different organs of *Nigella sativa*. *Physiol. Plant* **1975**, *33*, 185–187. [CrossRef]
39. Chand, S.; Roy, S.C. Study of callus tissues from different parts of *Nigella sativa* (Ranunculaceae). *Experientia* **1980**, *36*, 305–306. [CrossRef]

40. Klimek-Chodacka, M.; Kadluczka, D.; Lukasiewicz, A.; Malec-Pala, A.; Baranski, R.; Grzebelus, E. Effective callus induction and plant regeneration in callus and protoplast cultures of *Nigella damascena* L. *Plant Cell Tissue Organ. Cult.* **2020**, *143*, 693–707. [CrossRef]
41. Khan, M.A.; Abbasi, B.H.; Ali, H.; Ali, M.; Adil, M.; Hussain, I. Temporal variations in metabolite profiles at different growth phases during somatic embryogenesis of *Silybum marianum* L. *Plant Cell Tissue Organ. Cult.* **2015**, *120*, 127–139. [CrossRef]
42. Khan, M.S.; Tabrez, S.; Rabbani, N.; Oves, M.; Shah, A.; Alsenaidy, M.A.; Al-Senaidy, A.M. Physico-chemical stress induced amyloid formation in insulin: Amyloid characterization, cytotoxicity analysis against human neuroblastoma cell lines and its prevention using black seeds (*Nigella sativa* L.). *Chin. J. Integr. Med.* **2015**. [CrossRef]
43. Abouzid, S.F.; El-Bassuon, A.A.; Nasib, A.; Khan, S.; Qureshi, J.; Choudhary, M.I. Withaferin A production by root cultures of *Withania coagulans*. *Int. J. Appl. Res. Nat. Prod.* **2010**, *3*, 23–27.
44. Majid, A. The Chemical Constituents and Pharmacological Effects of *Nigella sativa*—A Review. *J. Biosci. Appl. Res.* **2018**, *4*, 389–400. [CrossRef]
45. Forouzanfar, F.; Bazzaz, B.S.; Hosseinzadeh, H. Black cumin (*Nigella sativa* L.) and its constituent (thymoquinone): A review on antimicrobial effects. *Iran. J. Basic. Med. Sci.* **2014**, *17*, 929–938. [PubMed]
46. Radusheva, P.; Pashev, A.; Uzunova, G.; Nikolova, K.; Gentscheva, G.; Perifanova, M.; Maria Marudova, M. Comparative physicochemical analysis of oils derived from *Nigella sativa* L. and *Coriandrum sativum* L. *J. Chem. Technol. Metall.* **2021**, *56*, 1175–1180.
47. Albakry, Z.; Karrar, E.; Ahmed, I.A.M.; Oz, E.; Proestos, C.; El Sheikha, A.F.; Oz, F.; Wu, G.; Wang, X. Nutritional Composition and Volatile Compounds of Black Cumin (*Nigella sativa* L.) Seed, Fatty Acid Composition and Tocopherols, Polyphenols, and Antioxidant Activity of Its Essential Oil. *Horticulturae* **2022**, *8*, 575. [CrossRef]
48. Hernandez, E.M. Specialty Oils: Functional and Nutraceutical Properties. In *Woodhead Publishing Series in Food Science, Technology and Nutrition, Functional Dietary Lipids*; Sanders, T.A.B., Ed.; Woodhead Publishing: Sawston, UK, 2016; pp. 69–101. [CrossRef]

Disclaimer/Publisher's Note: The statements, opinions and data contained in all publications are solely those of the individual author(s) and contributor(s) and not of MDPI and/or the editor(s). MDPI and/or the editor(s) disclaim responsibility for any injury to people or property resulting from any ideas, methods, instructions or products referred to in the content.

Review

Oat (*Avena sativa* L.) In Vitro Cultures: Prospects and Challenges for Breeding

Marzena Warchoł [1], Edyta Skrzypek [1,*], Katarzyna Juzoń-Sikora [1] and Dragana Jakovljević [2]

[1] Department of Biotechnology, The Franciszek Górski Institute of Plant Physiology, Polish Academy of Sciences, Niezapominajek 21, 30-239 Kraków, Poland; m.warchol@ifr-pan.edu.pl (M.W.); k.juzon@ifr-pan.edu.pl (K.J.-S.)

[2] Department of Biology and Ecology, Faculty of Science, University of Kragujevac, Radoja Domanovića 12, 34000 Kragujevac, Serbia; dragana.jakovljevic@pmf.kg.ac.rs

* Correspondence: e.skrzypek@ifr-pan.edu.pl

Abstract: Plant in vitro cultures have been a crucial component of efforts to enhance crops and advance plant biotechnology. Traditional plant breeding is a time-consuming process that, depending on the crop, might take up to 25 years before an improved cultivar is available to farmers. This is a problematic technique since both beneficial qualities (such as pest resistance) and negative ones (such as decreased yield) can be passed down from generation to generation. In vitro cultures provide various advantages over traditional methods, including the capacity to add desirable characteristics and speed up the development of new cultivars. When it comes to oat (*Avena sativa* L.), the efficient method of plant regeneration is still missing compared to the most common cereals, possibly because this cereal is known to be recalcitrant to in vitro culture. In this review, an effort has been made to provide a succinct overview of the various in vitro techniques utilized or potentially involved in the breeding of oat. The present work aims to summarize the crucial methods of *A. sativa* L. cultivation under tissue culture conditions with a focus on the progress that has been made in biotechnological techniques that are used in the breeding of this species.

Keywords: androgenesis; callus; doubled haploids; embryogenesis; organogenesis; wide crossing

Citation: Warchoł, M.; Skrzypek, E.; Juzoń-Sikora, K.; Jakovljević, D. Oat (*Avena sativa* L.) In Vitro Cultures: Prospects and Challenges for Breeding. *Agronomy* **2023**, *13*, 2604. https://doi.org/10.3390/agronomy13102604

Academic Editor: Dan Mullan

Received: 31 August 2023
Revised: 25 September 2023
Accepted: 11 October 2023
Published: 12 October 2023

Copyright: © 2023 by the authors. Licensee MDPI, Basel, Switzerland. This article is an open access article distributed under the terms and conditions of the Creative Commons Attribution (CC BY) license (https://creativecommons.org/licenses/by/4.0/).

1. Introduction

In crop breeding practice, the development of new cereal cultivars takes from several to several dozen years and is mainly based on generating plants with a high degree of homozygosity through inbreeding crosses, followed by the selection of individuals with desirable traits. The use of biotechnological methods allows us to shorten this procedure by up to several years and involves the obtaining of haploid plants, followed by the generation of doubled haploid (DH) lines through in vitro culture methods. The obtained homozygous lines guarantee that subsequent generations will be genetically and phenotypically identical. Therefore, they are increasingly utilized in breeding programs. It is also worth noting that DH lines have a larger percentage of plants with breeders' target genes than the F_2 generation and subsequent generations acquired using conventional techniques [1]. Additionally, DH lines find applications in studies involving molecular markers by accelerating the derivation of mapping populations and genetic transformations, estimating recombination fractions, and detecting recessive mutants. They are also an effective means of genetic enrichment in plants, introducing more favorable alleles into the genome, i.e., those that carry traits such as resistance to biotic and abiotic stresses [2].

More than 20 species belonging to the genus *Avena* L. exist at the diploid, tetraploid, and hexaploid ploidy levels. *Avena sativa* L., the most widely cultivated plant, is a hexaploid (2n = 6x = 42), having three genomes: AA, CC, and DD. Common oat is grown worldwide in agricultural regions with a temperate climate, and its grains are primarily used for feed and food production [3]. Despite having lesser economic and commercial significance compared to other grains, the scientifically proven health benefits of oat grains make it an interesting subject of both breeding and genetic research. However, using standard research methods, the large size and complexity of the oat genome are significant limitations, which has led to limited advancements in oat research [4].

Haploids are plants in the sporophytic development stage, but with the gametic number of chromosomes (n). They are produced without the involvement of fertilization by male gametophyte cells from in vitro cultures of anthers or isolated microspores (androgenesis) or via female gametophytic cells from in vitro cultures of ovaries or ovules (gynogenesis). Both methods involve the action of various factors leading to the reprogramming of the developmental pathway of haploid male and female cells, resulting in the formation of haploid androgenic or gynogenic embryos rather than gametes. Methods based on gynogenesis also include wide crosses, or wide hybridization, involving the forced pollination of plants belonging to different species or genera. The ground-breaking work on cereal haploidization was performed by Kasha and Kao [5], who pollinated *Hordeum vulgare* L. with the pollen of the wild species *Hordeum bulbosum* L. This technique, known as the "bulbosum method", proved to be highly efficient and found application in generating haploids not only of barley or common wheat but also of other plant species. Currently, the most common pollen donor in wide crosses of cereals is maize (*Zea mays* L.), followed by pearl millet (*Pennisetum glaucum* (L.) R. Br.), sorghum (*Sorghum bicolor* (L.) Moench), Job's tears (*Coix lacryma-jobi* L.), and cogon grass (*Imperata cylindrica*) [6]. These crosses result in hybrid embryos of wheat, oats, triticale, or barley, in which paternal chromosomes are eliminated during successive nuclear divisions, resulting in haploid embryos containing only maternal genetic material. Sometimes, however, elimination does not occur properly, and whole or fragments of donor chromosomes are incorporated into the recipient's genome. This most commonly occurs in crosses between plants belonging to the two subfamilies, *Pooideae* and *Panicoideae*, within the family Poaceae. Additional chromosomes from maize or pearl millet have been observed in both wheat haploids [7,8] and oat haploids [9]. Investigating the causes of this phenomenon, Mochida et al. [10] found incomplete attachment of the spindle apparatus to maize centromeres, while Ishii et al. [11] reported chromosome breaks in pearl millet. The presence of stable maize chromosomes in the oat genome was first described by Riera-Lizarazu et al. [8], and they were referred to as oat × maize addition (OMA) lines by Kynast et al. [12]. Since then, these OMAs have been used for physical mapping of the maize genome [13], studies on CENH3 centromere-specific histones [14], or gene expression analysis in the C_4 photosynthetic pathway [15].

The use of oat in vitro techniques still faces difficulties, even though many of the methodological issues have been resolved. Moreover, there is a lack of knowledge regarding the mechanism of oat haploid induction. This review paper focuses on oat in vitro organogenesis, embryogenesis, and haploidization via anther and microspores cultures, via wide crossing (chromosome elimination), and via the modification of centromere specific histone CENH3. The paper points out the recent advances in oat in vitro cultures which might be successfully incorporated in this crop breeding.

2. Callus Culture, Organogenesis, Somatic Embryogenesis, and Cell Suspension of Oat (*Avena sativa* L.)

The development of in vitro regeneration techniques is crucial for improving cereals biotechnologically. Like other Poaceae species, hexaploid oat (*Avena sativa* L.) can be regenerated from tissue culture via either organogenesis or somatic embryogenesis. Limited reports have been published on the development of effective plant regeneration systems from various tissues and organs in oat compared to the most common cereals, e.g., maize, rice, wheat, and barley.

2.1. Effect of Explant on Callus Production

Oat callus cultures might be induced from seeds, immature embryos, germination-stage seedling roots, and germination-stage embryo hypocotyls, as first described by Carter et al. [16]. Nine years later, Lörz et al. [17] reported successful plant regeneration from non-friable calluses that had structured, green primordia. In the same year, Cummings et al. [18] used germinating immature embryos from 25 oat genotypes as explants for initiation of callus cultures on B5 [19] or MS [20] media containing from 0.5 to 3.0 mg L^{-1} 2,4-Dichlorophenoxyacetic acid (2,4-D). These cultures were maintained by subculturing to B5 medium with 1 mg L^{-1} 2,4-D every 4 to 6 weeks. The callus from the cultivar "Lodi" has been maintained for the longest time and retained the regeneration ability through 18 months. Maddock [21] then notes that morphogenetic oat callus does not appear to develop from the scutellum, which becomes necrotic, especially when older embryos are cultivated, but rather from the entire embryo or the mesocotyl area.

The very efficient regeneration method from leaf tissue of six different oat cultivars named "Coolabah", "Cooba", "Blackbutt", "Mortlock", "Victorgrain", and "HVR" was reported by Chen et al. [22]. Callus was produced using leaf base segments from seedlings on MS medium with 2 mg L^{-1} 2,4-D; moreover, explants grown in the light or the dark responded to callus induction similarly. The two-to-five-day old seedlings and the callus from the first leaf segment demonstrated a comparatively high potential for regeneration. Therefore, seedling age must be considered as a key variable for in vitro regeneration from leaf explants in oat. Calluses had been proliferating for more than eight months without substantial reductions in regeneration capacity.

Shoot apical meristems of A. sativa were also used to establish an effective micropropagation method [23]. Explants were obtained from aseptically germinated oat seedlings and cultivated in vitro. After five weeks in culture on MS medium with various combinations of 2,4-D and N^6-benzyladenine (BA), the enlarged apical meristems and multiple adventitious shoots were produced. All tested oat cultivars formed seedlings at a high efficiency and fertile oat plants were produced. These in vitro multiplied shoots might serve as an alternate tissue of selection for oat genetic transformation. Cummings et al. [18] also obtained calluses from apical meristems capable of plant regeneration on B5 medium containing 2,4-D.

Nuutila et al. [24] enhanced the regeneration efficiency from oat leaf base cultures by altering the nitrogen composition and the concentrations of the sugar and auxins in the culture medium. The effectiveness of inducing embryogenic callus and plant growth was studied on MS and L3 [25] media. The callus-producing abilities of leaf base segments 1 through 6 were compared. Concerning all cultivars, the first three leaf base segments produced embryogenic callus, but segments 4–6 either produced very little or none. For both tested cultivars, "Aslak" and "Veli", the L3 medium turned out to be more efficient and produced more embryogenic calluses and plants, compared to the MS medium. Lower concentrations of ammonium (4.9 mM) and nitrate (29 mM) and high organic nitrogen (11 mM) in medium caused the highest regeneration of green plants in "Aslak". On the other hand, high ammonium (20.1 mM), high nitrate (46.8 mM), and low organic nitrogen (0.9 mM) concentrations resulted in the greatest green plant regeneration in "Veli". This study has additionally demonstrated that both sugar and auxins have a definite impact on the induction of embryogenesis. For oat leaf base in vitro culture, sucrose and maltose have been investigated as carbohydrate sources [22,26].

Establishing tissue cultures in oat has frequently run into complications because of the strong dependency of the donor plant cultivar. The variations in cultivars' susceptibilities to genetic programming and the external reprogramming of embryogenically capable cells might be the cause of the discrepancies. Although it may be reasonable, from the practical point of view, to choose a medium in routine systems that elicits an average response from the majority of cultivars, these responses do not always reflect how well or badly the cultivars generally respond in generally. This genotypic dependency was described by Cummings et al. [18]. In their investigation, from 25 tested genotypes, 2 failed to initiate calluses, and 9 of them were able to produce callus but had no regeneration ability. Next, Rines and McCoy [27] noted that the frequency of callus development varied between oat cultivars and ranged from 5% to 75%. Studies of Chen et al. [22] showed that it is possible to obtain the callus form of all tested genotypes in 100% or almost 100% frequency. However, the rate of regenerable callus formation obtained from leaf base segments was substantially higher in mature embryos [18,27]. Because they are less genotype dependent, leaf explants appear to be more appropriate donor materials to produce regenerable oat callus cultures.

2.2. Factors Affecting Organogenesis and Somatic Embryogenesis

In comparison to callus cultures only capable of organogenesis, those capable of somatic embryogenesis are more likely to show fast growth rates and very high levels of plant regeneration. Due to these, the production of friable and embryogenic callus has received most of the attention in the quest to create tissue cultures of monocotyledonous plants. As with all cereals, immature tissues must be used to initiate oat in vitro cultures since these differentiated tissues are typically unable to induce cell division and proliferation. Two different callus types, some of which can regenerate plants, are regularly generated during indirect somatic embryogenesis. Typically, the embryogenic callus is friable, compact, and yellowish-white, and this non-embryogenic is rough-looking, moist, non-friable, and transparent [28].

The formation of an embryogenic oat callus is reliant on the source of the explant, its physiological stage, genotype, and the composition of the culture initiation medium. Most of the work on this topic has utilized immature zygotic embryos as explants to initiate embryogenic oat cultures [21,27,29–32], cultured on MS medium and supplemented mostly with 2,4-D in different concentrations. Other explants exhibiting callus formation include mature seeds [16,17] and mesocotyl of germinated seedlings [30]. Embryogenic calluses from immature and mature embryos [33,34] and leaf segments [26] have also been used for gene transfer.

Avena sativa L., *Avena sterilis* L., and *Avena fatua* L. are three hexaploid oat species from which tissue cultures were started and plants were regenerated [27]. Immature embryos were used to start a variety of tissue cultures, with regenerable-type cultures distinguished by the presence of organized chlorophyllous primordia in a compact, yellowish-white, strongly lobed callus. The frequency of regenerable-type cultures was determined by the embryo size, species, genotype, growing conditions of the donor plants, and 2,4-D concentrations in the culture induction media. The highest rates of regenerable-type of cultures were consistently produced by the "Lodi" cultivar and 2 "Lodi"-related lines out of the 23 investigated *A. sativa* cultivars. For "Lodi", this frequency reached 80% in one test. Only 3 of the 16 *A. sterilis* lines produced regenerable-type cultures, but more than 20%. In 7 out of the 32 investigated *A. fatua* lines, regenerable-type cultures were generated at rates higher than 45%. The tissue cultures of all three species could regenerate plants after 9–10 subcultures and more than a year in in vitro culture.

Since genotype affects culture initiation frequency and culture type, genotype screening and selection, as well as the developmental stage of embryos, ought to be a successful strategy for enhancing oats' capacity for cell culture. The aim of King et al.'s [35] investigation was to determine the optimum size of immature embryos of 10 oat cultivars for callus induction and plant regeneration under in vitro culture. Plant regeneration was assessed after three months of culturing one hundred immature embryos on MS medium with different 2,4-D concentrations. No differences between cultivars were observed in the amount of callus produced, but the rate of regeneration from the different cultivars extended from 3 to 42%, suggesting that there are genotypic differences in the ability to regenerate plants from calluses. Scanning electron microscopy and light microscopy were used by Bregitzer et al. [30,31] to characterize the stages of development of somatic embryos in friable embryogenic callus. Following Bregitzer et al. [30], the cultivation of non-friable oat calluses produced from immature embryos on MS medium containing 20 mg L^{-1} sucrose and no hormones led to the production of separate somatic embryos that hatched into full-grown plants. During repeated culturing on a modified MS medium containing 2 mg L^{-1} 2,4-D, and 20 g L^{-1} sucrose, embryogenic sectors separated from non-friable calluses were visually selected to create friable callus. The maturation of somatic embryos was stimulated by transferring friable calluses to a modified MS medium containing 60 g L^{-1} sucrose and no hormones. Some of these embryos were able to germinate after being transferred from this friable callus to a modified MS medium that included 20 g L^{-1} sucrose and no hormones. According to Bregitzer et al. [30], the culture of non-friable oat calluses derived from immature embryos on MS medium containing 20 mg L^{-1} sucrose and no hormones resulted in the development of distinct somatic embryos that germinated to form complete plants. Embryogenic sectors isolated from non-friable calluses were visually selected during repeated subcultures on a modified MS medium containing 2 mg L^{-1}, 2,4-D and 20 g L^{-1} sucrose to produce friable callus. After the development of friable calluses, plants continued to grow from those callus lines for more than 78 weeks. Additionally, immature embryos of three genotypes and seedling mesocotyls of two genotypes were used to directly generate friable embryogenic callus. There was also evidence of genotypic heterogeneity in this reaction. For the first time, calluses produced from oat seed developed root primordia, and the meristems of these primordia were sites of somatic embryo production, according to Chen et al.'s study [36]. The callus that was kept on MS medium with 1, 2, or 4 mg L^{-1} 2,4-D with underlying root or shoot parts was transferred to a new supply of the same medium after one month of culture. The callus induction frequency among the five cultivars studied was 93% for cv. "Risto", and 76% for cvs. 'Victory', 'Sang', 'Sanna' and 'Vital' respectively. Between 30 and 200 mg per seed were produced as fresh weight of each callus. The proliferating callus cells caused the seedling roots to swell during callus induction. Epidermal and cortical cells of the roots tended to be expelled and fall off. Differentiated pericycle cells became meristematic. When these roots were sectioned longitudinally or transversely, numerous lateral root primordia were seen originating from the pericycle cells along the vascular strand. Additionally, callus induction took place in the shoot's basal regions. Meristematic cells and solitary xylem cells developed from parenchymatic cells. The meristematic cells could directly generate root primordia and contained noticeable nuclei. After callusing in the vicinity of the shoot bases, groups of root primordia were also inadvertently generated. On various media, somatic embryos connected to underlying callus cells grew. The MS medium that proved best for germination contained 6% sucrose and 0.01 mg L^{-1} of abscisic acid (ABA). Single plantlets or clumps with 2–5 mm green leaves, with or without roots, were produced after 30–40 days on this medium. Ten-month-old embryogenic tissue may produce 200 shoots or plantlets per gram. After being transferred to media devoid of hormones, plantlets grew stronger and had more developed roots. In the soil, more than 90% of the green plantlets survived and matured. More than 30 months have passed since the preservation of embryogenic tissue.

In the next study, young seedlings of five oat cultivars—"Fuchs", "Jumbo", "Gramena", "Bonus", and "Alfred"—were tested for their regenerative abilities [26]. Two different basal media—MS medium and L3 medium [25]—were enriched with phytohormones in different concentrations for the callus induction, shoot proliferation, and regeneration of plants. Four-week-old culture-produced calluses were transferred to induction medium, and one week later, somatic embryos began to germinate. To develop further, shoots were placed in hormone-free media, and developed plants were morphologically healthy and fertile. From the base of the oat leaves, a callus was induced on all the tested media. However, certain phytohormones had better effects on plant regeneration. The highest regeneration frequencies were attained on media with 2.5 mg L^{-1} 2,4-D. In five oat genotypes, 25 plants on average could be grown per explant, and for the most responsive Jumbo, more than 50 regenerants could be produced per explant. Hence, the oat leaf bases are very promising as primary explants for micropropagation due to their strong capacity for regeneration.

To examine the effects of polyamines on somatic embryogenesis and plant regeneration oat genotypes Tibor (*Avena nuda* L. with low regeneration factor), GP-1 (*Avena sativa* L. with high regeneration factor) and their crosses, GP-1 × Tibor and Tibor × GP-1, were grown in in vitro cultures [37]. Somatic embryos were produced in large amounts from mature embryos of Tibor and Tibor × GP-1 on MS medium supplemented with 2.0 mg L^{-1}

2,4-D and 0.5 mM putrescin. Putrescin treatments induced plant regeneration from other genotypes in most cases, compared with the results of Somers et al. [33] obtained with the same regeneration media. This suggests that media enriched with putrescin can be used to screen other oat lines for regeneration efficiency. Moreover, the shoot proliferation medium containing low concentration of putrescin induced significant numbers of plants from usually recalcitrant cultivars.

Bregitzer et al. [30] showed that immature embryos gave the highest and most repeatable callus initiation frequency; however, it was shown that, generally, the frequency of embryogenic callus initiation of some of the elite germplasm lines is still quite low [31]. A significant contribution to the formation of embryogenic calluses in oat and the subsequent demonstration of plant regeneration via somatic embryogenesis, as well as variables controlling plant regeneration, were reported by Somers et al. [33]. This paper includes the methods used to manipulate oat cells and tissues in tissue culture, the constraints on their usage, and both planned and actual uses for improving oats. The genotype of the oats, the explant utilized to start the development process, and the tissue choice made during subculture can all affect the oat callus structure and its ability to regenerate. However, according to these authors, by seeing and picking out extremely transparent to opaque, compact, highly lobed tissues within early cultures, oat cultures with long-term preservation of plant regeneration potential can be produced. Borji et al. [38] used mature caryopses as initial material for somatic embryogenesis from oat cultivar "Meliane". Longitudinal sections of caryopses were plated on MS medium supplemented with 3 mg L^{-1} 2,4-D. Primary calluses were removed from explants after four weeks of growth and placed into the proliferation medium (MS medium with 1.0 mg L^{-1} 2,4-D and 0.5 mg L^{-1} 6-Benzylaminopurine (BAP)). For germination, somatic embryos were transferred to MS medium without plant growth regulators and then to MS medium containing 0.5 mg L^{-1} indole-3-acetic acid (IAA) to promote root system. The regenerated seedlings were acclimated to ex vitro conditions and were grown until maturity in a greenhouse.

2.3. Oat (Avena sativa L.) Cell Suspension Culture

The experiment by Gana et al. [39], among others, set out to evaluate the relative adaptability and plant regeneration of four oat genotypes in suspension cultures and to examine plant regeneration in 19 genotypes from three different oat species in three successive callus subcultures. Highly significant differences were found between 19 Avena genotypes for callus initiation, germination, and rhizogenesis in this study, in which the ability of "88Ab3073" to regenerate plants in suspension culture and the highly regenerable capacity of "GAF/Park" in both agar and suspension culture systems were also described. A highly significant genotype impact (32.1% variation), genotype subculture interactions (9.9% variance), and a non-significant subculture effect (0.3% variance) were all seen in the analysis of variance for plant numbers for genotypes in three subcultures. Genotypes with the highest callus production were selected to initiate liquid cultures. Two-month-old calluses from "GAF/Park", "Tibor", "88Ab3073", and "87Ab5932" genotypes were used for the suspension culture initiation. To better assess the regeneration potential of small and big cell clusters, the suspensions were divided into fractions of <3 mm or >3 mm. The "GAF/Park" and "88Ab3073" clusters that were 3 mm and bigger generated yellow friable callus. Clusters of "87Ab5932" developed slowly and finally ceased to multiply, whereas "Tibor" clusters began to form a rhizomorphic callus, a propensity that was also seen in its suspension cultures. Within two weeks, clusters of "GAF/Park" had quadrupled in size on the solidified medium. Plant regeneration from clusters larger than 3 mm was observed after three weeks in three of the four tested genotypes. Additionally, on the "GAF/Park" callus, many globular somatic embryos were observed. A total of 42 plants were regenerated from suspension clusters larger than 3 mm, and 50 plants were derived from the agar-based callus culture.

Wise et al. [40] used suspension cultures of *Avena sativa* L. cv. "Belle" for the biosynthesis of avenanthramides. Calluses of oat were initiated from the shoot apical meristem on solid MS media in dark conditions. Liquid cultures were established from 1.0 g callus and 25 mL of MS media containing 2 mg L^{-1} 2,4-D. To stimulate avenanthramides production, chitin (poly-N-acetyl glucosamine) was added as elicitor, and two unique callus phenotypes, named "aggregate" and "friable", were identified. The more brittle aggregate tissue easily shed off and was easily separated from the friable tissue, which remained evenly scattered in the culture medium. Because the suspension cultures produced relatively large quantities of avenanthramides, these results point to the potential of oat suspension culture as a tool for future in-depth research into the processes that initiate their production, as well as the variables that determine the specific kinds of avenanthramides that are produced.

The summary of research on *A. sativa* L. callus induction, organogenesis, and somatic embryogenesis with the improved biotechnological potential of named species is presented in Table 1.

Table 1. Callus culture, organogenesis, and somatic embryogenesis of oat (*Avena sativa* L.).

A. sativa L. Genotype	Explant Used	Media/PGRs	Experimental Outcomes	References
cv. "Lodi", cv. "Moore", cv. "Lyon", cv. "Benson", cv. "Marathon", cv. "Dal", cv. "Stout", cv. "Tippecanoe", cv. "Lang", cv. "Victorgrain", cv. "Garry", cv. "Hudson", cv. "Terra", cv. "0A338", cv. "Victory", cv. "Black", cv. "Mesdag", cv. "Victoria", cv. "Selma", cv. "AJ10915", cv. "NP3/4", cv. "Karin", cv. "Rallus", cv. "Coolabah"	immature embryos	MS/B5 medium with 2 mg L^{-1} 2,4-D for initiation; MS/B5 medium with 1.0, 2.0 and 5.0 mg L^{-1} 2,4-D for embryo regeneration	Tissue cultures capable of plant regeneration after more than 12 months in culture	[27]
cv. "Victorgrain", cv. "Victoria" GAF (*A. sativa* L. cv. "Garland" × *A. fatua* L.) × *A. sativa* L. cv. "Victoria"	10- to 12-days old embryos	MS medium with 2 mg L^{-1} 2,4-D; MS medium with 1 mg L^{-1} 2,4-D and 5 units mL^{-1} victorin; MS medium with 2 mg L^{-1} NAA, and 0.2 mg L^{-1} BAP for regeneration	12 of 65 immature embryos of the cv. "Victorgrain" and 2 of 21 embryos of cv. "Victoria" developed regenerable callus; without tissue growth or survival on a victorin-containing medium	[29]
cv. "Trafalgar", cv. "Rollo", cv. "07408 in 111/2", cv. "Rhiannon", cv. "Dula", cv. "Avalanche", cv. "Caron", cv. "Pennal", cv. "Cabanna", cv. "Margam"	embryos	MS medium with 2 mg L^{-1} 2,4-D for callus initiation and growth; MS medium with 0.5 mg L^{-1} 2,4-D followed by PGRs free MS medium for regeneration	The highest level of regeneration from 4–4.5 mm long embryos with the genotyping differences of plant regeneration	[35]
GAF (*A. sativa* L. cv. "Garland" × *A. fatua* L.) × *A. sativa* L. cv. "Victoria" lines GAF-18, GAF-30, GAF-30, GAF-30/"Park" and GAF-30/Park//GAF-30	immature embryos	MS medium with 4 mg L^{-1} 2,4-D for callus initiation; MS medium with 2 mg L^{-1} 2,4-D for callus maintenance; MS medium with 2 mg L^{-1} NAA and 0.2 mg L^{-1} BAP for shoot differentiation; MS medium free of PGRs for rooting	Embryogenic cultures maintained the ability to regenerate plants for more than 78 weeks	[30]
cv. "Risto", cv. "Sang", cv. "Sanna", cv. "Vital", cv. "Sol"	Embryos	MS medium with 2 mg L^{-1} 2,4-D for embryos; MS medium with 0.01 mg L^{-1} ABA and 6% sucrose for germination	Suppressed root elongation, promoted secondary root initiation and proliferation of embriogenic cells with 2,4-D in the medium	[36]

Table 1. Cont.

A. sativa L. Genotype	Explant Used	Media/PGRs	Experimental Outcomes	References
cv. "Coolabah", cv. "Cooba", cv. "Blackbutt", cv. "Mortlock", cv. "Victorgrain", cv. "HVR"	Immature embryos, leaf segments	MS medium with 2 mg L^{-1} 2,4-D for callus induction and growth; N6 medium (Chu et al. 1975) with 2 mg L^{-1} KIN, and 2 mg L^{-1} NAA for shoot regeneration; MS medium with 0.3 mg L^{-1} KIN for root regeneration	callus formation from the leaf segments and plant regeneration are comparable to that of the immature embryos; plants were grown to maturity	[22]
line GAF, line GAF/Park	Immature zygotic embryos	MS medium with 2 mg L^{-1} 2,4-D for embryos; MS medium with 6% sucrose for embryo maturation, and sucrose reduction for bipolar plant development	Friable embryogenic callus inoculated into liquid medium will produce rapidly growing dedifferentiated suspension cultures	[31]
cv. "Corbit", cv. "Dark Husk", cv. "Winter Turf", cv. "Monida", cv. "SO87213", cv. "Dal"	Embryos	MS medium with 2 mg L^{-1} 2,4-D for callus initiation/proliferation; CIP medium with 0.5 mg L^{-1} picloram, and 5 mg L^{-1} KIN for plant regeneration	High level of plant regeneration	[39]
cv. "Prairie", cv. "Porter", cv. "Pacen", cv. "Ogle"	Apical meristems, leaf primordia, leaf bases	MS medium with 2,4-D (0 and 0.5 mg L^{-1}) and BA (0, 1.0, 2.0, 4.0, and 8.0 mg L^{-1})	Multiple shoot differentiation from shoot apical meristems on medium with 0.5 mg L^{-1} 2,4-D, and 2.0 or 4.0 L^{-1} BA	[23]
cv. "Fuchs", cv. "Jumbo", cv. "Gramena", cv. "Bonus", cv. "Alfred"	Leaf bases of young seedlings	L3 medium for callus induction; 2.5 mg L^{-1} 2,4-D for plant regeneration	for cv. "Jumbo" average of 50 regenerants per explant could be regenerated, whereas for cv. "Gramena", only 3–4 plants per explant could be regenerated	[26]
cv. "GP-1"	Mature embryos	MS medium with 2 mg L^{-1} 2,4-D for callus induction and shoot proliferation; after 6 weeks, 0.5 or 1.0 mM of putrescine was applied	Significant regeneration of plants in presence of 0.5 mM putrescine	[37]
cv. "Aslak", cv. "Velik"	Leaf based segments from 3- to 4-days old seedlings	L3 or MS medium for callus induction; L3 or MS medium with 0.2 mg L^{-1} for regeneration	Optimization of nitrogen, sugar, and auxin in media	[24]
cv. "Belle"	Shoot apical meristem	MS medium with 2 mg L^{-1} 2,4-D for liquid cultures	suspension cultures produced large quantities of aventhramides A and aventramides G in response to 0.25 mg mL^{-1} chitin (poly-N-acetyl glucosamine) elicitation	[40]
cv. "Meliane"	Mature caryopses	MS medium with 3 mg L^{-1} 2,4-D for callus induction; MS medium with 1 mg L^{-1} 2,4-D and 0.5 mg L^{-1} BAP for embryogenic callus induction and somatic embryos differentiation; MS medium with 0.5 mg L^{-1} IAA for rooting	Ultrastructural changes and cytological modifications of oat somatic embryogenesis	[38]

3. Androgenesis of Oat (*Avena sativa* L.)

In recent years, the production of doubled-haploid (DH) lines using methods involving male gametic lines for developing haploid plants has proven efficient for species belonging to the families Solanaceae, Brassicaceae, and Graminae. Consequently, in vitro induced androgenesis has become the most promising biotechnological method applied in breeding practice [41]. However, not all species respond equally to the induction of this process. There are model species that respond with high efficiency to the application of this method, but other species are more resistant to it. The largest group consists of species in which the induction of microspore embryogenesis is possible but not very efficient from a practical standpoint. Although about 250 protocols related to androgenesis have been described so far, only in a few species, such as barley (*Hordeum vulgare* L.), oilseed rape (*Brassica napus* L.), tobacco (*Nicotiana* spp.), wheat (*Triticum aestivum* L.), pepper (*Capsicum annum* L.), or rice (*Oryza sativa* L.) has this method have been applied in breeding programs due to the high regenerative efficiency of the obtained plants [42].

The process of androgenesis is defined as an alternative developmental pathway of microspores, involving redirecting their natural gametophytic development, which leads to pollen grain formation towards a sporophytic pathway, along with their reprogramming and the initiation of embryo development [43]. By inducing zygotic embryo-like structures (ELS), followed by their regeneration, androgenic embryos with a haploid number of chromosomes (n) are obtained. The literature indicates that microspores in the late uninucleate or early binucleate stage, directly after division, are most susceptible to androgenesis induction, and the process of microspore differentiation occurs under the influence of abiotic stress in the period preceding culture initiation [44]. Among the most used stress-inducing factors are exposure to low or high temperatures, application of sugar- or nitrogen-free media, and treatments with colchicine, heavy metal ions, or mannitol [44]. In cereals, storing spikes at a low temperature can disrupt cytoplasm polarity and impair the direction of spindle formation, leading to a change in the developmental pathway of microspores towards embryo formation [45]. Additionally, subjecting spikes to cold treatment prolongs the viability of anthers, which promotes synchronization of nuclear divisions and inactivates substances that inhibit androgenesis. In practice, a combination of two or three of these factors is applied, and depending on the method, they are used on whole donor plants, cut shoots with spikes, isolated spikes from leaf sheaths, or anthers alone [46]. The main factors determining the androgenic response in in vitro cultures include the genotype of the donor plants, the physiological state and growth conditions of the plants, in vitro medium composition, and physical factors at play during tissue culture and their interactions [47].

In cereals, the process of androgenesis is a more commonly used method for obtaining homozygous plants, and in vitro production of androgenic embryos is more efficient than methods based on gynogenesis [2]. In vitro anther cultures enable the rapid and efficient production of haploid plants, primarily due to the abundance of male reproductive cells. Thousands of microspores present in each anther can potentially give rise to androgenic embryos and, subsequently, haploid plants [48]. However, the main challenge associated with the anther culture technique is the strong dependence of androgenesis not only on the species but also on the genotype of the donor plant [47]. In addition, albinism, i.e., the formation of plants with disrupted chlorophyll production, is a serious problem in anther and isolated microspore cultures. Such plants significantly lower the efficiency of the applied method, expressed in the number of regenerated, green plants with a doubled chromosome number [49].

3.1. Effect of Panicle Pretreatment and Media Composition on ELS Formation

Despite the progress that has been made in improving the effectiveness of methods based on microspore embryogenesis in cereals, the common oat is still considered a recalcitrant species in this process. The first oat regenerants using androgenesis were obtained by Rines [50], who acquired one haploid ($n = 3x = 21$) and one diploid ($2n = 6x = 42$) plants of the cultivar "Stout" from around 65,000 isolated anthers. Before usual incubation at 22 °C and immediately following plating, these anthers were also heat-shocked at 35 °C for 24 h. MS medium without hormone, supplemented with 10% sucrose, had the highest anther callus initiation frequencies among all media tested. However, only from anthers which have been plated on a modified potato extract medium containing 2.0 mg L^{-1} 2,4-D and 0.5 mg L^{-1} kinetin (KIN) were seedlings produced. Subsequently, Kiviharju and Pehu [51] reported the unsuccessful regeneration of androgenic embryos in *Avena sativa* L. and the production of haploid plants in *Avena sterilis* L. Five days of heat pretreatment (32 °C) radically increased the embryos induction of *A. sterilis* L. (27.5 embryos/100 anthers), compared to three-day (3.8 embryos/100 anthers) and one-day (0.6 embryos/100 anthers) treatments. Embryo production of *A. sterilis* L. was better on high maltose concentrations than that of *A. sativa* L. The highest number of embryos was obtained on the medium with 14% maltose under both temperature pretreatments. For 10 weeks, 230 embryo-like structures were transferred onto differentiation media. Consequently, two haploid green plants survived transfer to the greenhouse, but these plants did not produce seeds. An attempt to induce androgenesis in Polish oat cultivars was made by Ślusarkiewicz-Jarzina [51,52], who tested the androgenic response of 15 genotypes on solid, liquid, and two-layer media. Oat panicles were harvested and cold-treated at 4 °C for a few days in an N6 mineral salt medium [53] with 2.0 mg L^{-1} 2,4-D. Of the 45,000 anthers plated in this experiment, 637 ELS (1.4%; in all three physical states) were generated on W14 media. Genotype had a significant impact on the frequencies of ELS and green plants production. Eight genotypes yielded ELS (average 1.4/100 anthers). Successful induction of ELS on W14 [54] and C17 media [55] from F3 generation of nine hexaploid oat hybrids was described by Ponitka and Ślusarkiewicz-Jarzina [56]. When compared to medium W14, which generated 137 ELS (from 0.6 to 3.3/100 anthers), medium C17 produced 409 ELS (from 0.6 to 12.1/100 anthers), achieving a greater induction efficiency for all genotypes. Crossing of Bohun × Deresz gave the best ELS induction rates on both media.

In the same year, Skrzypek et al. [57] analyzed the possibility of inducing androgenic ELSs depending on the genotype, the length of the panicle cooling period, the density of anthers in a Petri dish, and the type and physical properties of the media. These studies have shown that pretreatment of oat panicles at a low temperature (4 °C) for 1–2 weeks stimulated induction of ELS the most on W14 and C17 media. Thus far, the highest efficiency of this method has been achieved by Kiviharju et al. [58], resulting in 30 green plants per 100 anthers from the crossing of Aslak × Lisbeth. In this study, the cut tillers were pretreated for 7 days at 4 °C, and the isolated anthers were followed by treatment for 5 days at 32 °C on a double-layer induction medium. "Lisbeth" naked-type oat was used to examine the effects of cytokinins, amino acids, reducing and ethylene-increasing agents and light and temperature conditions. For cv. "Aslak" (2.1/100 anthers) and "Lisbeth" (5.3/100 anthers), the induction medium comprising 2,4-D, BAP, ethephon, cysteine, and myo-inositol produced noticeably higher rates of green plant regeneration than the media containing simply 2,4-D and KIN. In comparison to other treatments, the conversion rate of ELS to green plants was also noticeably greater for the cv. "Aslak" (33%) and much better for the cultivar "Lisbeth" (13%), demonstrating that the 2,4-D and KIN applied together enhance the quality of ELS. Regeneration rates between these two induction media did not significantly differ when weak light was utilized for induction, most likely because of a reduced entire response.

3.2. The Developmental Stage of Microspores Affects ELS Formation

Microspores' competence for androgenesis varies not only among species or cultivars; it is primarily limited temporally and has been referred to as the "developmental window" by Pechan and Smykal [59]. During this short period, it is possible to redirect microspore differentiation from the gametophytic to the sporophytic pathway by applying appropriate physicochemical factors known as stress factors. In addition, by manipulating the composition of the induction media, especially the content of auxins or their analogs, it is possible to effectively induce callus formation and subsequently embryogenic structures from microspores [60]. The architecture and morphology of oat panicles contribute to the non-linear maturation of anthers, which significantly complicates the identification of microspores at the appropriate developmental stage and likely accounts for the low efficiency of androgenesis in this species. Research conducted by De Cesaro et al. [44] has confirmed that the developmental stage of microspores depends not only on the genotype and age of the plant, but primarily on the position of the anthers in the inflorescence, which results in their uneven maturation. It has also been observed that microspores within a single anther often differ in their embryogenic competence due to slight differences in their developmental stage. The aim of the experiments presented in the work of Warchoł et al. [61] was to determine which external stimuli should be used to arrest the gametophytic pathway of the microspores and direct their development towards embryo formation. The optimization of media composition for the initiation of embryo-like structures was also performed. In addition, the distance from the base of the flag leaf to the penultimate leaf of the panicle was measured to correlate the developmental stage of microspores with shoot morphology. In this way, four distances were determined, i.e., (i) 0.0–4.0 cm, (ii) 4.1–8.0 cm, (iii) 8.1–12.0 cm, and (iv) 12.1–16.0 cm, thereby selecting panicles based on the competence of their microspores for androgenesis. In the first stage of the experiment, the cultivars "Akt", "Bingo", "Bajka", and "Chwat" were tested for their susceptibility to androgenesis induction. In the latter experiment, a significant impact of oat cultivar and the distance from the base of the flag leaf to the penultimate leaf of the inflorescence on the formation of ELS was observed. ELS formation was observed in all cultivars, but the highest number of structures was recorded in the cultivars "Chwat" and "Bingo" (3.6% and 1.6%, respectively). In addition, the highest ELS production was observed on anthers isolated from the youngest panicles, i.e., when the measured distance did not exceed 4.0 cm. The second stage of the experiment aimed to increase the efficiency of androgenesis in the cultivars "Bingo" and "Chwat" by changing the length and type of thermal stress, as well as modifying the composition of the induction media. For the first time, the induction of oat ELS was carried out using a combination of low temperature (4 °C) followed by high temperature (32 °C). The anthers were plated on C17 [62] and W14 [54] media, which were supplemented with the following auxins: 2,4-D, picloram, dicamba and NAA; and cytokinins: KIN and BAP. More ELS were obtained from the anthers of the cultivar "Chwat" compared to the cultivar "Bingo". Differences in androgenesis response depending on the hormones in the induction medium were manifested in the number of obtained haploid plants and DH lines. Based on the results, it was shown that treating oat panicles for 14 days with a low temperature of 4 °C and a high temperature of 32 °C for 24 h before anther isolation increased the efficiency of androgenesis in the cultivar "Chwat". The most susceptible to this process were anthers isolated from panicles where the distance from the base of the flag leaf to the penultimate leaf did not exceed 4 cm. The best medium for induction of ELS and haploid plants was W14 with the addition of 2.0 mg L^{-1} 2,4-D and 0.5 mg L^{-1} KIN.

3.3. Impact of Cu^{2+}, Zn^{2+} or Ag^+ Ions on ELS Formation

The literature suggests that increasing the concentration of Cu^{2+}, Zn^{2+}, or Ag^+ ions in the induction medium not only stimulates haploid embryogenesis of microspores but also regulates numerous physiological and biochemical cellular processes. Cu^{2+} and Zn^{2+} ions stimulate the normal division of chloroplasts, while Ag^+ ions act as an inhibitor of ethylene biosynthesis in in vitro cultures, preventing the aging of microspores [63–65]. Warchoł et al. [66] studied the efficiency of induction of embryonic structures in oat anther cultures depending on the concentration of $CuSO_4 \times 5\ H_2O$ (10 and 20 µM), $ZnSO_4 \times 7\ H_2O$ (90 and 180 µM), and $AgNO_3$ (25 and 50 µM). Copper, zinc, and silver ions were added to the media at two stages of androgenesis: during pretreatment of panicles of donor plants and as an addition to the induction medium. Ions added to the medium during the pretreatment of panicles had a significant effect on the formation of embryonic structures. The highest number of ELS was obtained when oat panicles were treated with 50% Hoagland medium supplemented with $CuSO_4 \times 5 H_2O$ at a concentration of 10 or 20 µM (2.1% and 1.8%, respectively). The introduction of Cu^{2+}, Zn^{2+}, or Ag^+ ions into the W14 induction medium had no significant statistical effect on the number of ELS. When comparing the cultivars, it was observed that the highest number of ELS (0.7%) was obtained from the cultivar "Chwat", resulting in the production of haploid plants only in this cultivar. The present results demonstrated that the treatment of panicles with $CuSO_4 \times 5 H_2O$ at a concentration of 10 or 20 µM increased the efficiency of androgenesis in the tested cultivars. Table 2 summarizes recent progress in androgenesis in various cultivars of A. sativa L.

Table 2. Androgenesis of oat (*Avena sativa* L.).

A. sativa L. Genotype	Culture Conditions	Experimental Outcomes	Reference
cv. "Clintford", cv. "Stout"	4 or 8 °C cold pretreatment	The highest anthers callusing initiation on MS medium with 10% saccharose and no hormones	[50]
Line WW 18019, cv. "Stout"	4 °C in the dark cold pretreatment for anthers from the main culm; 4 °C in the dark for cold pre-treatment for tillers, and MS medium with no PGRs; 32 °C heat pre-treatment for anther cultures	The pretreatment of isolated anthers for 5 days at 32 °C, before culture at 25 °C, is the key point	[51]
44 genotypes	4 °C in the dark cold pretreatment for anthers from the main culm; MS medium with or without 5 mg L^{-1} 2,4-D for ELS induction;	Callus growth, ELS * production rates and plant regeneration differed between naked oat, wild oat, and crosses	[67]
Line WW 18019, cv. 'Kolbu'	4 °C in the dark cold pretreatment for anthers from the main culm; MS medium with 2,4-D and KIN for anthers; MS medium with 1 mg L^{-1} KIN for embryo structures; 32 °C heat pre-treatment for anther cultures	High 2,4-D concentrations enhanced embryo induction with or without heat pre-treatment	[68]
cv. "Lisbeth", cv. "Virma", cv. "Cascade", cv. "Kolbu", cv. "WW 18019", cv. "OT 257", cv. "Stout", cv. "Sisu", cv. "Katri", cv. "Yty", cv. "Sisko", cv. "Talgai", cv. "Roope", cv. "Salo"	tillers pretreated at 4 °C for 7 days; double-layer induction medium MS or W14 with 10% maltose and PGRs; 32 °C heat pre-treatment for anther cultures	Regenerable-type embryos from heat-pretreated anthers on media containing 2, 3 or 5 mg L^{-1} mg 2,4-D and 0.2 or 0.5 mg L^{-1} KIN	[69]
cv. "Lisbeth"	4 °C for 7 days for the tillers; 32 °C heat pretreatment; W14 medium with 10% maltose and PGRs for anthers, W14 medium with 2 mg L^{-1} NAA, and 0.5 mg L^{-1} KIN for ELS and regeneration; MS with 0.2 L^{-1} NAA for rooting	Improved number of derived plants via application of W14	[58]
Oat hybrids 1705/05, 1717/05, 1725/05, 1780/05, 2038/05, 1889/05, 1893/05, 1903/05, 1944/05, 1954/05, 956/05, 1967/05, 1985/05, 1989/05, 1997/05	4 °C for 6–9 days for the tillers in N6 medium with 2 mg L^{-1} 2,4-D; liquid, solid or double-layer W14 salts and vitamins, 5.0 mg L^{-1} 2,4-D, and 0.5 mg L^{-1} BAP for ELS induction;	Development of ELS after 6 weeks of culture on liquid medium, and between the 7th and 8th weeks on solid and double-layer medium	[52]

Table 2. Cont.

A. sativa L. Genotype	Culture Conditions	Experimental Outcomes	Reference
cv. "UPF 7", cv. "UPF 18", cv. "UFRGS 14", cv. "Stout"	Samples were collected when the distance between the flag leaf and the last node was one third of the distance between the last node and flag leaf	The use of anther size for the identification of microspore developmental stage is inefficient selection criterion	[44]
Cross combination of hexaploid oat: Lisbeth × Bendicoot, Flämingsprofi × Rajtar, Scorpion × Deresz, Aragon × Deresz, Deresz × POB7219/03, Bohun × Deresz, Krezus × Flämingsprofi, Krezus × POB10440/01, Cwał × Bohun	4 °C for 6–9 days for the tillers in N6 medium with 2 mg L^{-1} 2,4-D; C17 induction medium with W14 salts and vitamins, 5.0 mg L^{-1} 2,4-D, and 0.5 mg L^{-1} BAP for ELS induction; 190-2 regeneration medium	The highest number of ELS on C17 medium; incubation at 22 °C in the dark for the first two weeks for the highest rate of green plants per 100 ELS	[56]
Genotype 2000QiON43 (LA9326E86)	0.3 M mannitol pretreatment of the tillers for 7 days; W14 medium and continuous incubation at 28 °C; W14 medium for embryos observed; 0.2% colchicine for 4 h for DH	Protocol for the production of microspore-derived embryos of oat, 80% of the plants were converted to DH	[46]
cv. "Akt", cv. "Bingo", cv. "Bajka", cv. "Chwat"	for tillers: 2 and 3 weeks at 4 °C, or 2 and 3 weeks at 4 °C followed by 32 °C for 24 h; for ELS induction: C17 medium with 0.5 mg L^{-1} picloram, 0.5 mg L^{-1} dicamba, and 0.5 mg L^{-1} KIN, or W14 medium with different concentrations of 2,4-D, NAA, and BAP	Cold pretreatment and high temperature enhanced the technique efficiency; W14 medium with 2 mg L^{-1} and 0.5 mg L^{-1} KIN for the highest number of ELS	[61]
cv. "Bingo", cv. "Chwat"	2 weeks at 4 °C for tillers pretreatment in liquid medium alone or with Cu^{2+}, Zn^{2+}, or Ag^{+} ions followed by 32 °C for 24 h	ELS formation depended on cold pretreatment combined with Cu^{2+}, Zn^{2+}, or Ag^{+}	[66]

* ELS—embryo-like structures.

4. Wide Crossing of Oat (*Avena sativa* L.) with Chosen Species from Poaceae Family

Obtaining DH lines of oat is very challenging compared to other cereals, and both breeding and biotechnological research conducted by research groups from Poland, the USA, Finland, or Japan unanimously confirm the recalcitrance of this species to haploidization. Since the techniques used successfully in other plants are still not very effective in oat, a commercially viable and efficient method of obtaining DH lines in this species has not yet been developed. The reasons for the low efficiency of the methods used, which typically yield between 0.5% and 10.0% of haploid embryos per emasculated floret [70], are attributed to the presence of numerous pre- and postzygotic barriers. Prezygotic barriers include all factors that hinder the successful fertilization of the ovum, i.e., the formation of a zygote. The most listed prezygotic barriers include the inability of pollen to germinate on a foreign stigma, inhibition of pollen tube growth, or rupture of the pollen tube [71]. On the other hand, postzygotic barriers impede the development of the zygote after fertilization [72] and are often a result of genetic incompatibility between the parental plants in wide crosses. Hence, developing a thorough understanding of these barriers and overcoming them can contribute to the development of an effective and universal method for obtaining oat haploids and subsequently DH lines. From a practical standpoint, this opens new possibilities for improving haploidization methods not only for oat but also for other plants recalcitrant to this process, such as legumes or woody plants. In addition, the production of new cultivars based on homozygous DH lines is becoming increasingly important in crop breeding programs and represents one of the key opportunities for adapting agriculture to ongoing climate change.

Pioneering work on obtaining DH oats through wide crosses with maize was conducted by Rines and Dahleen [73]. Pollen from maize (*Zea mays* L.) was applied to previously emasculated oat florets in a series of experiments. Extracted caryopses and the embryos formed from them were then cultured on an MS medium with 7% sucrose and amino acids supplements. Recovered plantlets were raised in soil-filled pots until they were fully developed. Following the pollination of maize pollen from around 3300 emasculated oat florets, 14 haploid oat seedlings were successfully produced via the embryo rescue technique. Subsequently, in studies conducted by Matzk [74], eastern gamagrass (*Tripsacum dactyloides* L.), pearl millet (*Pennisetum americanum* L.), and maize (*Zea mays* L.) were used to pollinate five varieties of oat. Postzygotic obstacles appeared while using early colchicine-mediated chromosomal doubling, exogenous auxins, and embryo rescue media. The embryo frequencies ranged from 0.4% in maize to 9.8% in pearl millet, depending on the type of pollinator. Although many plantlets in the embryo rescue process perished, the beginning of growth usually occurred. Four viable plants were formed overall, including hybrids with pearl millet and for the first time using eastern gamagrass. One to four chromosomes from pollinator species were discovered in oat root tip cells during the tillering stage. The authors stated that while the efficiency of haploid formation (0.1%) was too low to use in plant breeding programs, crossings of oat with maize and pearl millet looked promising for the transfer of genes or chromosomes. In 2015, Nowakowska et al. [75] conducted research aimed at developing an effective method for obtaining oat DH lines and demonstrated a significant influence of individual steps of the procedure on the efficiency of haploid production. In these experiments, the optimal timing between emasculation of florets, pollination with maize, treatment of ovaries with auxin, as well as the appropriate timing for the isolation of haploid embryos was determined. The highest number of haploid embryos and plants was obtained by pollinating donor plants with maize pollen 2 days after emasculation, when auxins were applied 2 days after pollination, and when embryos were isolated 3 weeks after pollination.

4.1. Induction of Haploid Embryos

The treatment of oat ovaries after pollination has also been the subject of many experiments. Initially, Rines et al. [76] negated the need for auxin application to oat ovaries to increase the efficiency of wide hybridization. However, Sidhu et al. [77] emphasized that growth regulators not only prevent the degeneration of ovaries; most importantly, they stimulate and sustain embryo development until its isolation from the ovary. Currently, to facilitate the formation of oat haploid embryos, pollinated flowers are most often treated with the following synthetic auxins: 2,4-D, dicamba, picloram, or gibberellic acid (GA_3) [70,75,77,78], or a combination of 2,4-D and GA_3 [12]. Only in a few cases were pollinated oat panicles cut and placed in a solution containing sucrose and 2,4-D [6]. Research by Smit and Weijers [79] has shown that auxins play a key role in the early stages of embryogenic plant development, mediating the formation of zygotic embryos. Exogenous 2,4-D application alters the levels of endogenous auxins, such as IAA, thereby modifying their intracellular metabolism, which leads to the establishment of proper embryonic symmetry [80]. Warchoł et al. [81] described the process of determining which auxin to apply at 100 mg L^{-1} to the ovary after removal of the anthers and pollination with maize pollen to induce the development of haploid embryos plants and the production of fertile DH lines. It was determined that the tested auxins did not affect the number of enlarged ovaries (83.4%—dicamba; 83.9%—2,4-D, calculated based on emasculated flowers), nor did they affect the number of resulting haploid embryos. However, the applied auxins significantly differentiated the capacity of embryos to germinate, thus affecting the production of haploid plants and DH lines. Nearly half of the generated embryos (48%) germinated when placed on 190-2 medium [62], but only 22% of them developed into haploid plants.

The final number of haploid plants was 45 (0.64%, based on emasculated florets) when using dicamba, and 104 plants (1.37%, based on emasculated florets) when 2,4-D was applied. The same concentration (100 mg L^{-1}) of the auxin analogues 2,4-D, dicamba, and picloram, as well as GA$_3$, were tested by Sidhu et al. [77]. A specific growth regulator was applied to emasculated oat florets of the AK-1 and F1 hybrid genotypes on the 2nd and 3rd post-pollination days after being pollinated with maize pollen. The ability of each hormone to promote caryopsis development varied significantly between the two genotypes. The largest proportion of caryopses were produced by the dicamba treatment, 94.5% for AK-1 and 94.1% for 01095, respectively. Following 2,4-D and GA$_3$, picloram stimulated caryopses development. There was no discernible difference between genotype and growth regulator interaction. Kynast et al. [12] used a phytohormone mixture (50 ppm 2,4-D + 50 ppm GA$_3$) and sprayed them 24 or 48 h after application of fresh pollen of maize Mo17 on emasculated oat panicles of Starter and Sun II to induce the growth of the haploid embryos.

As mentioned above, the application of synthetic auxins to pollinated ovaries is a required step in the process of oat haploidization because it leads to the proper distribution of endogenous auxins necessary for establishing embryogenic patterns. The studies conducted by Nowakowska et al. [75] and Mahato and Chaudhary [82] have emphasized that the efficiency of this process is influenced not only by the timing and method of hormone application but primarily by their concentration. However, it is important to remember that using high concentrations of 2,4-D in in vitro cultures is toxic to plants and can result in tissue necrosis [83] or the inhibition of embryo germination, as observed by Bronsema et al. [84] in maize. Considering that synthetic auxins applied at high concentrations exhibit strong toxic properties, which could consequently result in low survival rates of oat haploid embryos, the aim of the experiments published by Juzoń et al. [85] was to determine how two different 2,4-D concentrations affected the conversion of embryos into haploid plants and the subsequent development of fertile DH lines. Treating the ovaries with 50 mg L^{-1} 2,4-D yielded 27 haploid plants (8.5%, based on emasculated flowers), while using 100 mg L^{-1} of 2,4-D increased their number to 49 (16.3%, based on emasculated flowers). The higher concentration of 2,4-D led to the survival of all haploid plants from 17 genotypes after colchicine treatment (approx. 58% of obtained plants), resulting in twice as many DH lines (44 plants) compared to the lower concentration of 2,4-D (22 plants). Oat florets from genotype AK-1 which had been emasculated and pollinated with maize pollen were exposed to four different doses of dicamba (5, 25, 50, and 100 mg L^{-1}) [77]. With the increasing dicamba concentration, the proportion of caryopses per floret grew considerably, reaching a maximum at concentration of 50 mg L^{-1}. Caryopsis development and embryo formation at 50 and 100 mg L^{-1} did not significantly differ. Kynast et al.'s [12] studies, the phytohormone combination (50 ppm 2,4-D + 50 ppm GA$_3$) was proven to be more effective for embryo formation than the 100 ppm 2,4-D solution without GA$_3$.

4.2. Embryo Rescue Technique

Achieving approx. 10% of haploid embryos per emasculated flowers in the first stage of wide hybridization does not confirm a high efficiency of haploid plants or fertile DH lines production. This is because the embryos formed after fertilization have a very low viability, and most of them die in the early developmental stages. Rines [70] has reported that the rate of embryo germination and their regeneration into plants typically falls below 20%. This was also confirmed by other studies, e.g., Warchoł et al. [81] and Juzoń et al. [85], who isolated a relatively high number of embryos (683 and 619, respectively) but obtained only 149 and 76 haploid plants, respectively. In addition, as pointed out by Rines [70], the low regenerative capacity of embryos hampers the conducting of experiments that would allow for a statistical comparison of factors influencing their germination effectiveness; thus, in assessing the reproducibility of the applied method in oat, the haploid embryos resulting from wide crosses with maize are most often devoid of endosperm, or else this tissue is rudimentary. In consequence, the lack of access to nutrients leads to their death, and the in vitro culture stage where suitable conditions for their growth are provided is referred to as the embryo rescue technique [86,87]. The first attempts at cultivating plant embryos outside their maternal tissues to obtain an interspecific cross of *Linum perenne* × *Linum austriacum* were conducted by Laibach [88]. Analyzing his research, it can be seen that the smaller the embryo, the more complex the medium required to continue its growth and development. In practice, this means that the regeneration medium closely mimics the composition of the maternal endosperm, thereby providing the appropriate nutritional components for the specific developmental stage of the embryo. The literature data indicate that the concentration of carbohydrates depends on the developmental stage of the embryo; the younger the embryos, the higher the concentration of sugars in the medium should be (even up to 12%). Sugars added to the medium at the appropriate concentration serve not only as a carbon source for heterotrophic embryos; they also ensure a suitable level of osmotic pressure [89]. The necessity of overcoming postzygotic barriers, including the selection of an appropriate regeneration medium that serves as an endosperm substitute for developing embryos, makes germination of embryos a critical stage in the method for obtaining oat DH lines.

Since each species requires the development of a detailed procedure concerning both culture conditions and the appropriate selection of regeneration medium components, Warchoł et al. [90] optimized the composition of the medium for embryo germination under in vitro conditions. This experiment, for the first time, analyzed the germination capacity of embryos on media with varying maltose concentrations and pH values. The resulting haploid embryos were plated on 190-2 agar medium [91] enriched with KIN and NAA at a concentration of 0.5 mg L^{-1}. Maltose was added to the medium at two concentrations, 6% and 9%, and the pH was set at 5.5 and 6.0. The medium with a pH of 6.0, compared to pH 5.5, increased the efficiency of embryo germination, similarly to the increased maltose content (9%) in the medium. Previous studies on obtaining oat DH lines have demonstrated high efficiency in inducing haploid embryos but unsatisfactory conversion of these embryos into plants [75,77,78]. The most frequently indicated reasons involved not only the lack of endosperm but also disrupted hormonal balance and a range of deformations visible at various stages of their development [92]. Moreover, the development of haploid oat embryos is not synchronized in time. Despite the fact that maize ovary pollination and auxin treatment occurred at the same time, the embryos transferred onto regeneration media differ in size and level of differentiation. When establishing in vitro cultures of embryos isolated from immature seeds, known as pseudo-seeds [93], it is important to remember that the establishment of the axis of symmetry is possible only when the embryo's development, at least up to the early globular stage, occurs in its natural environment, i.e., in the ovary. On the contrary, the initiation of cultures must occur before the critical point of developmental arrest, namely, the cotyledon formation stage. Additionally, in immature embryos, a phenomenon called "premature germination" is observed, typically occurring before the embryo axis formation. Since this type of germination is characterized by the elongation of cells and low intensity of divisions, the resulting haploid plants are weak and usually die back [94].

The experiments conducted by Noga et al. [95] aimed to increase the conversion efficiency of haploid embryos into haploid plants and to analyze correlations between the germination capacity of oat embryos at different developmental stages and the type of growth regulators. Although the isolation was performed at the same time, the embryos plated on the media exhibited differences in morphological structure. As a result, they were divided into four size classes: <0.5 mm, 0.5–0.9 mm, 1.0–1.4 mm, and ≥1.5 mm. Subsequently, they were cultured on a 190-2 regeneration medium [91], containing 9% maltose, 0.6% agar, and the following growth regulators: KIN, NAA, zeatin (ZEA), dicamba, and picloram. Microscopic observations revealed that embryos smaller than 0.5 mm were spherical, those ranging from 0.5 to 1.4 mm were elongated without distinct basal and apical parts, while embryos larger than 1.5 mm had a visible coleoptile and embryonic root. The conducted analysis of oat embryo germination capacity concerning their developmental stage showed that the largest embryos germinated at nearly 80%, while the smallest ones lacked regenerative capacity and died after plating on the medium. Furthermore, it was observed that the size of haploid embryos and their germination capacity varied significantly among different oat genotypes. A similar observation was made by Sidhu et al. [77]. Compared to self-pollinated embryos, the white, embryo-like structures (ELS) of the four oat genotypes—AK-1, Carrolup, Dumont, Mortlock, and S093658 obtained by crossing oats with maize—differed in size and shape. The type of growth regulators added to the regeneration medium did not exert a significant effect on the regeneration of haploid embryos into plants. Nevertheless, the highest percentage of haploid embryos (19%) germinated on a medium with 0.5 mg L^{-1} NAA and 0.5 mg L^{-1} KIN, and the smallest (11%) on a medium with 1 mg L^{-1} dicamba, 1 mg L^{-1} picloram and 0.5 mg L^{-1} KIN.

Skrzypek et al. [96] examined the role of light intensity applied in vivo to initiate haploid embryos and in vitro to regulate their development. For the growth of donor plants, the light intensity of 800 µmol m^{-2} s^{-1} more effectively stimulated the formation of haploid embryos (9.4%) compared to the light intensity of 450 µmol m^{-2} s^{-1} (6.1%). Light intensity during in vitro cultures of embryos also had an impact on their conversion into plants. Light intensity of 110 µmol m^{-2} s^{-1} during culture most optimally stimulated embryo germination (38.9%) and plant development (36.4%) compared to light intensities of 20, 40, and 70 µmol m^{-2} s^{-1}. In a previous study by Sidhu et al. (2006), it was amply shown that temperature has no influence on the caryopsis development. Despite this, the authors observed higher embryos production at 24 °C. However, this difference was not statistically significant, most likely as a result of the few repetitions (one donor plant per treatment).

To understand the slow rate of oat embryo germination, research was conducted to investigate the phytohormone content in ovaries during embryo development. Additionally, the hormonal profiles of zygotic and haploid embryos were analyzed. Dziurka et al. [97] compared ovules with embryos (OE) and ovules without embryos (OWE). The latter study analyzed the phytohormone content and found significantly higher concentrations of IAA, trans-zeatin (tZ), and KIN in OE compared to OWE. It was also demonstrated that an excess of cytokinins in OE was detrimental to embryogenesis, while reduced cytokinin levels increased the efficiency of obtaining DH lines. The presence of IAA was detected only in OWE, indicating its role in plant aging processes. Although both haploid and zygotic oat embryos were isolated at the same time, the extremely low levels of endogenous auxins, larger amounts of cytokinins, and a ten-fold higher cytokinin/auxin ratio in the haploid embryos may indicate an earlier developmental stage for the former. It was also shown that inadequate germination of haploid embryos could rsult from an excess of reactive oxygen species, raising levels of low-molecular-weight osmoprotectants and stress hormones in addition to hormonal modulation of embryogenesis [92]. The summary of progress in the wide crossing of A. sativa with various species from the Poaceae family is presented in Table 3.

Table 3. Crossing of oat (*Avena sativa* L.) with chosen species from the Poaceae family.

Plant Material	Culture Conditions	Experimental Outcomes	Reference
Oat × maize Oat: cv. "Stout", cv. "Starter", cv. "Steele", cv. "Black Mesdag" Maize: A188, B73, Honeycomb, A619 × W64A	Haploid plants recovered via embryo rescue following field-grown maize pollen application to emasculated florets of growth chamber-grown oat	Recovered haploids were from a different oat cultivar and different source of maize pollen—the process is not genotype unique	[73]
Oat × maize Oat: genotypes AK-1, S093658, Carrolup, Dumont, Mortlock Maize: early extra sweet F1, and Kelvedon Glory F1 varieties	100 mg L^{-1} GA_3, 2,4-D, 3,6-dichloro-o-anisic acid (dicamba) or 4-amino-3,5,6,-trichloro-picolinic acid (picloram) applied after pollination; four different temperature regimes (32/24, 24/20, 21/17 and 17/14 °C day/night) applied before flowering	The highest number of caryopses produced with dicamba, but without effects on embryo production; genotype dependent temperature effects	[77]

Table 3. Cont.

Plant Material	Culture Conditions	Experimental Outcomes	Reference
Oat × pearl millet Oat: cv. "Best Enbaku" Pearl millet: *Pennisetum glaucum* cv. "Ugandi"	100 ppm 2,4-D dropped onto each floret 12 h after pollination; 100 ppm 2,4-D and 4% sucrose for the spike culture	Retention of all seven pearl millet chromosomes in embryos from the crosses with oat; oat haploid developed to a fertile adult plant	[6,11]
Oat × maize Oat: lines Black Mesdag, GAF-Park, Kanota, MN97201-1, Preakness, Starter, Steele, Stout, Sun II, and F1 (MN97201-1 × MN841801-1) oat hybrid Maize: lines Seneca60, *bz1-mum9*, A188, B73, Mo17, and the F1 (A188 9 W64A) maize hybrid	50 ppm 2,4-D and 50 ppm GA_3 for embryo formation delay of endosperm collapse	Euhaploid plants with complete oat chromosome complements without maize chromosomes; aneuhaploid plants with complete oat chromosome complements and different numbers of retained individual maize chromosomes; uniparental genome loss during early steps of embryogenesis causing the elimination of maize chromosomes in the hybrid embryo	[98]
Oat × maize Oat: genotypes 80022, 80031, 81711, 81350, 81384, 81524, 81559, 82072, 82091, 82230, 82266, 83200, 83207, 83213, 83421, 83430, 85924, and 85931 Maize: Waza, Dobosz, and Wania	Oat florets pollinated with maize pollen after 0, 1 or 2 days; 100 mg L^{-1} 2,4-D or 100 mg L^{-1} dicamba placed on the floret pistils 1, 2-, 3-, 4-, and 5-days following pollination	Genotype-dependent haploid embryo formation and plant regeneration; 2nd-day pollination together with auxin treatment was the most effective	[78]
Oat × maize Oat: 80031—(Deresz × Szakal), 81350 (Krezus × STH 454), 82072 (Bajka × STH 454), 82091 (Bajka × STH 7706), 83213 (Flamingstern × Chwat) Maize: *Zea mays* L. var. *saccharata*, Oat × sorghum Sorghum: *Sorghum bicolor* (L.) Moench Oat × common millet Common millet: *Panicum miliaceum* L.	100 mg L^{-1} dicamba one day after pollination; enlarged ovaries collected at 2, 3 and 4 weeks after pollination cultivated on 6 or 9% of maltose	2.5—6.9% of HE * for genotypes pollinated with maize, 1.3% for sorghum, and 1.2% for millet; the highest frequency of HE germination and number of plants 3 weeks after pollination; 9% maltose for embryo formation, germination, and haploid plants development	[75]
Oat × maize Oat: STH 4.8456/1, STH 4.8456/2, STH 4.8457/1, STH 4.8457/2, STH 5.8421, STH 5.8422, STH 5.8423, STH 5.8424, STH 5.8425, STH 5.8426, STH 5.8427, STH 5.8428, STH 5.8429, STH 5.8430, STH 5.8432, STH 5.8436, STH 5.8440, STH 5.8449, STH 5.8450, STH 5.8458, STH 5.8460 Maize: Waza	<0.5 mm HE, 0.5–0.9 mm HE, 1.0–1.4 mm HE, and ≥1.5 mm HE on 0.5 mg L^{-1} KIN and 0.5 mg L^{-1} NAA, or 1 mg L^{-1} ZEA and 0.5 mg L^{-1} NAA, or 1 mg L^{-1} dicamba, 1 mg L^{-1} picloram, and 0.5 mg L^{-1} KIN	Germination of HE ≥ 1.5 mm on medium with 0.5 mg L^{-1} NAA and 0.5 mg L^{-1} KIN	[95]
Oat × maize 32 oat genotypes were pollinated with *Zea mays* L. var. *saccharata* (maize) genotypes: MPC4, Dobosz and Wania	Different light intensity during the growing period of donor plants and in vitro cultures	9.4% HE formed in a greenhouse, 6.1% in a growth chamber; 38.9% of embryo germination, 36.4% conversion into plants, and 9.2% DH ** line production with 110 µmol m^{-2} s^{-1} light intensity	[96]
Oat × maize Oat: F1 progeny of thirty-three oat genotypes Maize: *Zea mays* L. var. *saccharata* (maize) genotypes MPC4, Dobosz and Wania	Immersion of haploid plants for 7.5 h in a 0.1% colchicine, 40 g L^{-1} DMSO, 0.025 g L^{-1} GA^3 at 25 °C and 80–100 µmol m^{-2} s^{-1} light intensity for chromosome doubling procedure	From 149 haploid plants 61 survived chromosome doubling procedure, 52 (85%) were fertile and produced seeds	[81]
Not specified	Colchicine solution with DMSO for chromosome doubling	Detailed description of a method for DHs generation	[1]

Table 3. Cont.

Plant Material	Culture Conditions	Experimental Outcomes	Reference
Oat × maize 80 oat genotypes pollinated with maize cv. "Waza"	Colchicine solution applied on HP roots for chromosome doubling	from 138 oat lines, the presence of maize chromatin was indicated in 66 lines from which 27 OMA lines were fertile and produced seeds	[99]
Oat × maize Oat: F1 progeny of twenty-two oat genotypes pollinated with *Zea mays* L. var. *saccharata* (maize) genotypes MPC4, Dobosz and Wania	For chromosome doubling HP roots were immersed in a 0.1% colchicine with 4% DMSO, 0.025 g L^{-1} GA^3, and 20 µL of Tween 20, left for 7.5 h at 25 °C and 80–100 µmol m^{-2} s^{-1} light intensity	591 HE formed, 48 fertile DH plants producing in all 4878 seeds	[90]
Oat × maize Oat: F1 progeny of twenty-nine oat genotypes pollinated with *Zea mays* L. var. *saccharata* (maize) genotypes MPC4, Dobosz and Wania	9465 florets were pollinated with maize pollen 2 days after emasculation and treated with 2,4-D at 50 mg L^{-1} and 100 mg L^{-1}; colchicine solution applied on HP roots for chromosome doubling	Higher 2,4-D concentration is more efficient in obtaining haploid/DH plants with better vitality and fertility	[85]

* HE—haploid embryos; ** DH—doubled haploids.

5. Conclusions and Future Directions

Many variables, including screening practices, tolerance bases and mechanisms, gene function and inheritance, and linkages to agronomical traits, all have an impact on the choice of an appropriate breeding strategy for the creation of cultivars that are of interest to us. As new cultivars have been created mostly using conventional breeding methods, the typical approach involves recombining DNA by distinct chromosomal assortment and crossing-over.

In conventional breeding, the number of generations is needed to produce stable variations via natural segregation from the heterozygous progeny of the original crosses. The in vitro methods could eliminate the necessity for back-crosses or repeated self-pollination. Although plant tissue culture techniques have developed significantly since the first publication on oat regeneration from calluses in 1967, the plant regeneration effectiveness is still low and strongly genotype dependent. Oat plants must be grown year-round for the isolation of immature embryos, which is costly, requires elaborate equipment, and may subject donor plants to physiological fluctuation that could influence how frequently tissue cultures are started. Furthermore, it takes a lot of effort to isolate immature embryos from oats since the panicle's fertilization is not as synchronized as it is, for example, in wheat or barley. As a substitute, mass-producing mature seeds is affordable and offers reliable explants for starting in vitro cultures, potentially removing the variability in cultures from different explants. On the other hand, by manipulating gamete development, it is possible to regenerate fully homozygous plants in just one generation.

Since the beginning of the 20th century, when the theories of totipotency and naturally occurring sporophytic haploids were discovered, the production of haploid and DH plants has been introduced in the breeding several crop plants. Research has resulted in a better understanding of the mechanisms of haploid formation, the identification of factors influencing haploid induction, and the increase in genetic benefits through the application of DH technology in plant breeding. The recent finding of a very efficient centromere-mediated genome deletion approach for haploid production has sparked intense interest in its application in plant breeding.

For oat species, the generation of DH has become an essential tool in advanced plant breeding. Combining the use of DH methods with applied genomics opens novel possibilities for maximizing genetic benefits in selection and for developing new, more cost-effective, and efficient massive techniques, as well as for minimizing the time needed for cultivar production. However, the study of oat molecular genetics remains substantially behind that of other grains mainly due to the genome size, and that the oat DNA sequence was not fully available. Since the publication of the first quantitative trait loci linkage map in oat, there have been constant attempts to enhance the density of the map via different kinds and numbers of markers. The recently sequenced hexaploid *Avena sativa* L. genome, although its size is expected to be over 11 Gb and it consists of two ploidy species: *Avena longiglumis* (AA) (3.7 Gb) and *Avena insularis* (CCDD) (7.3 Gb), should help to accelerate the process of enhancing oats for numerous features [100]. The most important objectives of oat breeding and modification of genes are to improve tolerance to diseases and environmental stressors, as well as yield and other important agronomical features [101].

Despite all these advancements, there are still DH line utilization difficulties that have yet to be resolved. There is a need for greater DH production efficiency (especially in anthers and microspore cultures compared with wide crossing), as well as improved germplasm control and a greater awareness of the molecular mechanisms that regulate the formation of haploid plants. Moreover, an important issue is the elaboration of the chromosome doubling method without using harmful chemicals. Recently, the production of haploids using centromere-mediated genetic engineering seems to have been of key importance. As described by Karimi-Ashtiyani et al. [102], point mutation in the histone H3 variant CENH3, which is specific to centromeres, may be utilized to create haploid plants. Plants with this single-point mutation in CENH3 are haploid inducers. Due to the high degree of conservation of the recognized mutation site and the fact that point mutation can be achieved via mutagenesis or genome editing, the disclosed method has the potential to be applicable to numerous crops.

The findings of this review may encourage the spread of this technology's application in accelerating and creating new oat breeding opportunities. We expect that this review will also assist molecular scientists in the construction of DH segregating populations in oat species, which are necessary in order to produce genetic maps employing molecular markers.

Author Contributions: Conceptualization, M.W. and E.S.; writing—original draft preparation, M.W., E.S. and D.J.; writing—review and editing, M.W., E.S., D.J. and K.J.-S.; visualization, D.J. All authors have read and agreed to the published version of the manuscript.

Funding: This research received no external funding.

Data Availability Statement: All data analyzed during this study are included in this published article.

Conflicts of Interest: The authors declare no conflict of interest.

References

1. Davies, P.A.; Sidhu, P.K. Oat doubled haploids following maize pollination. In *Oats: Methods in Molecular Biology*; Gasparis, S., Ed.; Humana Press: New York, NY, USA, 2017; Volume 1536, pp. 23–30.
2. Seguí-Simarro, J.M.; Jacquier, N.M.A.; Widiez, T. *Overview of in vitro and in vivo doubled haploid technologies*, In Doubled Haploid Technology, Vol. 1: General Topics, Alliaceae, Cereals, Methods in Molecular Biology; Segui-Simarro, J.M., Ed.; Humana Press: New York, NY, USA, 2021; Volume 2287, pp. 3–22.
3. Morikawa, T. Protocol for Producing Synthetic Polyploid Oats. In *Oat: Methods in Molecular Biology*; Gasparis, S., Ed.; Humana Press: New York, NY, USA, 2017; Volume 1536, pp. 43–52.
4. Gasparis, S. Agrobacterium-Mediated Transformation of Leaf Base Segments. In *Oat: Methods in Molecular Biology*; Gasparis, S., Ed.; Humana Press: New York, NY, USA, 2017; Volume 1536, pp. 95–111.
5. Kasha, K.J.; Kao, K.N. High frequency haploid production in barley (*Hordeum vulgare* L.). *Nature* **1970**, *225*, 874–876. [CrossRef]
6. Ishii, T.; Tanaka, H.; Eltayeb, A.E.; Tsujimoto, H. Wide hybridization between oat and pearl millet belonging to different subfamilies of Poaceae. *Plant Reprod.* **2013**, *26*, 25–32. [CrossRef]
7. Laurie, D.A.; Bennett, M.D. Cytological evidence for fertilization in hexaploid wheat × sorghum crosses. *Plant Breed.* **1988**, *100*, 73–82. [CrossRef]
8. Laurie, D.A.; Bennett, M.D. The timing of chromosome elimination in hexaploid wheat × maize crosses. *Genome* **1989**, *32*, 953–961. [CrossRef]
9. Riera-Lizarazu, O.; Rines, H.W.; Phillips, R.L. Cytological and molecular characterization of oat x maize partial hybrids. *Theor. Appl. Genet.* **1996**, *93*, 123–135. [CrossRef] [PubMed]
10. Mochida, K.; Tsujimoto, H.; Sasakuma, T. Confocal analysis of chromosome behavior in wheat × maize zygotes. *Genome* **2004**, *47*, 199–205. [CrossRef]

11. Ishii, T.; Ueda, T.; Tanaka, H.; Tsujimoto, H. Chromosome elimination by wide hybridization between Triticeae or oat plant and pearl millet: Pearl millet chromosome dynamics in hybrid embryo cells. *Chromosome Res.* **2010**, *18*, 821–831. [CrossRef]
12. Kynast, R.G.; Davis, D.W.; Phillips, R.L.; Rines, H.W. Gamete formation via meiotic nuclear restitution generates fertile amphiploid F1 (oat × maize) plants. *Sex. Plant Reprod.* **2012**, *25*, 111–122. [CrossRef]
13. Okagaki, R.J.; Kynast, R.G.; Livingston, S.M.; Russell, C.D.; Rines, H.W.; Phillips, R.L. Mapping maize sequences to chromosome using oat-maize chromosome addition materials. *Plant Phys.* **2001**, *125*, 1228–1235. [CrossRef]
14. Jin, W.; Melo, J.R.; Nagaki, K.; Talbert, P.B.; Henikoff, S.; Dawe, R.K.; Jiang, J. Maize centromeres: Organization and functional adaptation in the genetic background of oat. *Plant Cell* **2004**, *16*, 571–581. [CrossRef]
15. Kowles, R.V.; Walch, M.D.; Minnerath, J.M.; Bernacchi, C.J.; Stec, A.O.; Rines, H.W.; Phillips, R.L. Expression of C4 photosynthetic enzymes in oat-maize chromosome addition lines. *Maydica* **2008**, *53*, 69–78.
16. Carter, O.; Yamada, Y.; Takahashi, E. Tissue culture of oats. *Nature* **1967**, *214*, 1029–1030. [CrossRef]
17. Lörz, H.; Harms, C.T.; Potrykus, I. Regeneration of plants from callus in *Avena sativa* L. *Z. Pflanzenzuechtg* **1976**, *77*, 257–259.
18. Cummings, D.P.; Green, C.E.; Stuthman, D.D. Callus induction and plant regeneration in oats. *Crop Sci.* **1976**, *16*, 465–470. [CrossRef]
19. Gamborg, O.L.; Miller, R.A.; Ojima, K. Nutrient requirements of suspension cultures of soybean root cells. *Exp. Cell Res.* **1968**, *50*, 151–158. [CrossRef] [PubMed]
20. Murashige, T.; Skoog, F. A revised medium for rapid growth and bioassays with tobacco tissue cultures. *Physiol. Plant* **1962**, *15*, 473–497. [CrossRef]
21. Maddock, S.E. Cell culture, somatic embryogenesis and plant regeneration in wheat, barley, oats, rye and triticale. In *Cereal Tissue and Cell Culture. Advances in Agricultural Biotechnology*; Bright, S.W.J., Jones, M.G.K., Eds.; Springer: Dordrecht, The Netherlands, 1985; Volume 15. [CrossRef]
22. Chen, H.; Xu, G.; Loschke, D.C.; Tomaska, L.; Rolfe, B. Efficient callus formation and plant regeneration from leaves of oats (*Avena sativa* L.). *Plant Cell Rep.* **1995**, *14*, 393–397. [CrossRef]
23. Zhang, S.; Zhang, H.; Zhang, M.B. Production of multiple shoots from shoot apical meristems of oat (*Avena sativa* L.). *J. Plant Physiol.* **1996**, *148*, 667–671. [CrossRef]
24. Nuutila, A.M.; Villiger, C.; Oksman-Caldentey, K.M. Embryogenesis and regeneration of green plantlets from oat (*Avena sativa* L.) leaf-base segments: Influence of nitrogen balance, sugar and auxin. *Plant Cell Rep.* **2002**, *20*, 1156–1161. [CrossRef]
25. Jähne, A.; Lazzeri, P.A.; Jäger-Gussen, M.; Lörz, H. Plant regeneration from embryogenic suspensions derived from anther cultures of barley (*Hordeum vulgare* L.). *Theor. Appl. Genet.* **1991**, *82*, 74–80. [CrossRef]
26. Gless, C.; Lörz, H.; Jähne-Gärtner, A. Establishment of a highly efficient regeneration system from leaf base segments of oat (*Avena sativa* L.). *Plant Cell Rep.* **1998**, *17*, 441–445. [CrossRef]
27. Rines, H.W.; McCoy, T.J. Culture initiation and plant regeneration in hexaploid species of oats. *Crop Sci.* **1981**, *6*, 837–842. [CrossRef]
28. Silveira, V.; de Vita, A.M.; Macedo, A.F.; Dias, M.F.R.; Floh, E.S.; Santa-Catarina, C. Morphological and polyamine content changes in embryogenic and non-embryogenic callus of sugarcane. *Plant Cell Tissue Organ Cult.* **2013**, *114*, 351–364. [CrossRef]
29. Rines, H.W.; Luke, H.H. Selection and regeneration of toxin-insensitive plants from tissue cultures of oats (*Avena sativa*) susceptible to *Helminthosporium victoriae*. *Theor. Appl. Genet.* **1985**, *71*, 16–21. [CrossRef] [PubMed]
30. Bregitzer, P.; Somers, D.A.; Rines, H.W. Development and characterization of friable, embryogenic oat callus. *Crop Sci.* **1989**, *29*, 798–803. [CrossRef]
31. Bregitzer, P.P.; Milach, S.K.; Rines, H.W.; Somers, D.A. Somatic embryogenesis in oat (*Avena sativa* L.). In *Somatic Embryogenesis and Synthetic Seed II*; Bajaj, Y.P.S., Ed.; Springer: Berlin/Heidelberg, Germany, 1995; pp. 53–62.
32. Rines, H.W.; Phillips, R.L.; Somers, D.A. Application of tissue cultures to oat improvement. In *Oat Science and Technology Marshall*; Marshall, H.G., Sorrells, M.E., Eds.; American Society of Agronomy: Madison, WI, USA, 1992; pp. 777–791.
33. Somers, D.A. Transgenic cereals: *Avena sativa* (oat). In *Molecular Improvement of Cereal Crops. Advances in Cellular and Molecular Biology of Plants*; Vasil, I.K., Ed.; Springer: Dordrecht, The Netherlands, 1999; Volume 5, pp. 317–339. [CrossRef]
34. Torbert, K.A.; Rines, H.W.; Somers, D.A. Transformation of oat using mature embryo-derived tissue cultures. *Crop Sci.* **1998**, *38*, 226–231. [CrossRef]
35. King, I.P.; Thomas, H.; Dale, P.J. Callus induction and plant regeneration from oat cultivars. In Proceedings of the Second International Oats Conference: The University College of Wales, Welsh Plant Breeding Station, Aberystwyth, UK, 15–18 July 1985; Springer: Dordrecht, The Netherlands, 1985; pp. 46–47.
36. Chen, Z.; Klockare, R.; Sundqvist, C. Origin of somatic embryogenesis is proliferating root primordia in seed derived oat callus. *Hereditas* **1994**, *120*, 211–216. [CrossRef]
37. Kelley, R.Y.; Zipf, A.E.; Wesenberg, D.E.; Sharma, G.C. Putrescine-enhanced somatic embryos and plant numbers from elite oat (*Avena* spp. L.) and reciprocal crosses. *Vitr. Cell. Dev. Biol. Plant* **2002**, *38*, 508–512. [CrossRef]
38. Borji, M.; Bouamama-Gzara, B.; Chibani, F.; Teyssier, C.; Ammar, A.B.; Milki, A.; Zekri, S.; Ghorbel, A. Micromorphology, structural and ultrastructural changes during somatic embryogenesis of a Tunisian oat variety (*Avena sativa* L. var 'Meliane'). *Plant Cell Tissue Organ Cult.* **2018**, *132*, 329–342. [CrossRef]
39. Gana, J.A.; Sharma, G.C.; Zipf, A.; Saha, S.; Roberts, J.; Wesenberg, D.M. Genotype effects on plant regeneration in callus and suspension cultures of *Avena*. *Plant Cell Tissue Organ Cult.* **1995**, *40*, 217–224. [CrossRef]

40. Wise, M.L.; Sreenath, H.K.; Skadsen, R.W.; Kaeppler, H.F. Biosynthesis of avenanthramides in suspension cultures of oat (*Avena sativa*). *Plant Cell Tissue Organ Cult.* **2009**, *97*, 81–90. [CrossRef]
41. Islam, S.; Tuteja, N. Enhancement of androgenesis by abiotic stress and other pretreatments in major crop species. *Plant Sci.* **2012**, *182*, 134–144. [CrossRef]
42. Forster, B.P.; Heberle-Bors, E.; Kasha, K.J.; Touraev, A. The resurgence of haploids in higher plants. *Trends Plant Sci.* **2007**, *12*, 368–375. [CrossRef]
43. Testillano, P.S. Microspore embryogenesis: Targeting the determinant factors of stress induced cell reprogramming for crop improvement. *J. Exp. Bot.* **2019**, *70*, 2965–2978. [CrossRef] [PubMed]
44. De Cesaro, T.; Baggio, M.I.; Zanetti, S.A.; Suzin, M.; Augustin, L.; Brammer, S.P.; Iorczeski, E.J.; Milach, S.C.K. Haplodiploid androgenetic breeding in oat: Genotypic variation in anther size and microspore development stage. *Sci. Agric.* **2009**, *66*, 118–122. [CrossRef]
45. Oleszczuk, S.; Zimny, J. Mikrospory zbóż w kulturach in vitro. *Biotechnologia* **2001**, *2*, 142–161. (In Polish)
46. Ferrie, A.M.R.; Irmen, K.I.; Beattie, A.D.; Rossnagel, B.G. Isolated microspore culture of oat (*Avena sativa* L.) for the production of doubled haploids: Effect of pre-culture and post-culture conditions. *Plant Cell Tissue Organ Cult.* **2014**, *116*, 89–96. [CrossRef]
47. Murovec, J.; Bohanec, B. Haploids and doubled haploids in plant breeding. In *Plant Breeding*; Abdurakhmonov, I., Ed.; InTech Europe: Rijeka, Croatia, 2012; pp. 87–106.
48. Małuszyński, M.; Kasha, K.J.; Forster, B.P.; Szarejko, I. *Doubled Haploid Production in Crop Plants: A Manual*; Kluwer: Dordrecht, The Netherlands; Boston, MA, USA; London, UK, 2003; p. 428.
49. Makowska, K.; Oleszczuk, S.; Zimny, A.; Czaplicki, A.; Zimny, J. Androgenic capability among genotypes of winter and spring barley. *Plant Breed.* **2015**, *134*, 668–674. [CrossRef]
50. Rines, H.W. Oat anther culture: Genotype effects on callus initiation and the production of haploid plant. *Crop Sci.* **1983**, *23*, 268–272. [CrossRef]
51. Kiviharju, E.; Pehu, E. The effect of cold and heat pretreatments on anther culture response of *Avena sativa* and *A. sterilis*. *Plant Cell Tissue Organ Cult.* **1998**, *54*, 97–104. [CrossRef]
52. Ślusarkiewicz-Jarzina, A.; Ponitka, A. The effect of physical medium state on anther culture response in polish cultivated oat (*Avena sativa* L.). *Acta Biol. Crac. Ser. Bot.* **2007**, *49*, 27–31.
53. Chu, C.; Wang, C.; Sun, C.; Hsu, C.; Yin, K.; Chu, C.; Bi, F. Establishment of an efficient medium for another culture of rice through comparative experiments on the nitrogen sources. *Sci. Sin.* **1975**, *18*, 223–231.
54. Ouyang, T.W.; Jia, S.E.; Zhang, C.; Chen, X.; Feng, G. *A New Synthetic Medium (W14) for Wheat Anther Culture. Annual Report. 1987–1988*; Institute of Genetics Academia Sinica: Beijing, China, 1989; pp. 91–92.
55. Wang, P.; Chen, Y. Preliminary study on production of height of pollen H2 generation in winter wheat grown in the field. *Acta Agron. Sin.* **1983**, *9*, 283–284.
56. Ponitka, A.; Ślusarkiewicz-Jarzina, A. Regeneration of oat androgenic plants in relation to induction media and culture condition of embryo-like structures. *Acta Soc. Bot. Pol.* **2009**, *78*, 209213. [CrossRef]
57. Skrzypek, E.; Stawicka, A.; Czyczyło-Mysza, I.; Pilipowicz, M.; Marcińska, I. Wpływ wybranych czynników na indukcję androgenezy owsa (*Avena sativa* L.). *Zesz. Probl. Postępów Nauk. Rol.* **2009**, *534*, 273–281. (In Polish)
58. Kiviharju, E.; Moisander, S.; Laurila, J. Improved green plant regeneration rates from oat anther culture and the agronomic performance of some DH lines. *Plant Cell Tissue Organ Cult.* **2005**, *81*, 1–9. [CrossRef]
59. Pechan, P.M.; Smykal, P. Androgenesis: Affecting the fate of male gametophyte. *Physiol. Plant* **2001**, *111*, 1–8. [CrossRef]
60. Kruczkowska, H.; Pawłowska, H.; Skucińska, B. Próba indukcji androgenezy u polskich odmian owsa. *Zesz. Probl. Postępów Nauk. Rol.* **2007**, *523*, 137–142. (In Polish)
61. Warchoł, M.; Czyczyło-Mysza, I.; Marcińska, I.; Dziurka, K.; Noga, A.; Kapłoniak, K.; Pilipowicz, M.; Skrzypek, E. Factors inducing regeneration response in oat (*Avena sativa* L.) anther culture. *Vitr. Cell. Dev. Biol. Plant* **2019**, *55*, 595–604. [CrossRef]
62. Wang, X.Z.; Hu, H. The effect of potato II medium for 279 triticale anther culture. *Plant Sci. Lett.* **1984**, *36*, 237–239.
63. Dahleen, L.S. Improved plant regeneration from barley callus cultures by increased copper levels. *Plant Cell Tissue Organ Cult.* **1995**, *43*, 267–269. [CrossRef]
64. Echavarri, B.; Soriano, M.; Cistué, L.; Vallés, M.P.; Castillo, A.M. Zinc sulphate improved microspore embryogenesis in barley. *Plant Cell Tissue Organ Cult.* **2008**, *93*, 295–301. [CrossRef]
65. Makowska, K.; Oleszczuk, S.; Zimny, J. The effect of copper on plant regeneration in barley microspore culture. *Czech J. Genet. Plant Breed.* **2017**, *53*, 17–22. [CrossRef]
66. Warchoł, M.; Juzoń, K.; Dziurka, K.; Czyczyło-Mysza, I.; Kapłoniak, K.; Marcińska, I.; Skrzypek, E. The effect of zinc, copper and silver ions on oat (*Avena sativa* L.) androgenesis. *Plants* **2021**, *10*, 248. [CrossRef] [PubMed]
67. Kiviharju, E.; Puolimatka, M.; Saastamoinen, M.; Hovinen, S.; Pehu, E. The effect of genotype on anther culture response of cultivated and wild oats. *Agric. Food Sci.* **1998**, *7*, 409–422. [CrossRef]
68. Kiviharju, E.M.; Tauriainen, A.A. 2,4-Dichlorophenoxyacetic acid and kinetin in anther culture of cultivated and wild oats and their interspecific crosses: Plant regeneration from *A. sativa* L. *Plant Cell Rep.* **1999**, *18*, 582–588. [CrossRef]
69. Kiviharju, E.; Puolimatka, M.; Saastamoinen, M.; Pehu, E. Extension of anther culture to several genotypes of cultivated oats. *Plant Cell Rep.* **2000**, *19*, 674–679. [CrossRef]

70. Rines, H.W. Oat haploids from wide hybridization. In *Double Haploid Production in Crop Plants*; Małuszyński, M., Kasha, K.J., Forster, B.P., Szarejko, I., Eds.; Kluwer Acad Publishers: Dordrecht, The Netherlands, 2003; pp. 155–159.
71. Zenkteler, M. In vitro fertilization and wide hybridization in higher plants. *Crit. Rev. Plant Sci.* **1990**, *9*, 267–279. [CrossRef]
72. Mwangangi, I.M.; Muli, J.K.; Neondo, J.O. Plant hybridization as an alternative technique in plant breeding improvement. *Asian J. Crop Sci.* **2019**, *4*, 1–11. [CrossRef]
73. Rines, H.W.; Dahleen, L.S. Haploid oat plants produced by application of maize pollen to emasculated oat florets. *Crop Sci.* **1990**, *30*, 1073–1078. [CrossRef]
74. Matzk, F. Hybrids of crosses between oat and Andropogoneae or Paniceae species. *Crop Sci.* **1996**, *36*, 17–21. [CrossRef]
75. Nowakowska, A.; Skrzypek, E.; Marcińska, I.; Czyczyło-Mysza, I.; Dziurka, K.; Juzoń, K.; Cyganek, K.; Warchoł, M. Application of chosen factors in the wide crossing method for the production of oat doubled haploids. *Open Life Sci.* **2015**, *10*, 112–118. [CrossRef]
76. Rines, H.W.; Riera-Lizerazu, O.; Nunez, V.M.; Davis, D.W.; Phillips, R.L. Oat haploids from anther culture and from wide hybridizations. In *In Vitro Haploid Production in Higher Plants*; Jain, S.M., Sopory, S.K., Veilleux, R.E., Eds.; Kluwer Acad Publishers: Dordrecht, The Netherlands, 1997; pp. 205–221.
77. Sidhu, P.K.; Howes, N.K.; Aung, T.; Zwer, P.K.; Davies, P.A. Factors affecting haploid production following oat × maize hybridization. *Plant Breed.* **2006**, *125*, 243–247. [CrossRef]
78. Marcińska, I.; Nowakowska, A.; Skrzypek, E.; Czyczyło-Mysza, I. Production of double haploids in oat (*Avena sativa* L.) by pollination with maize (*Zea mays* L.). *Cent. Eur. J. Biol.* **2013**, *8*, 306–313. [CrossRef]
79. Smit, M.E.; Weijers, D. The role of auxin signaling in early embryo pattern formation. *Curr. Opin. Plant Biol.* **2015**, *28*, 99–105. [CrossRef]
80. Vondráková, Z.; Krajňáková, J.; Fischerová, L.; Vágner, M.; Eliášová, K. Physiology and role of plant growth regulators in somatic embryogenesis. In *Vegetative Propagation of Forest Trees*; Park, Y.S., Bonga, J., Moon, H.K., Eds.; National Institute of Forest Science: Seoul, Republic of Korea, 2016; pp. 123–169.
81. Warchoł, M.; Skrzypek, E.; Nowakowska, A.; Marcińska, I.; Czyczyło-Mysza, I.; Dziurka, K.; Juzoń, K.; Cyganek, K. The effect of auxin and genotype on the production of *Avena sativa* L. doubled haploid lines. *Plant Growth Regul.* **2016**, *78*, 155–165. [CrossRef]
82. Mahato, A.; Chaudhary, H.K. Auxin induced haploid induction in wide crosses of durum wheat. *Cereal Res. Commun.* **2019**, *47*, 552–565. [CrossRef]
83. Baklouti, E.; Beulé, T.; Nasri, A.; Romdhane, A.B.; Drira, R.; Doulbeau, S.; Rival, A.; Drira, N.; Fki, L. 2,4-D induction of somaclonal variations in in vitro grown date palm (*Phoenix dactylifera* L. cv Barhee). *Plant Cell Tissue Organ Cult.* **2022**, *150*, 1–15. [CrossRef]
84. Bronsema, F.B.F.; van Oostveen, W.J.F.; van Lammeren, A.A.M. Influence of 2,4-D, TIBA and 3,5-D on the growth response of cultured maize embryos. *Plant Cell Tissue Organ Cult.* **2001**, *65*, 45–56. [CrossRef]
85. Juzoń, K.; Warchoł, M.; Dziurka, K.; Czyczyło-Mysza, I.; Marcińska, I.; Skrzypek, E. The effect of 2,4-dichlorophenoxyacetic acid on the production of oat (*Avena sativa* L.) doubled haploid lines through wide hybridization. *PeerJ* **2022**, *10*, e12854. [CrossRef]
86. Sharma, D.R.; Kaur, R.; Kumar, K. Embryo rescue in plants—A review. *Euphytica* **1996**, *89*, 325–337. [CrossRef]
87. Lulsdorf, M.M.; Ferrie, A.; Slater, S.M.H.; Yuan, H.Y. Methods and role of embryo rescue technique in alien gene transfer. In *Alien Gene Transfer in Crop Plants, Innovations, Methods and Risk Assessment*; Pratap, A., Kumar, J., Eds.; Springer: New York, NY, USA, 2014; Volume 1, pp. 77–103.
88. Laibach, F. Das Taubwerden von Bastardsamen und die künstliche Aufzucht früh absterbender Bastardembryonen. *Z. Bot.* **1925**, *17*, 417–459.
89. Bridgen, M.P. A review of plant embryo culture. *Hort. Sci.* **1994**, *29*, 1243–1246. [CrossRef]
90. Warchoł, M.; Czyczyło-Mysza, I.; Marcińska, I.; Dziurka, K.; Noga, A.; Skrzypek, E. The effect of genotype, media composition, pH and sugar concentrations on oat (*Avena sativa* L.) doubled haploid production through oat × maize crosses. *Acta Physiol. Plant* **2018**, *40*, 93. [CrossRef]
91. Zhuang, J.J.; Xu, J. Increasing differentiation frequencies in wheat pollen callus. In *Cell and Tissue Culture Techniques for Cereal Crop Improvement*; Hu, H., Vega, M.R., Eds.; Science Press: Beijing, China, 1983; p. 431.
92. Dziurka, K.; Dziurka, M.; Muszyńska, E.; Czyczyło-Mysza, I.; Warchoł, M.; Juzoń, K.; Laskoś, K.; Skrzypek, E. Anatomical and hormonal factors determining the development of haploid and zygotic embryos of oat (*Avena sativa* L.). *Sci. Rep.* **2022**, *12*, 548. [CrossRef]
93. Gurtay, G.; Kutlu, I.; Avci, S. Production of haploids in ancient, local and modern wheat by anther culture and maize pollination. *Acta Biol. Crac. Ser. Bot.* **2021**, *63*, 43–53.
94. Orlikowska, T.; Chrząstek, M. Kultury zarodków roślinnych in vitro i możliwości ich wykorzystania w hodowli. *Postępy Nauk. Rol.* **1979**, *1*, 27–42. (In Polish)
95. Noga, A.; Skrzypek, E.; Warchoł, M.; Czyczyło-Mysza, I.; Dziurka, K.; Marcińska, I.; Juzoń, K.; Warzecha, T.; Sutkowska, A.; Nita, Z.; et al. Conversion of oat (*Avena sativa* L.) haploid embryos into plants in relation to embryo developmental stage and regeneration media. *Vitr. Cell. Dev. Biol.-Plant* **2016**, *52*, 590–597. [CrossRef]
96. Skrzypek, E.; Warchoł, M.; Czyczyło-Mysza, I.; Marcińska, I.; Nowakowska, A.; Dziurka, K.; Juzoń, K.; Noga, A. The effect of light intensity on the production of oat (*Avena sativa* L.) doubled haploids through oat × maize crosses. *Cereal Res. Comm.* **2016**, *44*, 490–500. [CrossRef]

97. Dziurka, K.; Dziurka, M.; Warchoł, M.; Czyczyło-Mysza, I.; Marcińska, I.; Noga, A.; Kapłoniak, K.; Skrzypek, E. Endogenous phytohormone profile during oat (*Avena sativa* L.) haploid embryo development. *Vitr. Cell. Dev. Biol.-Plant* **2019**, *55*, 221–229. [CrossRef]
98. Kynast, R.G.; Riera-Lizarazu, O.; Vales, M.I.; Okagaki, R.J.; Maquieira, S.B.; Chen, G.; Ananiev, E.V.; Odland, W.E.; Russel, C.D.; Stec, A.O.; et al. A complete set of maize individual chromosome additions of the oat genome. *Plant Phys.* **2001**, *125*, 1216–1227. [CrossRef]
99. Skrzypek, E.; Warzecha, T.; Noga, A.; Warchoł, M.; Czyczyło-Mysza, I.; Dziurka, K.; Marcińska, I.; Kapłoniak, K.; Sutkowska, A.; Nita, Z.; et al. Complex characterization of oat (*Avena sativa* L.) lines obtained by wide crossing with maize (*Zea mays* L.). *PeerJ* **2018**, *6*, e5107. [CrossRef]
100. Hilli, H.J.; Kapoor, R. An overview of breeding objectives to improve the economically important traits in oat. *Curr. Agric. Res. J.* **2023**, *11*, 18–27. [CrossRef]
101. Gasparis, S.; Nadolska-Orczyk, A. Oat (*Avena sativa* L.). In *Agrobacterium Protocols. Methods in Molecular Biology*; Wang, K., Ed.; Springer: New York, NY, USA, 2015; Volume 1223. [CrossRef]
102. Karimi-Ashtiyani, R.; Ishii, T.; Niessen, M.; Stein, N.; Heckmann, S.; Gurushidze, M.; Banaei-Moghaddam, A.M.; Fuchs, J.; Schubert, V.; Koch, K.; et al. Point mutation impairs centromeric CENH3 loading and induces haploid plants. *Proc. Natl. Acad. Sci. USA* **2015**, *112*, 11211–11216. [CrossRef]

Disclaimer/Publisher's Note: The statements, opinions and data contained in all publications are solely those of the individual author(s) and contributor(s) and not of MDPI and/or the editor(s). MDPI and/or the editor(s) disclaim responsibility for any injury to people or property resulting from any ideas, methods, instructions or products referred to in the content.

Article

Antioxidant Response in the Salt-Acclimated Red Beet (*Beta vulgaris*) Callus

Jarosław Tyburski * and Natalia Mucha

Department of Plant Physiology and Biotechnology, Nicolaus Copernicus University, Lwowska 1, 87-100 Torun, Poland; mucha_nata13@umk.pl
* Correspondence: tybr@umk.pl

Abstract: Callus cultures initiated from red beet tubers were acclimated to 75 or 100 mM NaCl salinity by exposing them to gradually increasing NaCl concentrations. The acclimated callus lines displayed growth rates comparable to the control culture cultivated on the NaCl-free medium. Several antioxidant system components were analyzed to assess the role of the antioxidant defense in the acclimated callus's ability to proliferate on salt-supplemented media. It was found that proline and ascorbate concentrations were increased in salt-acclimated callus lines with respect to the control line. On the other hand, glutathione concentration was unchanged in all tested callus lines. Total activities of the antioxidant enzymes, namely superoxide dismutase (SOD, EC 1.15.1.1), catalase (CAT, EC 1.11.1.6), ascorbate peroxidase (APX, EC 1.11.1.11), and class III peroxidase (POX, EC 1.11.1.7) were increased in salt-acclimated cultures. The enzymatic components of the antioxidant systems were upregulated in a coordinated manner during the initial phases of the culture cycle when the increase in callus fresh mass occurs.

Keywords: antioxidant; ascorbate; ascorbate peroxidase; beet; callus; class III peroxidase; proline; salt stress; superoxide dismutase

Citation: Tyburski, J.; Mucha, N. Antioxidant Response in the Salt-Acclimated Red Beet (*Beta vulgaris*) Callus. *Agronomy* **2023**, *13*, 2284. https://doi.org/10.3390/agronomy13092284

Academic Editors: Justyna Lema-Rumińska, Danuta Kulpa and Alina Trejgell

Received: 15 August 2023
Revised: 25 August 2023
Accepted: 28 August 2023
Published: 30 August 2023

Copyright: © 2023 by the authors. Licensee MDPI, Basel, Switzerland. This article is an open access article distributed under the terms and conditions of the Creative Commons Attribution (CC BY) license (https://creativecommons.org/licenses/by/4.0/).

1. Introduction

Currently, it is estimated that about 20% of the world's agricultural land and 33% of all irrigated land is exposed to excessive salinization. Moreover, the soil areas threatened with excessive salinity are increasing at an alarming rate of 10% per year. According to estimates, by 2050, over 50% of arable land will be considered saline [1]. From year to year, largely as a result of human activity, the problem of soil salt contamination is becoming more and more important [2,3]. The salinity, due to changes in the osmotic pressure between the cell and its environment, and the toxic effects of ions, mainly Na^+ and Cl^-, triggers both the osmotic and the ionic stress. This is manifested by water deficit, oxidative stress, disorganization of cell membranes, disturbances in the mineral nutrient homeostasis, reduced cell division rate and photosynthetic rates, genotoxicity, and sometimes cell death [2,4,5].

In response to the reductions in the osmotic potential of the salt-contaminated soil solution, plants developed a mechanism of osmotic adjustment. It consists of the synthesis of osmoprotective compounds and their accumulation in the cell sap. Several organic compounds of different chemical properties, such as proline, mannitol, sorbitol, betaine, or spermine, belong to this category. These molecules allow for maintaining an appropriate osmotic balance between the vacuole, the cytoplasm, and the external environment of the cell. In addition, many of them act as antioxidants [6].

Exposure to various biotic and abiotic agents, including salinity, is followed by increased rates of reactive oxygen species (ROS) formation in plant cells [7]. Due to their high and nonspecific reactivity, ROS may provoke oxidative damage to the cell structures and macromolecules, such as lipids, proteins, and nucleic acids. In order to prevent oxidative damage, the antioxidant systems are activated in plant cells. Their action consists of restoring the oxidative balance by scavenging excess ROS molecules. The antioxidant systems include compounds with enzymatic activity and non-enzymatic antioxidants. Many studies have confirmed that the activation of antioxidant enzymes such as superoxide dismutase (SOD), glutathione and ascorbate peroxidase (GPX and APX), catalase (CAT), or glutathione reductase (GR) and the accumulation of non-enzymatic antioxidants such as ascorbate (ASC), glutathione (GSH), tocopherol, flavonoids, and carotenoids contributes to the formation of salt stress tolerance in plants [8,9].

Several studies demonstrated that callus cultures may serve as a model system for studies on the cellular dimension of salt stress tolerance mechanism [10]. The callus consists, in its major part, of rapidly dividing cells. Therefore, it may be particularly advantageous to use callus for studies on salt stress effects on proliferating plant cell populations. Due to the relatively large scale of the culture, high growth rates and susceptibility to phytohormonal stimuli, and ability to induce different developmental pathways, the callus offers an easily accessible alternative or useful complement to studies performed using intact plant meristems [11–13]. Furthermore, tissue culture techniques, including callus cultures, were frequently proposed, either as a tool to identify markers of salt tolerance [14] or for breeding salt-tolerating crops. The latter may be accomplished by regenerating plants from the salt-tolerant somaclonal callus variants selected for salt tolerance [10,15].

Furthermore, callus culture may be exploited as a source of useful metabolites. Red beet callus is particularly promising in this regard due to its capacity to synthesize betalains, a class of nitrogen-containing plant pigments present in a vacuole of most plant families classified in the order *Caryophyllales* [16]. Betalains are subdivided into red-violet betacyanins (betains) and yellow-orange betaxanthins. Red beet tubers, which contain two main pigments, betanin (red betacyanin) and vulgaxanthin I (yellow betaxanthin), are the most common source of betalains [17]. Due to their strong antioxidant properties, these pigments are considered a desirable component of the everyday diet. Possible health benefits of betalains and betalain-containing red beet juice were frequently reported. The effectiveness of betanins in the long-term inhibition of the development of skin and liver tumors has been demonstrated in mice [18]. It has also been found that betanins and betanidines, acting at very low concentrations, have the ability to inhibit the process of lipid oxidation and heme breakdown in vitro [19]. The ability of betalains to prevent the oxidation of endothelial cells has also been observed [20]. It has also been proven that betalains have the ability to increase the activity of quinone reductase, the enzyme involved in detoxification processes associated with cancer chemoprevention [21]. Red beet juice ingestion has been suggested to have beneficial effects on endothelial functions due to its high nitrate content. After ingestion, the nitrate from the beet juice is first reduced to nitrite and then easily converted to nitric oxide. The latter promotes artery vasodilation and thus may contribute to cardiovascular disease prevention. Therefore, red beet ingestion was particularly recommended during the recent COVID pandemic for periods of home confinement, when changes in lifestyle behavior, such as unhealthy diet, emotional disorders, and reduced physical activity, may increase the risk of cardiovascular disease. It was also speculated that the boost of NO production following red beet juice ingestion may prevent the release of a large amount of pro-inflammatory cytokines (called the "cytokine storm"), which is involved in the development of respiratory distress and multiple organ failure under severe COVID-19 [22]. Red beet juice is the most common source of beneficial beet metabolites, but plant tissue cultures offer several advantages over raw plant material, such as the possibility of manipulating the betalain composition and synthesis rate [23,24].

Here, we show that the callus cultures derived from red beet tubers and acclimated to either 100 or 75 mM NaCl display growth increments comparable to the control culture grown on the NaCl-free medium. Then, we analyzed several enzymatic and non-enzymatic components of the antioxidant defense system to assess their role in sustaining callus growth on salt-supplemented media.

2. Material and Methods

2.1. Red Beet Tuber Callus and Its Acclimation to Salinity

The plant material used for the study was a callus culture derived from red beet (*Beta vulgaris*) hypocotylar tuber explants. Red beet tubers were purchased at a local market, rinsed in distilled water, surface sterilized with 70% ethanol for 5 min, and then, for 20 min, in a 50% solution of commercial bleach (about 2% Cl_2). Next, the tubers were washed several times for 5 min in sterile distilled water. Afterward, cuboid fragments of storage parenchyma were excised under aseptic conditions and placed on a solid medium containing mineral salts and vitamins according to Murashige and Skoog's medium [25] and 3% sucrose, 0.75% agar, 1 mg/L 6-benzylaminopurine (BAP), and 0.1 mg/L 2,4-dichlorophenoxyacetic acid (2,4-D). The pH of the medium was adjusted to 5.7 before autoclaving the medium. Phytohormones were added to the medium from filter-sterilized stock solutions after autoclaving. Callus culture was carried out in a culture room under continuous irradiation and at 25 °C.

The callus was subcultured at 4-week intervals into the fresh medium. During subsequent subcultures, three callus lines were established. The control callus line was grown on the medium of the composition mentioned above. Simultaneously, in order to produce salt-acclimated callus lines, distinct callus cultures were subjected to gradual acclimation to salinity by transferring tissue onto the NaCl-supplemented media. For this purpose, the NaCl content in the media was increased by 5–10 mM NaCl with each subsequent subculture until final concentrations of 75 or 100 mM NaCl in the medium were achieved. The callus lines generated in this way were referred to as 75 mM NaCl-acclimated callus or 100 mM NaCl-acclimated callus. Then, the acclimated lines were cultivated for two years on media containing target NaCl concentrations. Subcultures were initiated using a callus inoculum of 1 g. The samples of callus tissue were collected in weekly intervals for fresh mass and biochemical parameter analysis.

2.2. Total Ascorbate Content

The assay was based on the reduction of Fe^{3+} to Fe^{2+} by ASC after reducing dehydroascorbic acid to ascorbic acid and the spectrophotometric detection of Fe^{2+} complexed with 2.2'-dipyridyl. In order to isolate ascorbate, samples consisting of 250 mg of callus were homogenized with 5 volumes of 5% TCA at 4 °C in a porcelain mortar. The homogenate was centrifuged at $10,000 \times g$ for 10 min at 4 °C. The supernatant was collected for analysis of ascorbate. In order to reduce dehydroascorbic acid to ascorbic acid, 135 µL of supernatant was mixed with 16.87 µL of 10 mM dithiothreitol (DTT) and 16.87 µL of 80 mM K_2HPO_4 and incubated 5 min at room temperature. Afterwards, the following reagents were added in sequence: 80 µL of 85% H_3PO_4, 1.37 mL of 0.5% 2,2'-dipiridyl, and 280 µL of 1% ferric chloride. The samples were allowed to stay for 30 min at room temperature for the color to develop. Then, the absorbance at 525 nm (A_{525}) was measured. The values of absorbance were compared with a standard curve based on ASC in the range of 0–50 µg mL^{-1}.

2.3. Total Glutathione Content

The measurement principle was based on the measurement of the rate of reduction of 5,5'-dithiobis(2-nitrobenzoic acid) (DTNB) to 2-nitro-5-thiobenzoic acid (TNB) by glutathione, which was oxidized to GSSG. The resulting GSSG was converted back to GSH by glutathione reductase at the expense of NADPH oxidation. To extract glutathione, 0.2 g of callus tissue was frozen in liquid nitrogen and ground to a powder using a porcelain mortar. After that, the powder was extracted with 2 mL of 5% sulfosalicylic acid. The resulting homogenate was centrifuged for 10 min at $10,000 \times g$ at 4 °C. The glutathione content was assayed with a Glutathione Assay Kit (Sigma-Aldrich, Burlington, VT, USA) according to the producer's protocol.

2.4. Proline Content

Proline was measured following the methods of Abrahám et al. [26]. The proline was extracted with 3% sulfosalicylic acid (5 µL/mg fresh weight). Next, 100 µL of the extract was mixed with 100 µL of 3% sulfosalicylic acid, 200 µL glacial acetic acid, and 200 µL acidic ninhydrin. The mixture was incubated at 96 °C for 60 min, cooled on ice, and extracted with 1 mL toluene. The absorbance of the chromophore phase was measured at 520 nm against toluene as a reference.

2.5. Enzyme Extraction, Assays, and Isoenzyme Profiling

The protein extract for superoxide dismutase, catalase, and peroxidase activity assay was prepared by homogenizing 0.5 g callus tissue at 4 °C in 1.5 mL of the homogenization buffer composed of 100 mM phosphate buffer, pH 7.8, 1 mM EDTA, 8 mM $MgCl_2$, 4 mM DTT, and 0.1% Triton X-100. The extraction of ascorbate peroxidase was performed in the 100 mM phosphate buffer, pH 7.5, supplemented with 1 mM EDTA and 5 mM ascorbic acid. Ascorbate oxidase was extracted with 100 mM phosphate buffer (pH 6.1) supplemented with 0.5 mM EDTA. The homogenates were centrifuged at 4 °C for 10 min at 10,000× g, and supernatants containing the extracted enzymes were used for the activity assays and isoenzyme analysis. Total protein was assayed according to Bradford [27] with BSA as a standard.

The activity of superoxide dismutase (SOD, EC 1.15.1.1) was assayed by its ability to inhibit photochemical reduction in NBT at 560 nm [28]. The assays were carried out in the 1.6 mL of the reaction mixture composed of 50 mM sodium phosphate buffer (pH 7.8), 33 mM NBT, 10 mM L-methionine, 0.66 mM EDTA, 0.0033 mM riboflavin, and 50 µL of enzyme extract. The concentrations of the reagents are their final concentrations in the reaction mixture. Simultaneously, control samples were prepared, where enzyme extract was replaced by homogenization buffer. Riboflavin was added last to the reaction mixtures. Thereafter, the test and control tubes were irradiated under constant white light for 10 min. Following irradiation, the A_{560} was measured against the nonirradiated reaction mixture. One unit of SOD was defined as the amount of enzyme that inhibits 50% NBT photoreduction. Cyanide-resistant SOD activity was assayed in the presence of 10 mM KCN. Peroxide-resistant SOD activity was assayed in the presence of 5 mM H_2O_2.

The activity of the SOD isoforms was determined after the electrophoretic separation of protein extract in the polyacrylamide gel. Samples containing 200 µg of total protein were subjected to discontinuous PAGE under nondenaturating, nonreducing conditions as described by Laemmli [29] using a stacking gel containing 4% (w/v) acrylamide and a separating gel containing 12% (w/v) acrylamide. After the completion of electrophoresis, the gels were stained for the activities of SOD. To visualize the SOD activity, the method of Rao et al. [30] was applied. Firstly, the gels were incubated for 25 min with 100 mM phosphate buffer, pH 7.8, containing 2.5 mM NBT. Next, the gels were transferred to the solution containing 28 mM tetramethyl ethylene diamine and 28 µM riboflavin in 100 mM phosphate buffer, pH 7.8, and kept for 25 min. in the darkness with gentle agitation. After that, the gels were placed in distilled water and exposed to white light for 10–15 min. until the bands were visible. SOD isoenzymes appeared as colorless bands on a deep-blue background. The identification of SOD isoforms was based on the inhibitory effect of KCN and H_2O_2 on the activity of SOD isoenzymes. After the accomplishment of the electrophoresis, the gels were incubated in the NBT-containing buffer, prepared as described above but supplemented with either 5 mM KCN or 5 mM H_2O_2. Following the 25-min-long incubation, the procedure of the visualization of the SOD activity was performed. Different sensitivity of given isoforms to inhibitors—KCN and H_2O_2—was used. Namely, the activity of FeSOD is inhibited by hydrogen peroxide, while KCN remains insensitive. The manganese isoform is not inhibited by KCN or inactivated by H_2O_2. On the other hand, CuZnSOD is sensitive to both KCN and H_2O_2 [31–33].

The activity of catalase (CAT, EC 1.11.1.6) was assayed according to Rao et al. [30]. The measurement of CAT activity was based on the following decrease in absorbance at

240 nm in 1 mL of the reaction mixture composed of 100 mM phosphate buffer, pH 7.0, 30 µL enzymatic extract, and 1.5 µL of 30% (v/v) H_2O_2. The decrease in A_{240} was followed at 25 °C against a plant extract-free blank. The activity of CAT was expressed as the number of µmol of H_2O_2 decomposed during 1 min by 1 mg of total protein [30].

The assay for ascorbate peroxidase (APX, EC 1.11.1.11) activity was performed according to Rao et al. [30]. The APX activity was determined by following the decrease in A_{290} for 3 min in 1 mL of reaction mixture composed of 100 mM K^+-phosphate buffer, pH 7.5, 0.5 mM ASC, and 0.2 mM H_2O_2 [30,34]. The concentrations of the reagents are their final concentrations in the reaction mixture. A slight decrease in absorbance observed in the absence of enzyme extract was subtracted. No decrease in absorbance occurred unless H_2O_2 was present in the reaction mixture. The activity of APX was expressed as a number of µmol ASC oxidized by 1 mg of total protein in 1 min [30,34].

Gels for APX staining were run in carrier buffer containing 2 mM ascorbate. These gels were prerun for 30 min to allow ascorbate present in the buffer to enter the gel prior to the application of samples [30,35]. Staining the gels for APX activity was performed according to Rao et al. [30]. Firstly, the gels were equilibrated with 50 mM K^+-phosphate buffer, pH 7.0, containing 2 mM ASC for 30 min. Then, the gels were incubated for 20 min in a reaction mixture composed of 4 mM ASC and 2 mM H_2O_2 in 50 mM K^+-phosphate buffer, pH 7.0. Then, the gels were briefly washed with buffer and stained in the solution of 28 mM tetramethyl ethylene diamine and 2.45 mM nitroblue tetrazolium in 50 mM K^+-phosphate buffer, pH 7.8. The APX bounds remained colorless on the deep blue background [30].

The activity of soluble class III peroxidases (POX, EC 1.11.1.7) was determined by recording changes in absorbance in a 1 mL reaction mixture containing 100 mM phosphate buffer, pH 6.0, 60 mM pyrogallol, 0.66 mM H_2O_2, and 5 µL of protein extract. The concentrations of the reagents are their final concentrations in the reaction mixture. The reaction was started by adding H_2O_2 to the mixture. Absorbance measurement at λ = 420 nm was recorded for 60 s against phosphate buffer using a U-1800 spectrophotometer (Hitachi, Tokyo, Japan). In parallel, absorbance changes in enzyme-free reaction mixtures were determined to correct for non-enzymatic oxidation of pyrogallol.

In order to separate peroxidase isoenzymes, 50 µg of total protein was subjected to native electrophoresis on polyacrylamide gel. The separating gel containing 7.5% (w/v) acrylamide and a stacking gel containing 4% (w/v) acrylamide were used. After the electrophoresis, the gels were immersed in the solution containing 0.63 mM o-dianisidine and 30 mM H_2O_2 in 50 mM phosphate buffer, pH 6.0, for 15 min. Peroxidase isoenzymes stained red.

2.6. Statistics

Statistical significance of differences between mean values of biochemical parameters were determined with two-way ANOVA to evaluate the effect of two factors (callus line and sampling point) and their interaction. The significance of differences between experimental variants was assessed in the course of a multiple comparison analysis employing Tukey's HSD. Differences at the level of $p < 0.05$ were considered significant. All statistical tests were performed in SigmaPlot 11.0 (Systat Software, San Jose, CA, USA). Callus growth and the biochemical parameter analysis were repeated three times during three independent culture periods. At each sampling point, five independent cultures, representing 75 mM NaCl-acclimated callus, 100 mM NaCl-acclimated callus, or control callus, were analyzed. The mean and standard deviation were calculated for each of the three callus lines separately. Error bars shown in all figures represent the standard deviation calculated from all repetitions representing a given callus line in each experiment.

3. Results

3.1. Callus Morphology and Its Fresh Mass (FM)

The callus cultures, derived from red beet tubers, were characterized by intensive red pigmentation. The control line was quite uniformly red-pigmented (Figure 1A). The salt-acclimated calli were also predominantly composed of red-pigmented tissue, but the bright-red, yellowish, and greenish clumps of the callus were discernible (Figure 1B,C).

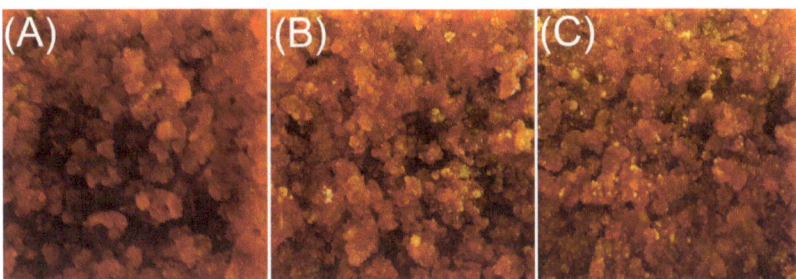

Figure 1. Photographs of the representative cultures of the red beet tuber—derived callus, representing the control callus line, grown on the NaCl—free medium (**A**), 75 mM NaCl—acclimated, (**B**) and 100 mM NaCl—acclimated callus line (**C**). The photographs were taken 28 days after subculture.

Callus FM gradually increased in all lines over the culture period. The results of the two-way ANOVA indicated a significant difference in FM between the sampling points (F = 64.025, $p < 0.001$). However, the largest FM increments occurred during the first 14 days of the culture period, resulting in significant differences in the FM between the values determined at the two first sampling points. Measurements performed 21 and 28 days after subculture show that growth increments, occurring during the second half of the culture period, were smaller, with negligible differences between the sampling points. The tissues acclimated to salinity were characterized by comparable FM to one of the control line when analyzed during subsequent sampling points. No significant differences in the FM were detected between the tested lines at all sampling points. At the first sampling point, i.e., 7 days after subculture, the acclimated lines displayed smaller FM than the control one, but as mentioned above, the difference did not pass the significance test (Figure 2). Comparing the difference in the FW between the 7th and 14th day of culture shows a higher increase in the FW in the salt-acclimated lines (115% and 136%, for 75 mM NaCl- and 100 mM NaCl-acclimated callus, respectively), as compared to the control line (49%). However, no statistically significant differences were detected between the tested lines (F = 3.217, $p = 0.052$), and no interaction between the callus line and sampling points was detected (F = 1.302, $p = 0.281$). These data show that the 75 mM or 100 mM NaCl-acclimated lines are not deficient in their growth rate with respect to the control line grown on the salt-free medium.

3.2. Ascorbate, Glutathione, Proline

The outcome of the two-way ANOVA analysis indicated the significant effect of the callus line (F = 60,170, $p < 0.001$) and the length of culture (F = 26.1, $p < 0.001$) on the ascorbate level, and there was an interaction between the two factors (F = 8.985, $p < 0.001$). After the first week of the growth cycle, the ascorbate content was similar in the control line and the salt-acclimated lines. The ascorbate level increased by the end of the second week of the culture period in all callus lines. However, in the control line, the antioxidant content decreased over the second half of the culture, whereas it remained relatively stable in the salt-acclimated lines. Consequently, the majority of the culture period was marked by a significantly increased ascorbate content in the lines acclimated to either 75 mM or 100 mM NaCl when compared to the control line. The highest ascorbate levels were detected in the 100 mM NaCl-acclimated line. In this line, the ascorbate level was 53% and 77% higher than in the control line after 21 and 28 days of culture, respectively. (Figure 3A). In contrast to ascorbate, the glutathione content was stable across the growth cycle (F = 2.267, $p = 0.097$), and it did not differ significantly between the salt-acclimated and the control callus line (F = 1.338, $p = 0.275$, Figure 3B). The proline content in callus differed significantly in function of both the callus line (F = 77.140, $p < 0.001$) and the sampling time (F = 40.299, $p < 0.001$). Furthermore, there was a statistically significant interaction between the two factors (F = 3.886, $p = 0.008$). The concentration of proline in the control line and the line adapted to the salinity of 75 mM NaCl was similar over the entire culture period. On the other hand, the line acclimated 100 mM NaCl accumulated significantly higher proline levels. Irrespective of the callus line, the proline levels slightly decreased in all lines over the first three weeks of the growth cycle. However, during this period, proline concentrations in the 100 mM NaCl-acclimated line were significantly higher when compared to other callus lines. When compared to the control line, the proline levels in the 100 mM NaCl-acclimated callus were 46%, 56%, and 80% higher after 7, 14, and 21 days of culture, respectively. After 28 days of the growth cycle, the proline content was similar in the tested callus lines (Figure 3C).

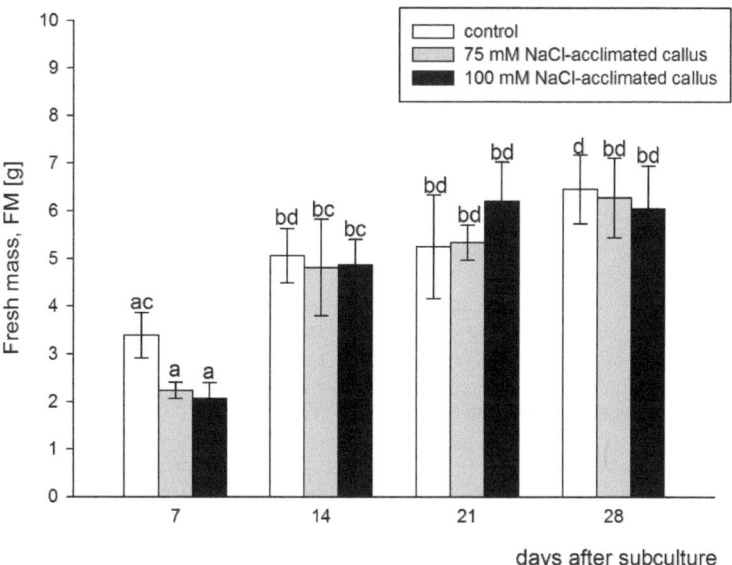

Figure 2. Time course of the fresh mass increments in the red beet tuber–derived callus acclimated to 75 (grey bars) or 100 mM NaCl (black bars) or cultured on basal medium (control, white bars). Fresh mass was assessed after 7, 14, 21, and 28 days of the four–week–long culture period. Different letters denote significant differences at $p < 0.05$.

3.3. Superoxide Dismutase (SOD) and Catalase (CAT)

Irrespective of the callus line, the highest SOD activity was observed 7 days after subculture. After that, the SOD activity gradually decreased until the lowest levels were attained after 28 days of culture. Over the entire culture period, a clear correlation of dismutase activity with the presence of NaCl in the culture medium was observed. A significantly higher level of SOD activity was noted in tissues acclimated to salinity, compared to the callus cultured on medium without the addition of NaCl. After 7 days of culture, the lowest SOD activity was detected in the control tissue, intermediate one in the tissue adapted to 100 mM NaCl (47% higher than control), and the highest activity was found in the tissue adapted to 75 mM NaCl (62% higher than control). When analyzed after 14, 21, or 28 days of culture, a dose–response relationship was observed between the salt concentration and the SOD activity, with the highest SOD activities revealed in a 100 mM NaCl-acclimated line (Figure 4A). Compared to the control line, SOD activity increased by 31%, 36%, and 16% in the 75 mM NaCl-acclimated line and 48%, 53%, and 50% in the 100 mM NaCl-acclimated line on the 14th, 21st, and 28th day, respectively. The results of the two-way ANOVA show that the SOD activity was strongly dependent on the type of the callus line (F = 533.902, $p < 0.001$) and on the stage of the culture period (F = 1116, $p < 0.001$). There was also a significant interaction between the two factors (43.194, $p < 0.001$).

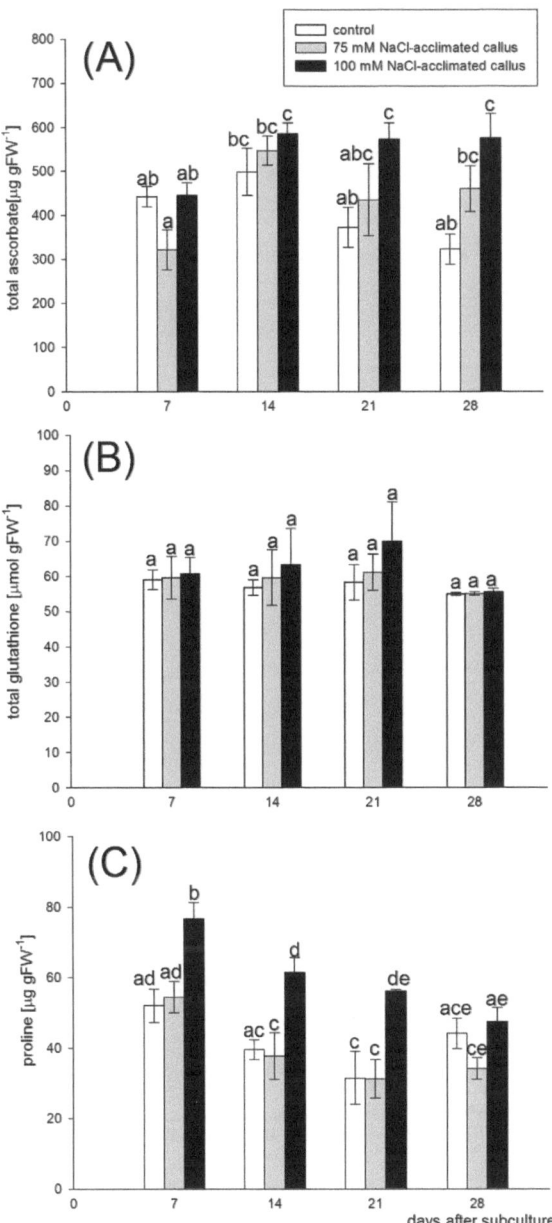

Figure 3. Ascorbate (**A**), glutathione (**B**), and proline (**C**) content in the red beet tuber–derived callus acclimated to 75 (grey bars) or 100 mM NaCl (black bars) or cultured on basal medium (control, white bars). The metabolites were assessed after 7, 14, 21, and 28 days of the four–week–long culture period. Different letters denote significant differences at $p < 0.05$.

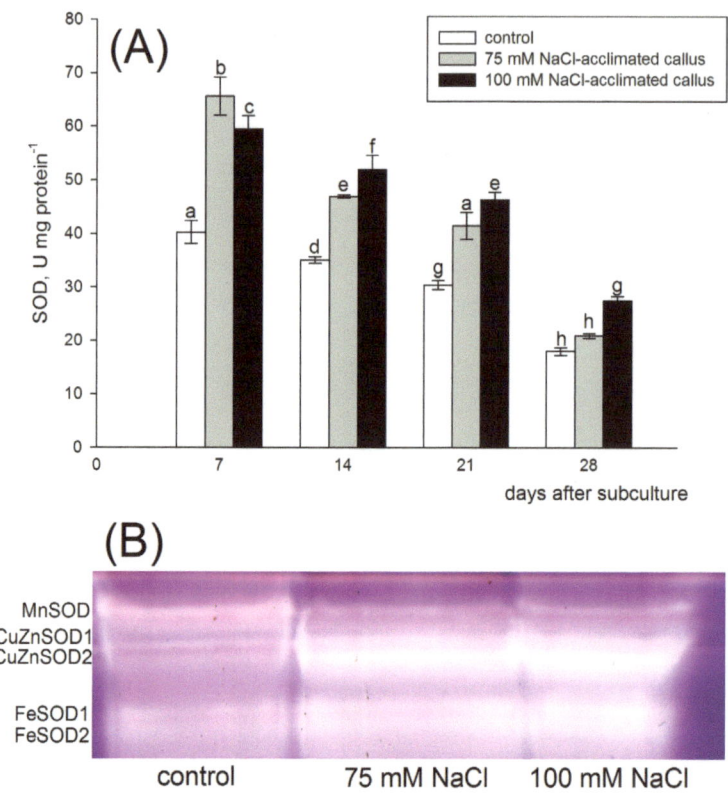

Figure 4. Total superoxide dismutase (SOD) activity (**A**) and zymogram analysis of SOD isoenzymes (**B**) in the red beet tuber−derived callus acclimated to 75 (grey bars) or 100 mM NaCl (black bars) or cultured on basal medium (control, white bars). SOD activity was assessed after 7, 14, 21, and 28 days of the four−week−long culture period. Different letters denote significant differences at $p < 0.05$. SOD isoenzyme analysis was performed 14 days after subculture.

Superoxide dismutase isoenzyme pattern analysis in the red beet callus revealed the presence of one MnSOD, two CuZnSOD, and two FeSOD isoenzymes. Total protein preparations were analyzed for SOD isoenzyme patterns 14 days after subculture. The bands representing CuZnSOD and FeSOD displayed visibly higher intensity in salt-acclimated tissues than in the control line. Contrastingly, the MnSOD activity was higher in the control line (Figure 4B).

All lines displayed the highest CAT activities in the first week of the growth cycle. A dose–response dependence of the CAT activity on the salt concentration in the medium was observed at this sampling point. In the 75 mM NaCl-acclimated callus, CAT activity was higher than in the control line, but the difference was not statistically significant. Significantly higher enzyme activity, as compared to the control, was detected in the tissues acclimated to 100 mM NaCl (80% increase). In the following weeks, CAT activity strongly decreased in all callus lines, and there were also no significant differences between them, except the 100 mM NaCl-acclimated callus, determined 21 days after subculture. At this sampling point, the CAT activity was significantly lower than in the control and the one acclimated to 75 mM NaCl (Figure 5). The two-way ANOVA analysis did not indicate the significant effect of the type of the callus line on CAT activity (F = 0.128, $p < 0.880$).

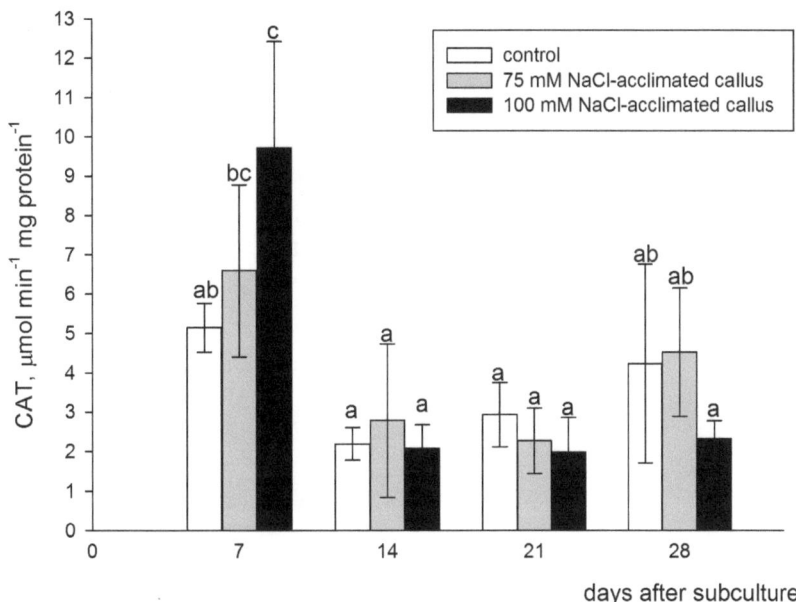

Figure 5. Catalase (CAT) activity in the red beet tuber–derived callus acclimated to 75 (grey bars) or 100 mM NaCl (black bars) or cultured on basal medium (control, white bars). CAT activity was assessed after 7, 14, 21, and 28 days of the four–week–long culture period. Different letters denote significant differences at $p < 0.05$.

3.4. Ascorbate Peroxidase

APX activity was relatively stable over the entire culture period in all callus lines, except the 28 days after subculture when the activity in the salt-acclimated lines significantly decreased when compared to the previous part of the culture period. However, during the first three weeks of the culture, APX activity in the salt-acclimated lines was maintained at significantly higher levels than in the control line. Compared to the control line, APX activity increased by 26%, 73%, and 62% in the 75 mM NaCl-acclimated line and 45%, 90%, and 64% in the 100 mM NaCl-acclimated line on the 7th, 14th and 21st day, respectively. Contrastingly, after 4 weeks of the culture, APX activities in the salt-acclimated lines dropped below the level detected in the control line, with a 25% and 34% decrease, in the 75 mM NaCl- and 100 mM NaCl-acclimated line, respectively (Figure 6A). The results of the two-way ANOVA show that both factors, i.e., the type of callus line and the sampling point, during the culture period (F = 147.107, $p < 0.001$ and F = 335.955, $p < 0.001$, respectively) affected APX activity and a statistically significant interaction between the APX activity and the stage of the culture period (F = 49.342, $p < 0.001$) occurred.

Analysis of the APX isoenzyme pattern performed 14 days after subculture revealed the presence of one APX isoenzyme, which was common for callus lines (APX1). The band was most intensely stained in the 75 mM NaCl-acclimated line. In contrast to the control line and the 75 mM NaCl-acclimated line, the 100 mM NaCl-acclimated callus was distinguished by the second APX isoenzyme (APX2). The band representing this isoenzyme was characterized by strong intensity in the 100 mM NaCl-acclimated but was absent in other lines (Figure 6B).

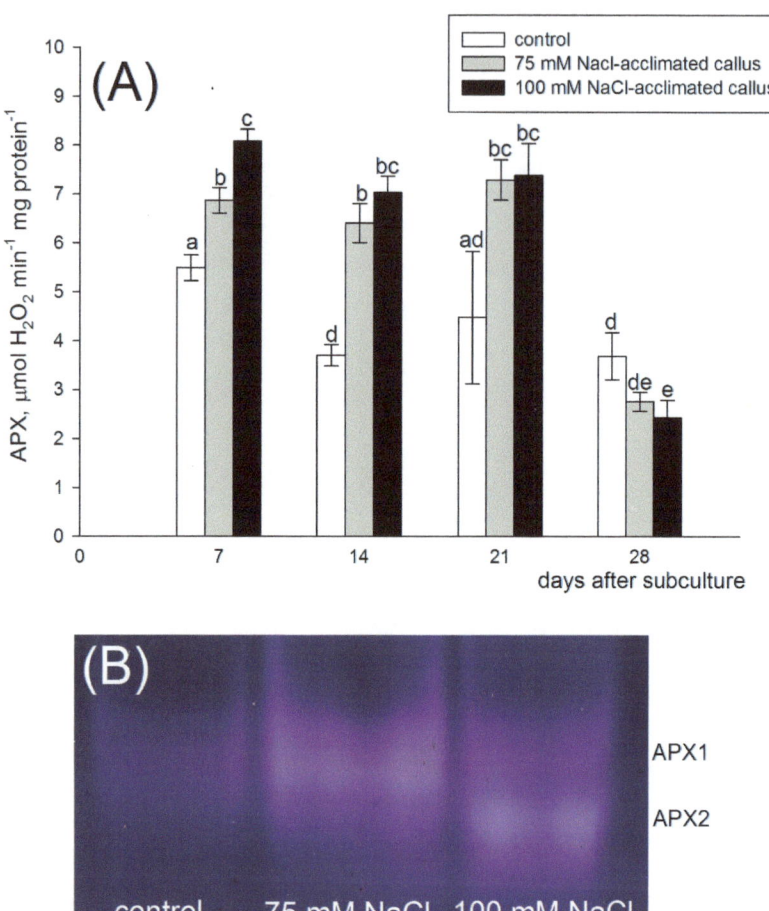

Figure 6. Ascorbate peroxidase (APX) activity (**A**) and zymogram analysis of APX isoenzymes (**B**) in the red beet tuber–derived callus acclimated to 75 (grey bars) or 100 mM NaCl (black bars) or cultured on basal medium (control, white bars). APX activity was assessed after 7, 14, 21, and 28 days of the four–week–long culture period. Different letters denote significant differences at $p < 0.05$. APX isoenzyme analysis was performed 14 days after subculture.

3.5. Class III Peroxidases

POX activity was dependent on the type of the callus culture (F = 142.870, $p < 0.001$) and the sampling time (F = 117.949, $p < 0.001$). There was also a significant interaction between these factors (F = 36.632, $p < 0.001$). During the first three weeks of the culture period, the POX activity was significantly increased in the salt-acclimated lines in a dose–response manner. Compared to the control line, POX activity increased by 46%, 49%, and 146% in the 75 mM NaCl-acclimated line and 155%, 153%, and 175% in the 100 mM NaCl-acclimated line on the 7th, 14th, and 21st day, respectively. The control line displayed a relatively stable level of POX activity over the first three weeks of the culture period. However, the enzyme's activity increased in the fourth week. The 75 mM NaCl-acclimated line was marked by a gradual increase in the POX activity till the third week after subculture. On the other hand, the POX activity in 100 mM NaCl-acclimated callus reached its maximal level when determined 14 days after subculture. Then, a gradual decrease in its activity occurred (Figure 7A).

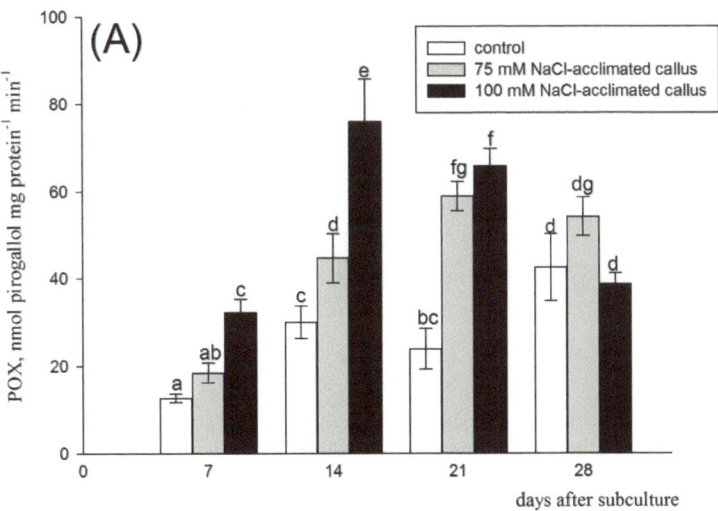
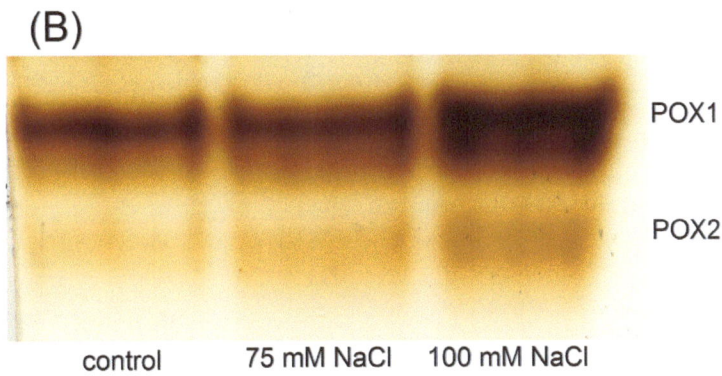

Figure 7. Class III peroxidase (POX) activity (**A**) and zymogram analysis of POX isoenzymes (**B**) in the red beet tuber−derived callus acclimated to 75 (grey bars) or 100 mM NaCl (black bars) or cultured on basal medium (control, white bars). POX activity was assessed after 7, 14, 21, and 28 days of the four−week−long culture period. Different letters denote significant differences at $p < 0.05$. POX isoenzyme analysis was performed 14 days after subculture.

Soluble peroxidase isoenzyme analysis revealed the presence of two peroxidase isoenzymes in the control line (POX1 and 2). The same soluble peroxidase isoenzymes were present in the salt-acclimated lines, but they were more intensive in these lines, suggesting higher peroxidase activity (Figure 7B).

4. Discussion

Beets are highly salt-tolerant crops. In spite of the recent progress in the field of elucidating the salt-tolerance traits in beets at the morphological, physiological, and molecular levels [36], the complete view of the mechanisms underlying the outstanding salt tolerance in this crop is still far from being accomplished. New experimental systems are required to provide a deeper understanding of the salt tolerance at different levels of the plant's body. Callus is an unorganized or loosely organized structure [11]. Therefore, we assume that studies on salt-acclimated calli may contribute to deciphering the salt-tolerance mechanism acting solely at the cellular level in the absence of correlative interactions occurring in the highly structured organs and tissues of the intact plant body. Thereafter, the analyses of the salt-acclimated calli may complement the salt tolerance studies performed using entire plants.

Here, we selected salt-tolerant red beet callus lines by gradually increasing salt concentration in the culture medium over subsequent subcultures until 75 or 100 mM NaCl was reached. Then, the culture media supplemented with final salt concentrations was used for a long-term culture. Red pigmentation of cultures proves that they retained the red beet tuber tissue's capacity for betalain synthesis. However, some differences in pigmentation between the lines suggest that salt treatment may, to some extent, affect pigment composition in the callus (Figure 1). Clarifying the mechanism behind this finding needs further research. Callus growth rate, analyzed over the four-week-long culture period, shows that the growth of salt-acclimated calli was slightly retarded, with respect to the control line, at the beginning of the culture period. However, in subsequent phases of the culture period, the growth increments of the salt-acclimated lines were comparable to the control line (Figure 2). Therefore, we conclude that the NaCl-treated lines achieved tolerance to the salt stress. Several studies were conducted with different plant tissue cultures in order to obtain salinity-resistant cell lines [37]. Similar to our results, the NaCl-tolerant callus, derived from sunflower cotyledons, showed no reductions in growth when grown in the presence of 175 mM NaCl [38]. Previous salt acclimation seems to be mandatory for callus growth on salt-supplemented media. For example, the mung bean callus culture, not adapted to growth on a medium supplemented with 150 mM sodium chloride, showed no signs of growth and died within two weeks. In contrast, the salt-adapted cell lines, transferred to media containing 150 mM NaCl concentration, showed a marked growth of the tissue. However, in this case, the tissue mass was lower compared to the callus cultured in a salt-free medium [39]. Similar results were obtained by comparing the growth of the potato calli, either exposed directly to 50–200 mM NaCl or gradually acclimated to increased NaCl concentrations. Both treatments were effective in selecting tissues acclimated to salinity, but better results were obtained when concentrations of NaCl were gradually added to the medium. Only in the case of callus culture carried on a medium with 200 mM NaCl did the cells not show a clear growth and many of them died after two weeks of the experiment, irrespective of the mode of treatment [40].

Since oxidative stress is a common consequence of salinity, the antioxidant metabolites, ascorbate and glutathione, were investigated in the salt-acclimated and control calli. Data shown in Figure 3A show that negligible differences in ascorbate content between the salt-acclimated and control lines were present in the first half of the cycle period. However, in the second half of the cycle period, the ascorbate content was maintained at elevated levels (especially in the 100 mM NaCl-acclimated callus), whereas the antioxidant content decreased in the control line. This finding suggests that increased ascorbate contents may contribute to salt tolerance in the acclimated lines. Furthermore, ascorbate levels may affect betalain levels in the callus since ascorbic acid is a cofactor of DOPA-dioxygenase, a key enzyme involved in betalain synthesis [41]. It was also suggested that the degradation of betalains during storage is suppressed in the presence of ascorbate [42]. On the other hand, the control line and salt-acclimated lines did not differ in the glutathione content throughout the culture period (Figure 3B). The results of previous studies show that GSH and/or ASC usually increase in plant callus tissues subjected to salinity stress. Studies on calli derived from potato tissues showed a 30% increase in ascorbate concentration in calli adapted to 50 mM NaCl medium compared to the control line, whereas a twofold increase in ASC content was observed in tissues cultured in the presence of 100 mM NaCl [14,40]. Cotton callus accumulated higher concentrations of ascorbate compared to the control line when cultured in a medium with 150 mM NaCl. An increase in glutathione levels relative to the control line was also observed [43]. On the other hand, the salt-tolerant line of sunflower callus contained much less GSH as compared to the callus cultured in the absence of salinity [38].

Proline is a versatile molecule being involved in mitigating salt stress due to its osmoprotectant, antioxidant, and redox-buffering properties [44]. Based on the results of the proline measurement, which show that solely the 100 mM NaCl-acclimated callus accumulates its increased levels (Figure 3C), it can be concluded that the ability to increase proline content under salinity conditions is dependent primarily on the stress intensity, and not on the presence of salt itself. Noteworthy, the proline content attained its highest levels at the beginning of the culture period. Then, its content gradually declines to the minimal values attained at the final phase of the culture cycle (Figure 3C). Possibly, the osmoprotective effect of proline is important at the initial phase of the growth cycle to ensure efficient water and nutrient uptake for intensively growing tissue. The decline of proline concentration in the later phase may be beneficial for callus growth and viability since the high proline may impart toxic effects if over-accumulated or applied exogenously at excessive concentrations [45]. Proline accumulation is a common phenomenon in salt-adapted calli. Salt-adapted and salt-stressed calli of potato accumulated about 10 times more proline than the control [14]. Under salt stress, callus developed from rice seeds accumulated higher levels of proline as compared to the unstressed control conditions [46]. Proline content was strongly enhanced in the soybean callus line selected for salt tolerance [47]. Proline was also accumulated in salt-stressed calli of *Salicornia* sp. and *Guizotia abyssinica* [48,49].

The activities of intracellular antioxidant enzymes, namely CAT, SOD, and APX, were markedly enhanced in the salt-acclimated lines, especially in the initial and middle phases of the culture period (Figures 4–6). This finding is particularly striking for CAT. The activity of this enzyme, as well as the ones of SOD and APX, was increased, in the salt-acclimated lines, solely in the first week of the culture period. Since CAT has a relatively low affinity for H_2O_2, the enzyme is an efficient H_2O_2 scavenger only at high oxidant concentrations [50,51]. High oxidant levels may be present in cells at the beginning of the cycle period, during the first days following subculture onto salt-supplemented medium, when the rapid growth increments occur, whereas CAT activity quickly declines, elevated SOD and APX activities are maintained until the third week of the culture period (Figures 4–6). The activities of these enzymes decline in the final phase of the culture period (Figures 4–6), when no significant increment in the callus growth occurs (Figure 2).

The rapid growth is promoted by a high metabolic rate. Its side effect is increased ROS production, which is further accelerated if callus cells are grown on salt-containing media [52]. Therefore, the increased activities of the antioxidant enzymes may promote growth on salt-supplemented media by safeguarding normal cell function under salinity, and thus, they contribute to the salt-acclimation mechanism in the red beet callus. In line with our results, increased SOD and APX activities were noticed in mung bean callus tissue, tolerant to 150 mM NaCl. Contrastingly, the activities of these enzymes were decreased and highly inhibited by salt in the callus nonselected for salt tolerance [30]. NaCl-tolerant lines of the eggplant callus showed significantly higher SOD and CAT activities compared to control [15]. In contrast to our results, SOD and APX activities were lower in salt-adapted calli of potato in relation to the control, cultured on salt-free medium [14].

Zymogram analysis of SOD activity revealed the presence of five superoxide dismutase isoforms in red beet calli. Namely, the one representing MnSOD, FeSOD, and two CuZnSOD isoenzymes were distinguished. MnSOD activity was reduced, but FeSOD and CuZnSOD activities were markedly enhanced in the salt-acclimated lines (Figure 4B). Therefore, the increased total SOD activity in the salt-acclimated lines may be attributed to the increase in the activity of FeSOD and CuZnSOD isoenzymes (Figure 4B). The acclimation to salinity enhanced the activity of isoenzymes which were present in the control line, i.e., the salt-acclimation did not change the SOD isoenzyme pattern. A similar situation was observed in the potato callus, where the same SOD isoenzymes were present in the salt-tolerant and the control calli. However, they differed in staining intensity between the adapted and control calli [14]. In contrast to SOD, the APX zymogram pattern analysis revealed a separate isoenzyme (APX2), which was expressed solely in the 100 mM NaCl-acclimated callus. On the other hand, the 75 mM NaCl-acclimated line showed an increased activity of the isoenzyme, represented by the band of the same mobility as in the control line (APX1) (Figure 6B). Therefore, the APX isoenzyme pattern in the red beet salt-acclimated callus is regulated by salt in a salt dose manner. Contrastingly, the native PAGE of the soluble fraction from both control and salt-adapted potato calli showed two bands with the same mobility [14].

Soluble POX activity strongly increased over three weeks following callus subculture in the cell lines acclimated to the saline medium. On the other hand, the POX activity in the control was relatively stable during the culture period. Significant differences in peroxidase activity were found both between the control line and both salinity acclimated lines, as well as between tissues cultured in the presence of 75 and 100 mM NaCl. Enzymatic activity in 100 mM NaCl-acclimated callus was the highest. Analysis of the peroxidase activity of proteins separated by native PAGE was in line with total activity measurements. Two bands corresponding to soluble peroxidase isoenzymes were present in all callus lines, and both appeared with higher staining intensity if extracted from salt-acclimated lines (Figure 7B). High and maintained over the major part of the culture period, peroxidase activity may largely contribute to antioxidant safeguard against salt toxicity in the salt-acclimated calli. The mechanism underlying the role of POX in mitigating oxidative stress in the salt-acclimated red beet callus requires further studies. Possibly, the betalains might be involved in the POX-dependent H_2O_2 detoxification. Previous research indicates that betalains can act as electron donors for POX to detoxify H_2O_2 [42,53]. Being abundant in the red beet callus (Figure 1), the betalains may provide the reducing force for the POX-catalyzed H_2O_2 scavenging parallel to one of the ascorbic acids in the APX-catalyzed reaction. Similar to other antioxidant enzymes, a decrease in POX activity, especially in the 100 mM NaCl-acclimated line, coincided with the transition of the culture to the stationary phase (Figure 7A). Stimulatory effects of salinity on POX activity in plant tissue cultures were usually reported, but POX activity frequently differed, depending on the culture technique and plant species. Salt-adapted tomato suspension culture was characterized by elevated POX activity, but in contrast to our results, maximal enzyme activities were reached during the last days of the growth cycle [54]. The salt-tolerant line of sunflower callus displayed increased POX activity during a 28-day-long culture period [38]. The glutathione peroxidase activity was increased under NaCl treatments in the alcaligrass (*Puccinellia tenuifora*) callus. However, in this study, the nonsalinity acclimated tissue was used [55]. In the calli derived from halophytic species, such as *Salicornia* sp., the salt treatments may either increase or decrease POX activities, depending on the plant species [48].

5. Conclusions

We developed two lines of the salt-acclimated red beet callus and determined its growth and several antioxidative parameters after a long-term culture on salt-supplemented media. The study revealed that SOD, CAT, APX, and POX activities were characterized by higher activity values in salinity-acclimated tissues when compared to the control callus cultured on salt-free media. From among the low molecular substances determined in the tested lines, the ascorbate and proline concentrations were increased in the acclimated lines, whereas the glutathione content was at the control level. The enzymatic components of the antioxidant systems and proline content were more pronounced during the initial phases of the culture cycle when the increase in callus fresh mass occurs. An increased level of antioxidant protection contributes to sustaining cell division and cell expansion underlying the red beet callus growth increments under continuous exposition to salinity.

Author Contributions: Conceptualization, J.T.; Methodology, J.T.; Software, N.M.; Validation, N.M.; Formal analysis, N.M.; Investigation, J.T.; Writing—original draft, J.T.; Writing—review & editing, N.M.; Visualization, N.M. All authors have read and agreed to the published version of the manuscript.

Funding: This research received no external funding. The APC was funded by Cells as EXperimental platforms and bioFACTories (CExFact) program running in the Nicolaus Copernicus University in Torun.

Data Availability Statement: Not applicable.

Conflicts of Interest: The authors declare no conflict of interest.

References

1. Shrivastava, P.; Kumar, R. Soil salinity: A serious environmental issue and plant growth promoting bacteria as one of the tools for its alleviation. *Saudi J. Biol. Sci.* **2015**, *22*, 123–131. [CrossRef]
2. Dajic, Z. Salt stress. In *Physiology and Molecular Biology of Stress Tolerance in Plants*; Madhava, R.K., Raghavebdra, A., Reddy, J., Eds.; Springer: Berlin/Heidelberg, Germany, 2006; pp. 41–99.
3. Munns, R.; Tester, M. Mechanisms of salinity tolerance. *Plant Biol.* **2008**, *59*, 651–681. [CrossRef] [PubMed]
4. Munns, R. Comparative physiology of salt and water stress. *Plant Cell Environ.* **2002**, *25*, 239–250. [CrossRef] [PubMed]
5. Zhu, J. *Plant Salt Stress*; Wiley: Hoboken, NJ, USA, 2007; pp. 1–3.
6. Xiong, L.; Schumaker, K.; Zhu, J. Cell signaling during cold, drought, and salt stress. *Plant Cell* **2002**, *14*, S165–S183. [CrossRef] [PubMed]
7. Hasanuzzaman, M.; Nahar, K.; Fujita, M. Plant response to salt stress and role of exogenous protectants to mitigate salt-induced damages. In *Ecophysiology and Responses of Plants under Salt Stress*; Ahmad, P., Azooz, M.M., Prasad, M.N.V., Eds.; Springer Science & Business Media: Berlin/Heidelberg, Germany, 2013.
8. Hasegawa, P.M.; Bressan, R.A.; Zhu, J.K.; Bohnert, H.J. Plant cellular and molecular responses to high salinity. *Ann. Rev. Plant Physiol. Plant Mol. Biol.* **2000**, *51*, 463–499. [CrossRef]
9. Gupta, B.; Huang, B. Mechanism of Salinity Tolerance in Plants: Physiological, Biochemical, and Molecular Characterization. *Int. J. Genom.* **2014**, *2014*, 701596. [CrossRef]
10. Rai, M.; Kalia, R.; Singh, R.; Gangola, M.P.; Dhawan, A. Developing stress tolerant plants through in vitro selection—An overview of the recent progress. *Environ. Exp. Bot.* **2011**, *71*, 89–98. [CrossRef]
11. Ikeuchi, M.; Sugimoto, K.; Iwase, A. Plant Callus: Mechanisms of Induction and Repression. *Plant Cell* **2013**, *25*, 3159–3173. [CrossRef]
12. Efferth, T. Biotechnology Applications of Plant Callus Cultures. *Engineering* **2019**, *5*, 50–59. [CrossRef]
13. Fehér, A. Callus, Dedifferentiation, Totipotency, Somatic Embryogenesis: What These Terms Mean in the Era of Molecular Plant Biology? *Front. Plant Sci.* **2019**, *10*, 536. [CrossRef]
14. Queirós, F.; Rodrigues, J.A.; Almeida, J.M.; Almeida, D.P.F.; Fidalgo, F. Differential responses of the antioxidant defense system and ultrastructure in a salt-adapted potato cell line. *Plant Physiol. Biochem.* **2011**, *49*, 1410–1419. [CrossRef]
15. Hannachi, S.; Werbrouck, S.; Bahrini, I.; Abdelgadir, A.; Siddiqui, H.A.; Van Labeke, M.C. Obtaining Salt Stress-Tolerant Eggplant Somaclonal Variants from In Vitro Selection. *Plants* **2021**, *10*, 2539. [CrossRef] [PubMed]
16. Mabry, T.J. Selected topics from forty years of natural products research: Betalains to flavonoids, antiviral proteins, and neurotoxic nonprotein amino acids. *J. Nat. Prod.* **2001**, *64*, 1596–1604. [CrossRef] [PubMed]
17. Fu, Y.; Shi, J.; Xie, S.Y.; Zhang, T.Y.; Soladoye, O.P.; Aluko, R.E. Red Beetroot Betalains: Perspectives on Extraction, Processing, and Potential Health Benefits. *J. Agric. Food Chem.* **2020**, *68*, 11595–11611. [CrossRef] [PubMed]
18. Kapadia, G.J.; Tokuda, H.; Konoshima, T.; Nishino, H. Chemoprevention of lung and skin cancer by *Beta vulgaris* (beet) root extract. *Cancer Lett.* **1996**, *100*, 211–214. [CrossRef]
19. Kanner, J.; Harel, S.; Granit, R. Betalains—A new class of dietary cationized antioxidants. *J. Agric. Food Chem.* **2001**, *49*, 5178–5185. [CrossRef]
20. Gentile, C.; Tesoriere, L.; Allegra, M.; Livrea, M.A.; D'Alessio, P. Antioxidant betalains from cactus pear (*Opuntia ficus-indica*) inhibit endothelial ICAM-1 expression. *Ann. N. Y. Acad. Sci.* **2004**, *1028*, 481–486. [CrossRef] [PubMed]
21. Lee, C.H.; Wettasinghe, M.; Bolling, B.W.; Ji, L.L.; Parkin, K.L. Betalains, phase-II enzyme-inducing components from red beetroot (*Beta vulgaris* L.) extracts. *Nutr. Cancer* **2005**, *53*, 91–103. [CrossRef]
22. Volino-Souza, M.; de Oliveira, G.V.; Conte-Junior, C.A.; Alvares, T.S. COVID-19 Quarantine: Impact of Lifestyle Behaviors Changes on Endothelial Function and Possible Protective Effect of Beetroot Juice. *Front. Nutr.* **2020**, *7*, 582210. [CrossRef]
23. Girod, P.A.; Zryd, J.P. Clonal variability and light induction of betalain synthesis in red beet cell cultures. *Plant Cell Rep.* **1987**, *6*, 27–30. [CrossRef]
24. Girod, P.A.; Zryd, J.P. Secondary metabolism in cultured red beet (*Beta vulgaris* L.) cells: Differential regulation of betaxanthin and betacyanin biosynthesis. *Plant Cell Tissue Organ Cult.* **1991**, *25*, 1–12. [CrossRef]
25. Murashige, T.; Skoog, F. A Revised Medium for Rapid Growth and BioAssays with Tobacco Tissue Cultures. *Physiol. Plant.* **1962**, *15*, 473–497. [CrossRef]
26. Abraham, E.; Hourton-Cabassa, C.; Erdei, L.; Szabados, L. Methods for Determination of Proline in Plants. *Methods Mol Biol.* **2010**, *20*, 317–331.

27. Bradford, M.M. A rapid and sensitive method for the quantitation of microgram quantities of protein utilizing the principle of protein-dye binding. *Anal Biochem.* **1976**, *7*, 248–254. [CrossRef]
28. Beauchamp, C.; Fridovich, I. Superoxide dismutase: Improved assays and an assay applicable to acrylamide gel. *Anal. Biochem.* **1971**, *44*, 276–287. [CrossRef]
29. Laemmli, U.K. Cleavage of structural proteins during the assembly of the head of bacteriophage T4. *Nature* **1970**, *227*, 680–685. [CrossRef]
30. Rao, M.V.; Paliyath, G.; Ormrod, D.P. Ultraviolet-B- and ozone-induced biochemical changes in antioxidant enzymes of *Arabidopsis thaliana*. *Plant Physiol.* **1996**, *110*, 125–136. [CrossRef]
31. Guzik, T.J.; Olszanecki, R.; Sadowski, J.; Kapelak, B.; Rudziński, P.; Jopek, A.; Kawczynska, A.; Ryszawa, N.; Loster, J.; Jawien, J.; et al. Superoxide dismutase activity and expression in human venous and arterial bypass graft vessels. *J. Physiol. Pharmacol.* **2005**, *56*, 313–323.
32. Veljovic-Jovanovic, S.; Kukavica, B.; Stevanovic, B.; Navari-Izzo, F. Senescence- and drought-related changes in peroxidase and superoxide dismutase isoforms in leaves of *Ramonda serbica*. *J. Exp. Bot.* **2006**, *57*, 1759–1768. [CrossRef]
33. Miszalski, Z.; Ślesak, I.; Niewiadomska, E.; Baczek-Kwinta, R.; Lüttge, U.; Ratajczak, R. Subcellular localization and stress responses of superoxide dismutase isoforms from leaves in the C3-CAM intermediate halophyte *Mesembryanthemum crystallinum* L. *Plant Cell Environ.* **1998**, *21*, 169–179. [CrossRef]
34. Chen, G.-X.; Asada, K. Ascorbate Peroxidase in Tea Leaves: Occurrence of Two Isozymes and the Differences in Their Enzymatic and Molecular Properties. *Plant Cell Physiol.* **1989**, *30*, 987–998.
35. Mittler, R.; Zilinskas, B.A. Detection of ascorbate peroxidase activity in native gels by inhibition of the ascorbate-dependent reduction of nitroblue tetrazolium. *Anal. Biochem.* **1993**, *212*, 540–546. [CrossRef] [PubMed]
36. Yolcu, S.; Alavilli, H.; Ganesh, P.; Panigrahy, M.; Song, K. Salt and Drought Stress Responses in Cultivated Beets (*Beta vulgaris* L.) and Wild Beet (*Beta maritima* L.). *Plants* **2021**, *10*, 1843. [CrossRef] [PubMed]
37. Gandonou, C.B.; Errabii, T.; Abrini, J. Selection of callus cultures of sugarcane (*Saccharum* sp.) tolerant to NaCl and their response to salt stress. *Plant Cell Tissue Organ Cult.* **2006**, *87*, 9–16. [CrossRef]
38. Davenport, S.B.; Gallego, S.M.; Benavides, M.P. Behaviour of antioxidant defense system in the adaptive response to salt stress in *Helianthus annuus* L. cells. *Plant Growth Regul.* **2003**, *40*, 81–88. [CrossRef]
39. Rao, S.; Patil, P. In Vitro Selection of Salt Tolerant Calli Lines and Regeneration of Salt Tolerant Plantlets in Mung Bean (*Vigna radiata* L. Wilczek). In *Biotechnology—Molecular Studies and Novel Applications for Improved Quality of Human Life*; Sammour, R., Ed.; IntechOpen: London, UK, 2012.
40. Queirós, F.; Fidalgo, F.; Santos, I. In vitro selection of salt tolerant cell lines in *Solanum tuberosum* L. *Biol. Plant.* **2007**, *51*, 728–734. [CrossRef]
41. Sasaki, N.; Abe, Y.; Goda, Y.; Adachi, T.; Kasahara, K.; Ozeki, Y. Detection of DOPA 4,5-Dioxygenase (DOD) Activity Using Recombinant Protein Prepared from *Escherichia coli* Cells Harboring cDNA Encoding DOD from *Mirabilis jalapa*. *Plant Cell Physiol.* **2009**, *50*, 1012–1016. [CrossRef]
42. Sakihama, Y.; Yamasaki, H. Phytochemical Antioxidants: Past, Present and Future. *IntechOpen* **2021**. [CrossRef]
43. Gossett, D.R.; Millhollon, E.P.; Lucas, M.C.; Banks, S.W.; Marney, M.M. The effects of NaCl on antioxidant enzyme activities in callus tissue of salt-tolerant and salt-sensitive cotton cultivars (*Gossypium hirsutum* L.). *Plant Cell Rep.* **1994**, *13*, 498–503. [CrossRef]
44. Szabados, L.; Savoure, A. Proline: A Multifunctional Amino Acid. *Trends Plant Sci.* **2009**, *15*, 89–97. [CrossRef]
45. Hayat, S.; Hayat, Q.; Alyemeni, M.N.; Wani, A.S.; Pichtel, J.; Ahmad, A. Role of proline under changing environments: A review. *Plant Signal. Behav.* **2012**, *11*, 1456–1466. [CrossRef] [PubMed]
46. Summart, J.; Thanonkeo, P.; Panichajakul, S.; Prathepa, P. Effect of salt stress on growth, inorganic ion and proline accumulation in Thai aromatic rice, Khao Dawk Mali 105, callus culture. *Afr. J. Biotechnol.* **2010**, *9*, 145–152.
47. Liu, T.; van Staden, J. Selection and characterization of sodium chloride-tolerant callus of *Glycine max* (L.) Merr cv. Acme. *Plant Growth Regul.* **2000**, *31*, 195–207. [CrossRef]
48. Torabi, S.; Niknam, V. Effects of Iso-osmotic Concentrations of NaCl and Mannitol on some Metabolic Activity in Calluses of Two *Salicornia* species. *In Vitro Cell. Dev. Biol.-Plant* **2011**, *47*, 734–742. [CrossRef]
49. Ghane, S.G.; Lokhande, V.; Nikam, T. Growth, physiological, and biochemical responses in relation to salinity tolerance for In Vitro selection in oil seed crop *Guizotia abyssinica* Cass. *J. Crop Sci. Biotech.* **2014**, *17*, 11–20. [CrossRef]
50. Mittler, R.; Zilinskas, B.A. Purification and characterization of pea cytosolic ascorbate peroxidase. *Plant Physiol.* **1991**, *97*, 962–968. [CrossRef]
51. König, J.; Baier, M.; Horling, F.; Kahmann, U.; Harris, G.; Schürmann, P.; Dietz, K.J. The plant-specific function of 2-Cys peroxiredoxin-mediated detoxification of peroxides in the redox-hierarchy of photosynthetic electron flux. *Proc. Natl. Acad. Sci. USA* **2002**, *99*, 5738–5743. [CrossRef]
52. Niknam, V.; Meratan, A.A.; Ghaffari, S.M. The effect of salt stress on lipid peroxidation and antioxidative enzymes in callus of two *Acanthophyllum* species. *In Vitro Cell. Dev. Biol.-Plant* **2011**, *47*, 297–308. [CrossRef]
53. Allegra, M.; Tesoriere, L.; Livrea, M.A. Betanin inhibits the myeloperoxidase/nitrite-induced oxidation of human low-density lipoproteins. *Free Rad. Res.* **2007**, *41*, 335–341. [CrossRef]

54. Sancho, M.A.; de Forchetti, S.M.; Pliego, F.; Valpuesta, V.; Quesada, M.A. Peroxidase activity and isoenzymes in the culture medium of NaCl adapted tomato suspension cells. *Plant Cell Tissue Organ Cult.* **1996**, *44*, 161–167. [CrossRef]
55. Zhang, Y.; Zhang, Y.; Yu, J.; Zheng, H.; Wang, L.; Wang, S.; Guo, S.; Miao, Y.; Chen, S.; Li, Y.; et al. NaCl-responsive ROS scavenging and energy supply in alkaligrass callus revealed from proteomic analysis. *BMC Genom.* **2019**, *20*, 990. [CrossRef] [PubMed]

Disclaimer/Publisher's Note: The statements, opinions and data contained in all publications are solely those of the individual author(s) and contributor(s) and not of MDPI and/or the editor(s). MDPI and/or the editor(s) disclaim responsibility for any injury to people or property resulting from any ideas, methods, instructions or products referred to in the content.

Article

Effect of Different Culture Conditions on Anthocyanins and Related Genes in Red Pear Callus

Wantian Yao, Diya Lei, Xuan Zhou, Haiyan Wang, Jiayu Lu, Yuanxiu Lin, Yunting Zhang, Yan Wang, Wen He, Mengyao Li, Qing Chen, Ya Luo, Xiaorong Wang, Haoru Tang and Yong Zhang *

College of Horticulture, Sichuan Agricultural University, Chengdu 611130, China; 2021305082@stu.sicau.edu.cn (W.Y.); 20152539@stu.sicau.edu.cn (D.L.); 2021205001@stu.sicau.edu.cn (X.Z.); 2022205002@stu.sicau.edu.cn (H.W.); jiayulu@stu.sicau.edu.cn (J.L.); linyx@sicau.edu.cn (Y.L.); asyunting@gmail.com (Y.Z.); wangyanwxy@163.com (Y.W.); hewen0724@gmail.com (W.H.); limy@sicau.edu.cn (M.L.); supnovel@gmail.com (Q.C.); luoya945@163.com (Y.L.); wangxr@sicau.edu.cn (X.W.); htang@sicau.edu.cn (H.T.)
* Correspondence: zhyong@sicau.edu.cn

Abstract: Red pears are appreciated for their abundant nutritional benefits and visually striking red hue, rendering them a favored option among consumers and stimulating substantial market demand. The present study employs the flesh of a red pear as the explant, subjecting the flesh callus to varying sugar sources, MS concentrations, light qualities, and temperatures to investigate the alterations in secondary metabolites, including anthocyanins, within the callus. It was found that sucrose can induce more anthocyanins, and its related metabolites and genes also increase as the sucrose and MS concentrations increase. Under the conditions of red-blue light and a temperature of 15 °C, it can further induce the production of more anthocyanins and secondary metabolites and can also upregulate the synthesis of anthocyanin-related genes. As such, this investigation serves to elucidate the factors that contribute to anthocyanin accumulation in red pears, thereby providing a theoretical foundation for understanding the mechanisms underlying color change.

Keywords: callus; pear; anthocyanin; sucrose; light; temperature

Citation: Yao, W.; Lei, D.; Zhou, X.; Wang, H.; Lu, J.; Lin, Y.; Zhang, Y.; Wang, Y.; He, W.; Li, M.; et al. Effect of Different Culture Conditions on Anthocyanins and Related Genes in Red Pear Callus. *Agronomy* 2023, 13, 2032. https://doi.org/10.3390/agronomy13082032

Academic Editors: Justyna Lema-Rumińska, Danuta Kulpa and Alina Trejgell

Received: 25 June 2023
Revised: 23 July 2023
Accepted: 25 July 2023
Published: 31 July 2023

Copyright: © 2023 by the authors. Licensee MDPI, Basel, Switzerland. This article is an open access article distributed under the terms and conditions of the Creative Commons Attribution (CC BY) license (https://creativecommons.org/licenses/by/4.0/).

1. Introduction

The process of coloration in most fruits involves the degradation of chlorophyll and the accumulation of anthocyanins, which are naturally occurring water-soluble pigments that are found in various plant tissues and organs [1,2]. Anthocyanins contribute to the red, purple, and blue colors that are observed in plants [3]. The specific coloration of a plant is influenced by various factors, including the type of anthocyanin present, the plant species, and environmental conditions such as light exposure [4]. Anthocyanins can play a crucial role in promoting fruit coloration, enhancing fruit quality, providing resistance to UV damage and pathogens, and improving resistance to adverse environmental conditions and challenges [5]. Furthermore, anthocyanins exhibit antioxidant properties and are capable of scavenging free radicals, thus slowing down the aging process [6]. Available evidence suggests that anthocyanins may also decrease the incidence of several chronic diseases, such as cardiovascular disease and type 2 diabetes, while enhancing visual function, memory, and preventing Alzheimer's disease [7–10]. Consequently, increasing the anthocyanin content in fruits can effectively enhance their commercial value.

Researchers have demonstrated that the concentration and type of anthocyanins significantly influence the red coloration of pears [11]. There is a relatively limited number of red-skinned pear varieties in China, with even fewer cultivars exhibiting bright coloring, superior quality, and large fruit sizes [12]. Due to their vibrant coloration and diverse nutritional benefits, red pears have gained immense popularity and tremendous demand among consumers, thereby boosting the market value of this fruit [13]. According to the available literature on fruit germplasm resources, there are records of 652 pear varieties, only 10% of which comprise red-skinned cultivars, accounting for a total of 89 varieties [14]. However, abiotic and biotic stresses pose significant threats to the synthesis and stability of anthocyanins in pears, leading to their degradation at any point during plant growth and development even before harvest. In red pears, high temperatures can substantially contribute to anthocyanin degradation, which often results in the loss of pre-harvest reddening [15]. Thus, it is crucial to explore and investigate various factors that impact the synthesis and degradation of anthocyanins to better understand the underlying mechanisms and implications for improving the commercial value of red pears.

Anthocyanins possess superior antioxidant activity in comparison to other flavonoids, owing to the presence of oxygen atoms carrying a positive charge [16]. Wang et al. conducted a comparative analysis of the active components and antioxidant activity in the juices of four small berries including raspberry, blueberry, lingonberry, and Lonicera cerulea. The study revealed a positive correlation between the levels of polyphenols, anthocyanins, SOD (superoxide dismutase), total flavonoids, and the corresponding clearance rates [17]. The biosynthesis of floral pigments initiates from phenylalanine in the cytoplasm and undergoes sequential hydroxylation, glycosylation, methylation, and acylation modifications, ultimately leading to their accumulation in vacuoles [18]. This intricate process is tightly regulated by a combination of structural and regulatory genes. The biosynthesis of flavonoid glycosides, on the other hand, is controlled by several key structural genes encoding specific enzymes, including CHI (Chalcone Isomerase), CHS (Chalcone Synthase), F3H (Flavanone 3-Hydroxylase), PAL (Phenylalanine Ammonia Lyase), DFR (Dihydroflavonol-4-Reductase), ANS (Anthocyanidin Synthase), and UFGT (UDP-glucose: flavonoid 3-O-glucosyltransferase) [19,20].

To improve the biosynthesis and accumulation of anthocyanins, it is crucial to develop biotechnologies that are capable of regulating experimental environmental conditions effectively [21]. In this study, we utilize callus material to investigate the changes in anthocyanin content in red pear. Notably, tissue culture can be initiated under controlled conditions, irrespective of seasonal variations [22]. Therefore, we conducted an experiment with the aim of exploring how different conditions, including varied sugar sources, light sources, and temperatures, influence anthocyanin synthesis. Our findings can serve as a theoretical foundation for enhancing the anthocyanin accumulation in red pears, and hold potential for applications in agricultural production, such as the promotion of fruit coloration and the genetic variations of fruit color.

2. Materials and Methods

2.1. Different Treatments of Induced Callus Anthocyanins

Mature 90 d 'Red Zao Su' red pear, sourced from Chongzhou base of Sichuan Agricultural University, were thoroughly rinsed with running water for two hours. Subsequently, they were disinfected with 75% ethanol on a culture bench for 30 s, followed by a 20 min treatment with 2% sodium hypochlorite solution. Finally, the fruits were rinsed three times with sterile water. Subsequently, the mature 'Red Zao Su' red pear pulp was placed in MS (Murashige and Skoog, 1962) medium containing 1.0 mg·L^{-1} BA (6-Benzyladenine), 1.0 mg·L^{-1} 2,4-D (2,4-Dichlorophenoxyacetic acid), agar 7.0 g·L^{-1}, pH = 6.0. After 30 d, the grown callus was cut off and transferred to 1.0 mg·L^{-1} BA and 0.5 mg·L^{-1} NAA (Naphthaleneacetic acid) medium for 3 weeks in dark culture. An amount of 1.0 g of callus was transferred to MS medium with the same hormone ratios. In the context of callus culture, the following five treatments were performed, with all treatments using MS medium supplemented with 7 g·L^{-1}, pH = 6.0, and a light cycle of 16/8 h:

1. Treatment with different sugars (sucrose, glucose, fructose, maltose, mannitol) at a concentration of 3% in MS medium, under white light conditions at 25 °C;
2. Treatment with different concentrations of sucrose (1%, 3%, 5%, 7%, 9%) in MS medium, under white light conditions at 25 °C;

3. Treatment with different concentrations of MS medium (MS, 1/2MS, 1/3MS, 1/5MS, 0MS) in the presence of 3% sucrose, under white light conditions at 25 °C;
4. Treatment with MS medium supplemented with 3% sucrose, under different light qualities (white, blue at 461 nm, red at 620 nm, and 50% red + 50% blue) at 25 °C;
5. Treatment with MS medium supplemented with 3% sucrose, under white light conditions at different temperatures and lighting conditions (15 °C light, 25 °C light, 15 °C dark, 25 °C dark).

The control treatment consisted of MS medium supplemented with 3% sucrose, exposed to white light at 25 °C with a light cycle of 16/8 h and an intensity of 2500 Lux. All callus tissues were sampled on the 15th day of treatment for subsequent analysis of secondary metabolites and RNA extraction. Each treatment had three biological repetitions and three technical repetitions. For each repetition, 1g of callus tissue was placed in a single Petri dish, resulting in a total of nine Petri dishes used for each treatment.

2.2. Determination of the Anthocyanin Content

To prepare the sample for analysis, the fresh red pear callus was pulverized using liquid nitrogen and a mortar and pestle. Subsequently, 0.1 g of the sample was transferred to a 1.5 mL centrifuge tube and blended with 1 mL of 0.1 mol·L^{-1} hydrochloric ethanol solution. The mixture was incubated in a water bath at 65 °C for 30 min, centrifuged at 12,000× g for 10 min at 4 °C, and the supernatant was collected for further use. The experiment was conducted with three replicates.

The absorbance values of anthocyanin extracts at 530 nm, 620 nm, and 650 nm were measured, with 0.1 mol ethanolic hydrochloric acid solution used as a blank control. To accurately calculate the anthocyanin absorbance values, Greey's formula OD = (OD_{530} − OD_{620}) − 0.1 × (OD_{650} − OD_{620}) was used. The anthocyanin content was then determined using the following formula: anthocyanin content [μg·g^{-1}(FW)] = OD/ε*V/m*M*10^3. Here, OD represents the anthocyanin absorbance value, whereas ε denotes the molar extinction coefficient of anthocyanin, measured as 4.62 × 10^4. V represents the total volume of the extract (mL), m denotes the sample mass (g), and M represents the molar molecular weight of anthocyanin, measured as 287.24. Lastly, 10^3 signifies the conversion of the calculated result into μg [23].

2.3. Determination of Total Phenol and Total Flavonoids

The method by Singleton et al. [24] was used to determine the total phenols. A total of 200 μL of sample extract was combined with 1.8 mL of distilled water and 0.5 mL of Folin (Solarbio, Chengdu, China). After allowing the mixture to stand for 5 min at room temperature, 4 mL of a 15% sodium carbonate solution was added, and the absorbance of the extract was measured at a wavelength of 765 nm after standing for 1.5 h. The method by Kim et al. [25] was employed to determine the total flavonoids. An amount of 2 mL of sample extract was blended into 3.8 mL of distilled water, 0.6 mL of 5% $NaNO_2$ solution, and 0.6 mL of 10% $AlCl_3$ solution after standing for 5 min at ambient temperature. After another rest period of 6 min, 3 mL of 1 mol·L^{-1} NaOH solution was added. After a 15 min stand at ambient temperature, the absorbance was measured at 510 nm.

2.4. Determination of Antioxidant Activity

This analysis was based on the method by Barreca et al. [26]. To prepare the sample mixture, 0.1 mL of each extract was combined with 63 μmol·L^{-1} DPPH and 80% methanol, and the final volume was adjusted to 4.0 mL. The absorbance was determined at a wavelength of 517 nm following a 30 min incubation period. By following the method by Almeida et al. [27], the ABTS solution was prepared by combining 25 mL of 7 mol·L^{-1} ABTS solution with 440 μL of 140 mol·L^{-1} potassium persulfate solution. The solution was allowed to react for 12–16 h. The ABTS solution was subsequently diluted with ethanol to an absorbency value of 0.7 ± 0.02. A sample volume of 0.1 mol·L^{-1} was collected, and 4.9 m·L^{-1} ABTS was mixed with it for 10 min prior to detection at 734 nm. The method by

Jang et al. [28] was used to conduct the FRAP assay. Specifically, 0.1 mL of callus extract was mixed with the FRAP reagent and deionized water. After a 30 min incubation period, the absorbance was measured at 593 nm. Results were compared to a standard curve generated using a Trolox standard solution.

2.5. RNA Extraction

The RNA extraction was carried out using the modified CTAB method by Chen et al. [29]. The extracted RNA solution was taken, its concentration and OD value were determined via protein-nucleic acid assay, and the purity of RNA was judged based on the ratio of $OD_{260/280}$ as well as $OD_{260/230}$. The integrity of RNA was detected via 1.0% agarose gel electrophoresis.

2.6. Design and Synthesis of the Primers

Based on the cloned gene sequences, a pair of specific quantitative primers (Table 1) was designed using SnapGene® 1.3.2 software while using the *Pyrus bretschneideri* Rehd *Actin* gene as the internal reference gene, and the primer sequences were sent to Sangong Biotech (Shanghai, China). 'Red Zao Su' is a red bud sport variety derived from 'Zao Su' and belongs to the white pear (*Pyrus bretschneideri* Rehd). Therefore, the genetic sequence of white pear was utilized.

Table 1. Design of primers for gene quantification.

Target Gene	Primer Sequences (5'-3')
Actin	Actin-F: CCATCCAGGCTGTTCTCTC Actin-R: GCAAGGTCCAGACGAAGG
PbANS	PbANS-F: TGGTAAGATTCAAGGCTATGGAAGC PbANS-R: TCACGCTTGTCCTCTGGGTATAC
PbCHS	PbCHS-F: ACCCAACTGTGTGCGAGTAC PbCHS-R: TGGGTGATTTTGGACTTGGGC
PbCHI	PbCHI-F: TCGGAGTGTACTTGGAGGAAAACG PbCHI-R: TCTCAAACGGACCTGTAACGATG
PbDFR	PbDFR-F: CAGGAACTGTGAACGTGGAGG PbDFR-R: GAGACGAAGTACATCCAACCAGTC
PbF3H	PbF3H-F: TCGCTAGAGAGTTCTTTGCTTTGC PbF3H-R: TTTCACGCCAATCTTGCACAG
PbPAL	PbPAL-F: ATCGCTACGCTCTCCGAAC PbPAL-R: GTGCAAGGCCTTGTTCCTC
PbUFGT	PbUFGT-F: ACACTCTCTTCTCGTTCTTCAGC PbUFGT-R: CATCGTACACCCTTAGGTTAGGC
PbMYB10	PbMYB10-F: CAGGAAGAACAGCGAATGATGTG PbMYB10-R: GGGCTGAGGTCTTATCACATTGG

2.7. cDNA Compose

The synthesis of the first strand of cDNA was performed according to the *TakaRa PrimeScxipe*^{TM} (China, Shanghai) *reagent kit with gDNA Erasex (PRT) for PCR* instruction manual. The PCR procedure was as follows: 42 °C for 2 min (or 5–30 min at ambient temperature); lower the temperature to 40 °C and remove for reverse transcription (operated on ice).

2.8. Related Gene Expression Determination

In a 0.5 mL thin-walled PCR reaction tube, qPCR reaction solution was added sequentially, gently mixed, transiently centrifuged and then placed in a fluorescence quantification instrument for amplification. The reaction procedure consisted of pre-denaturation at 94 °C for 5 min, followed by 35 cycles of denaturation at 94 °C for 40 s, annealing at 58 °C for 40 s, and extension at 72 °C for 1 min. The reaction was completed with a final extension at 72 °C for 10 min.

2.9. Statistics and Analysis

Proliferation rate = (final fresh weight-initial fresh weight)/initial fresh weight.

IBM SPSS Statistics 20.0 was used to identify significant differences between treatments ($p < 0.05$). Subsequently, the data were processed and graphed using Microsoft Excel software and Origin 2023.

3. Results

3.1. Effect of Different Sugar Sources on the Callus

The sucrose treatment resulted in the highest anthocyanin content of 52.97 $\mu g \cdot g^{-1}$, followed by glucose and fructose, which induced similar levels of anthocyanin content. The anthocyanin content induced by the mannitol treatment was the lowest at only 34.80 $\mu g \cdot g^{-1}$ (Figure 1a). The sucrose treatment had the most significant impact on the callus proliferation rate, which was observed to be lower under maltose and mannitol treatments (Figure A1). The highest total phenol and total flavonoid contents of the callus were recorded under the sucrose, glucose, and fructose treatments, with no significant differences observed among them. Conversely, the lowest total phenol and total flavonoid contents were observed under the maltose and mannitol treatments (Figure 1b,c). In terms of the antioxidant capacity, the DPPH, ABTS, and FRAP values were highest for the sucrose treatment, followed by fructose and glucose, and then the mannitol and maltose treatments (Figure 1d–f).

Carbohydrates are known to be the primary drivers of plant metabolism, growth, and development [30]. Sugar uptake is critical for plant growth and development in tissue culture [31]. The results of this study are consistent with those of previous findings [32], demonstrating a strong correlation between all phenolics and antioxidant activity. The hydroxyl groups of phenolics are known to combat free radicals, making it a valuable antioxidant [33]. Different sugars have varying roles in anthocyanin biosynthesis, with sucrose playing the most significant regulatory role [34]. In a study by Kumar et al. [35], cotton hypocotyls were cultured in different sugar sources, and it was found that the sucrose and fructose cultures had a high phenolic content in the callus, while maltose reduced the phenolic content of the callus. Similarly, radish sprouts were also treated with different types of sugars, and the results show that sucrose treatment resulted in the highest anthocyanin content, followed by fructose, mannitol, and glucose [36]. Some sugars, such as sucrose, glucose, or mannitol, were shown to promote anthocyanin accumulation [37].

Through an analysis of the genes involved in anthocyanin synthesis, it was observed that PAL expression was lower in sucrose, fructose, and glucose compared to mannitol and maltose. Notably, PAL expression was the highest in the mannitol treatment, being ten times greater than that observed in the fructose treatment, which showed an opposing trend to the anthocyanin content (Figure 2a). The expressions of both CHS and DFR were similar, exhibiting little difference in the sucrose, fructose, and glucose, but both were significantly higher than the maltose and mannitol treatments (Figure 2b,e). Among all treatments, fructose exhibited the highest expression of CHI (Figure 2b). Both F3H and ANS displayed similar trends, with the highest expression observed under sucrose treatment, followed by glucose and fructose, with the first three treatments being significantly higher than maltose and mannitol (Figure 2d,f). The expressions of UFGT and MYB10 were the highest under mannitol treatment, being roughly twice as high as those observed for other treatments (Figure 2g,h). These results suggest that saccharides may increase anthocyanin levels by upregulating the expression levels of CHS, F3H, DFR, and ANS. Sugars are known to regulate the majority of the structural and regulatory genes involved in flavonoid metabolic pathways, including PAL, CHS, DFR, and UFGT [38]. Dai et al. [39] found that saccharides, acting as a carbon source, can enhance the accumulation of anthocyanin in cultures by directly affecting both anthocyanin regulation and the expression of various structural genes (such as CHS, CHI, F3H, DFR, LAR, LDOX, and ANR), while simultaneously decreasing the phenylalanine content.

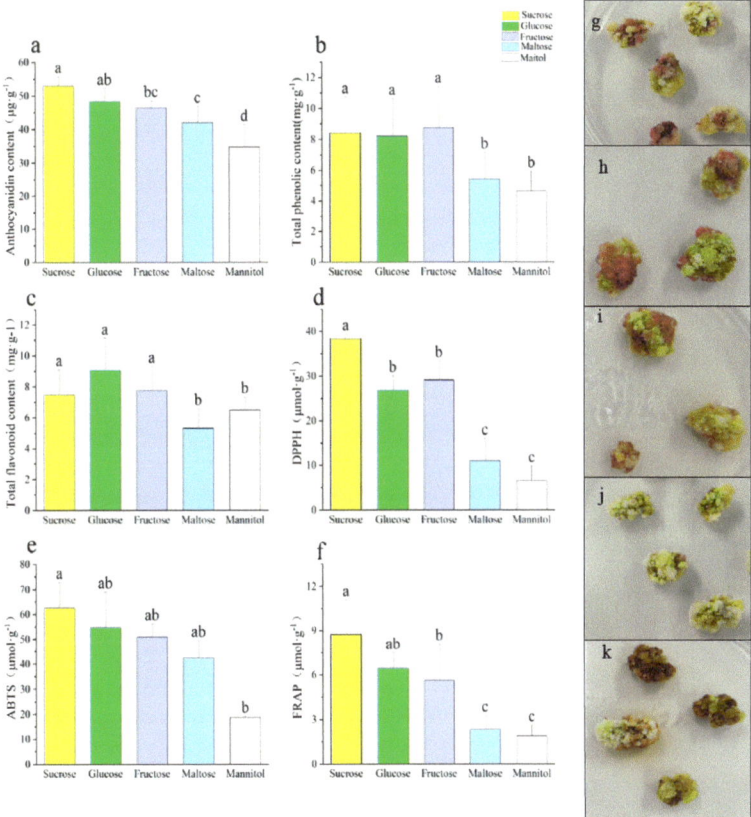

Figure 1. Effects of different sugar sources on callus secondary metabolites ((**a**) anthocyanin content, (**b**) total phenolic content, (**c**) total flavonoid content, (**d**) DPPH, (**e**) ABTS, (**f**) FRAP, (**g**) sucrose, (**h**) glucose, (**i**) fructose, (**j**) maltose, (**k**) mannitol). Note: Normal letters in every column indicate significant differences at 0.05 level by Duncan's multiple range test.

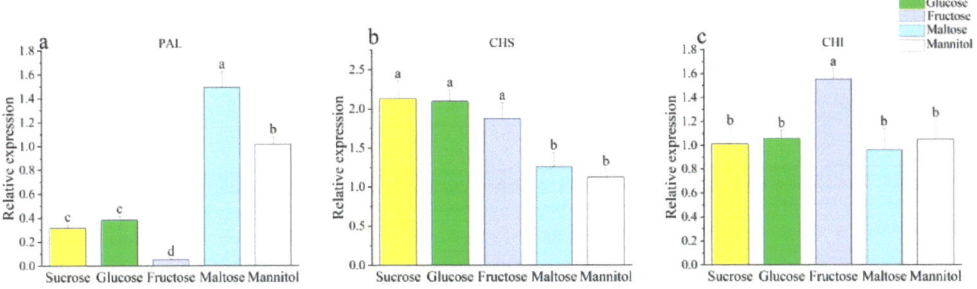

Figure 2. *Cont.*

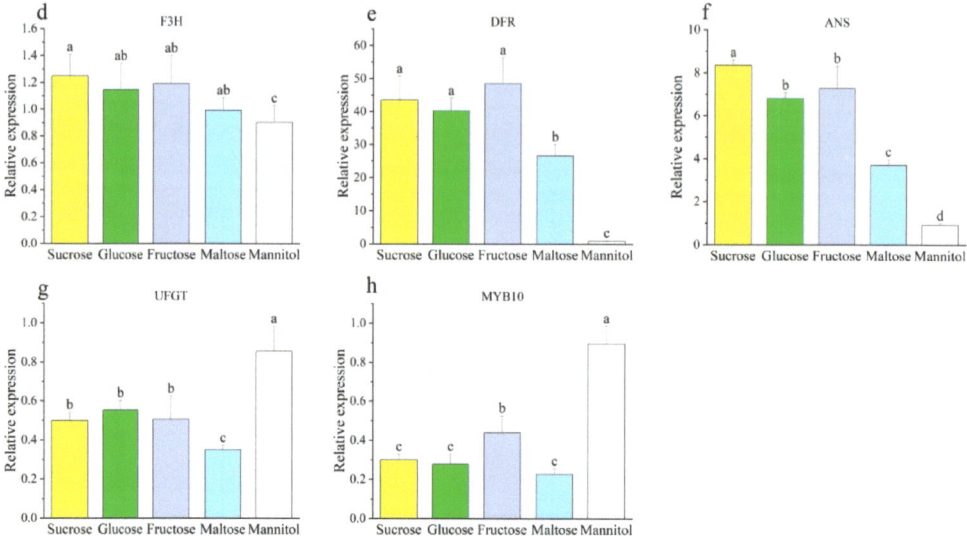

Figure 2. Effect of different sugar sources on the callus anthocyanin gene ((**a**) PAL, (**b**) CHS, (**c**) CHI, (**d**) F3H, (**e**) DFR, (**f**) ANS, (**g**) UFGT, (**h**) MYB10). Note: Normal letters in every column indicate significant differences at 0.05 level by Duncan's multiple range test.

3.2. Effect of Different Sucrose Concentrations on the Callus

We chose to use sucrose for our sugar concentration studies because it induced both anthocyanins and maximized the growth of the callus. An increase in the sucrose concentration led to an increase in anthocyanin content, with the highest anthocyanin content of 75.54 µg·g^{-1} being achieved with 9% sucrose treatment (Figure 3a). The callus proliferation rate increased initially and then decreased with the increase in the sucrose concentration, with the callus proliferation rate reaching its highest value under 5% sucrose treatment (Figure A2). Similar patterns were observed in the total phenol and total flavonoid contents as well as the antioxidant indicators, compared to the trends observed for the anthocyanin levels under different sucrose concentration treatments (Figure 3b–f). Specifically, an increase in the sucrose concentration resulted in a gradual rise in both the total phenol and total flavonoid contents, as well as in the antioxidant capacity. Sarmadi et al. [40] treated yew callus with different concentrations of glucose and found that the callus browning became more severe with an increasing glucose concentration, while the phenols and flavonoids also increased significantly. An increase in the sucrose concentration also resulted in an increased phenolic content in willow and camptotheca [41,42]. According to Ram et al. [43], rose callus was found to exhibit a rich reddish hue and a significant accumulation of anthocyanins when cultured with a sucrose content of 6–7% in the medium. It was also found that anthocyanins were induced in dandelion callus under 5.5% sucrose culture [22].

A gradual upregulation in the expression of PAL, CHS, CHI, F3H, DFR, and ANS was observed with the increasing sucrose concentration (Figure 4a–f). Studies conducted on plant species such as petunia and chrysanthemum revealed that the expression of relevant genes was found to be positively correlated with increasing concentrations of sucrose in the medium [44,45]. It was proposed that sucrose plays a key role in regulating the expression of biosynthetic genes involved in anthocyanin production, thereby leading to an increase in the production of these pigments [46]. Studies have demonstrated that the addition of exogenous sucrose to the growth medium can lead to a marked increase in the transcript levels of DFR and LDOX genes [47,48].

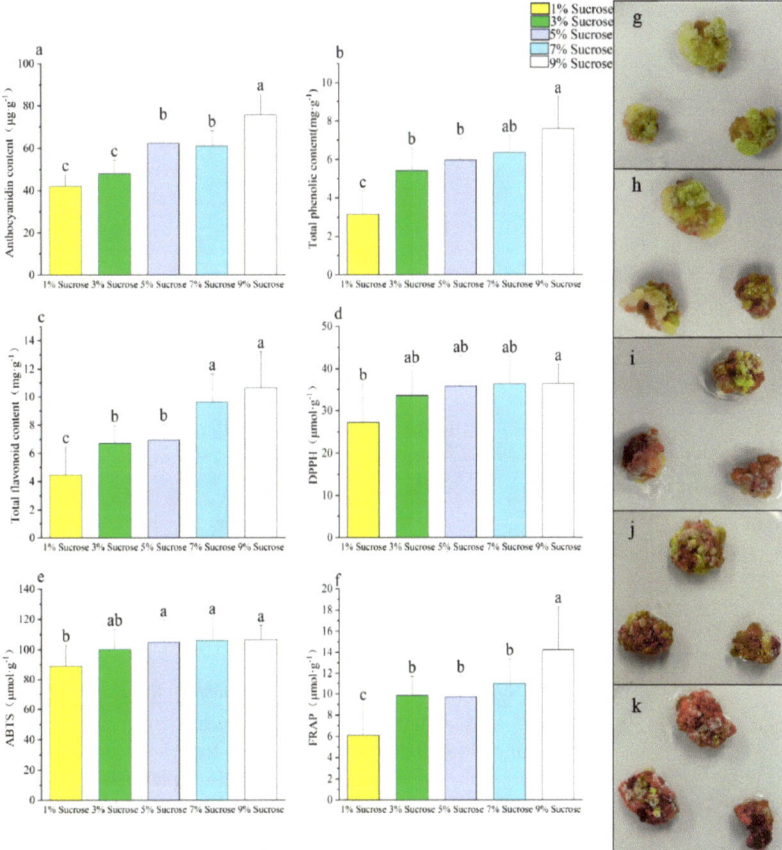

Figure 3. Effect of different sucrose concentrations on callus secondary metabolites ((**a**) anthocyanin content, (**b**) total phenolic content, (**c**) total flavonoid content, (**d**) DPPH, (**e**) ABTS, (**f**) FRAP, (**g**) 1% sucrose, (**h**) 3% sucrose, (**i**) 5% sucrose, (**j**) 7% sucrose, (**k**) 9% sucrose). Note: Normal letters in every column indicate significant differences at 0.05 level by Duncan's multiple range test.

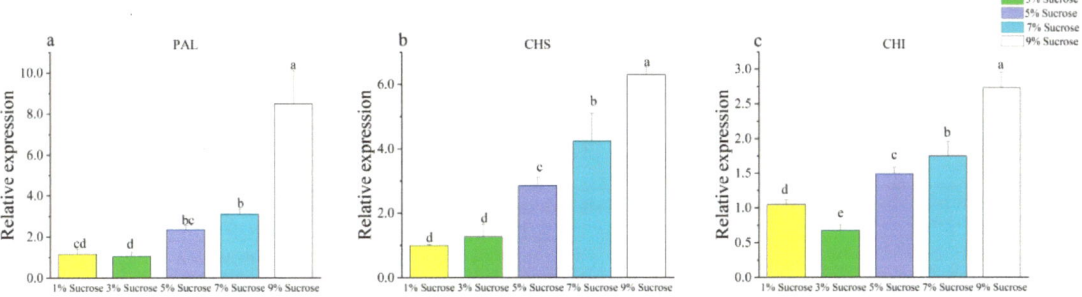

Figure 4. *Cont.*

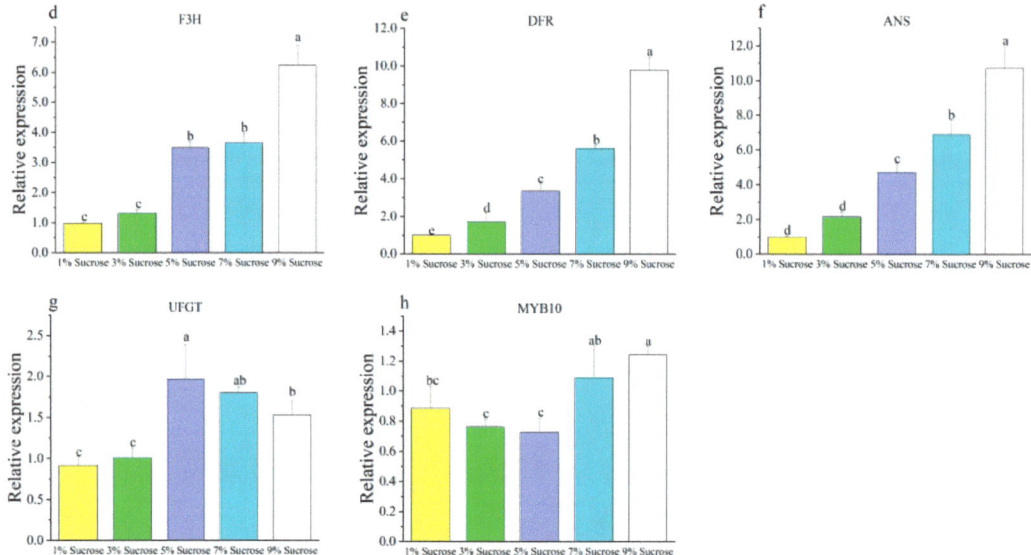

Figure 4. Effect of different sucrose concentrations on callus anthocyanin genes ((**a**) PAL, (**b**) CHS, (**c**) CHI, (**d**) F3H, (**e**) DFR, (**f**) ANS, (**g**) UFGT, (**h**) MYB10). Note: Normal letters in every column indicate significant differences at 0.05 level by Duncan's multiple range test.

3.3. Effect of Different MS Concentrations on the Callus

As depicted in Figure 5a, a decrease in the concentration of MS in the growth medium resulted in a gradual increase in the anthocyanin content, with the highest level of 86.42 µg·g^{-1} being achieved in the absence of MS supplementation. The total phenolic content was the lowest under MS and 1/2 MS treatment, and as the MS concentration decreased, the total phenolic content gradually increased, with the highest value of 13.79 mg·g^{-1} observed in the medium without the addition of MS (Figure 5b). The highest level of total flavonoids was observed under 0MS treatment, followed by 1/3MS and 1/5MS, and the lowest level was observed under MS treatment (Figure 5c). Similar trends were observed for the three indicators of antioxidant capacity (Figure 5d–f). The results indicate that the deficiency of certain nutrients can stimulate the production of secondary metabolites and enhance the antioxidant capacity in callus cultures, as evidenced by the data obtained in this study. Despite the increase in the anthocyanin content achieved through nutrient deficiency, it was observed that the growth of callus was negatively impacted, as depicted in Figure A3, indicating the need to find a relatively reasonable concentration to increase anthocyanin production when nutrient deficiency is used. According to previous studies [49], nutrient deficiencies, particularly in nitrogen, phosphorus, and sulfur, can cause an accumulation of anthocyanins as a mechanism to prevent physiological disorders that arise from the excessive accumulation of carbohydrates in tissues. Furthermore, Landi et al. [50] reported that anthocyanins play a pivotal role in preventing premature aging in plants that are subjected to mineral deficiencies like nitrogen or phosphorus deficit. Simões et al. [51] lowered the MS salt concentration to a quarter, which increased the anthocyanin content of rose calli. Studies conducted on various plant species, including Arabidopsis, tomato, tobacco, rose, and grape plants, have demonstrated that a nitrogen deficiency can trigger a substantial increase in the anthocyanin accumulation [43,52]. At lower concentrations of growth compounds (NH_4NO_3 and KNO_3), inhibitory effects on anthocyanin production can be effectively reversed [53]. When there is a nitrogen deficiency, the amount of sugar used to form amino acids decreases, and more sugars are used to form more anthocyanins, resulting in a red coloration [54].

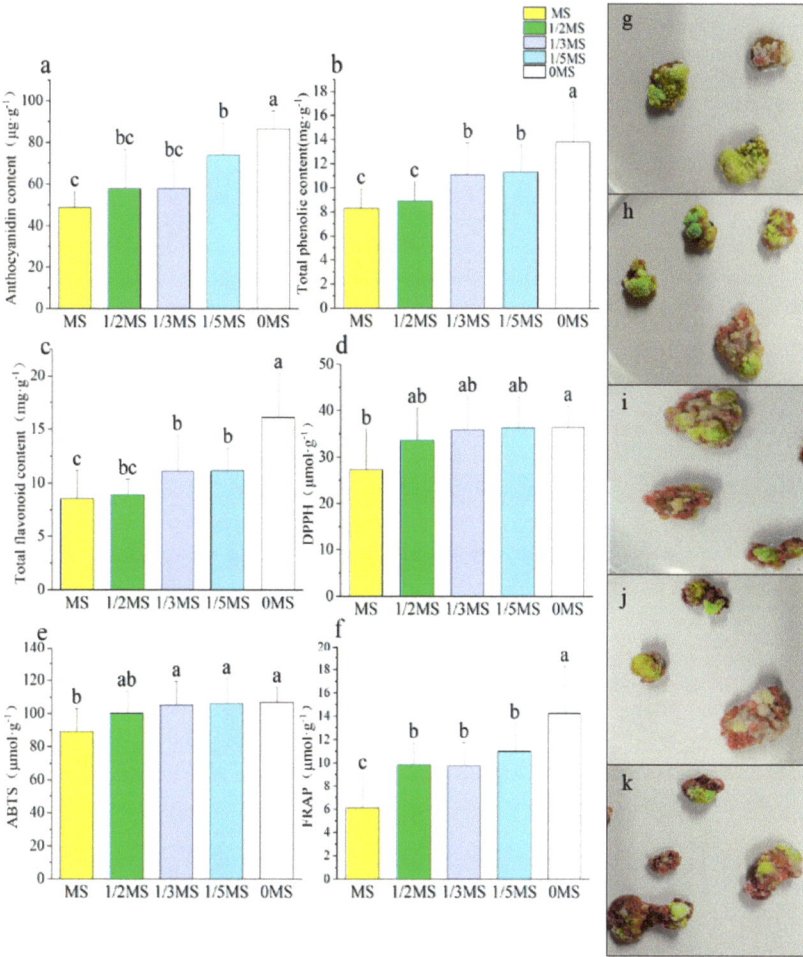

Figure 5. Effect of different MS concentrations on callus secondary metabolites ((**a**) anthocyanin content, (**b**) total phenolic content, (**c**) total flavonoid content, (**d**) DPPH, (**e**) ABTS, (**f**) FRAP, (**g**) MS, (**h**) 1/2MS, (**i**) 1/3MS, (**j**) 1/5MS, (**k**) 0MS). Note: Normal letters in every column indicate significant differences at 0.05 level by Duncan's multiple range test.

The expressions of PAL, CHS, CHI, and DFR all showed a progressive increase with the decreasing MS concentration (Figure 6a–c,e), while the expressions of F3H and ANS were the highest at 1/2MS and 0MS treatment (Figure 6d,f). The expression of UFGT was the highest at 0MS (Figure 6g), and the other treatments had little effect on UFGT. It was established that nutrient deficiency is a key factor that can induce significant anthocyanin accumulation in plants, with the concomitant upregulation of genes being responsible for anthocyanin synthesis, as reported by Li et al. [55]. In this experiment, the expressions of PAL, CHS, CHI, and DFR showed similar trends to the anthocyanin content. Huang et al. [56] suggested that PAL is critical in the synthesis of anthocyanins starting from deamination.

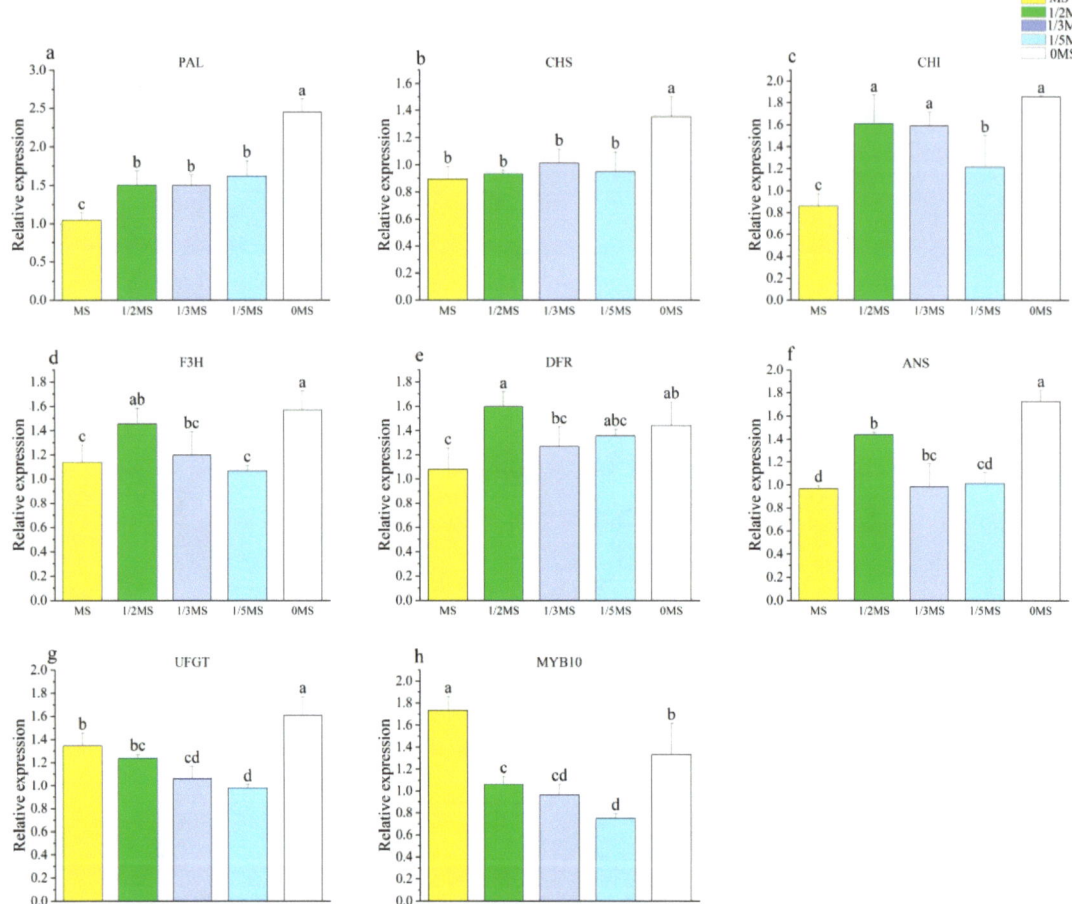

Figure 6. Effect of different MS source concentrations on the callus anthocyanin gene ((**a**) PAL, (**b**) CHS, (**c**) CHI, (**d**) F3H, (**e**) DFR, (**f**) ANS, (**g**) UFGT, (**h**) MYB10). Note: Normal letters in every column indicate significant differences at 0.05 level by Duncan's multiple range test.

3.4. Effect of Different Light Qualities on the Callus

Under red-blue light, the callus showed the highest content of anthocyanins at 83.21 µg·g^{-1}, while the lowest content was 43.95 µg·g^{-1} under white light (Figure 7a). The callus growth was faster under white and red-blue lights, but slower under red and blue lights (Figure A4). The total phenolic content was also the highest under red-blue light and the lowest under white light (Figure 7b). The flavonoid content was the highest under red-blue light, while the remaining differences in the flavonoid content between red, white, and blue lights were not significant (Figure 7c). Similar trends were also found in the detection of antioxidant capacity, as with substances such as anthocyanins (Figure 7d–f). Overall, a correlation between anthocyanins and total phenols, total flavonoids, and antioxidant capacity was observed.

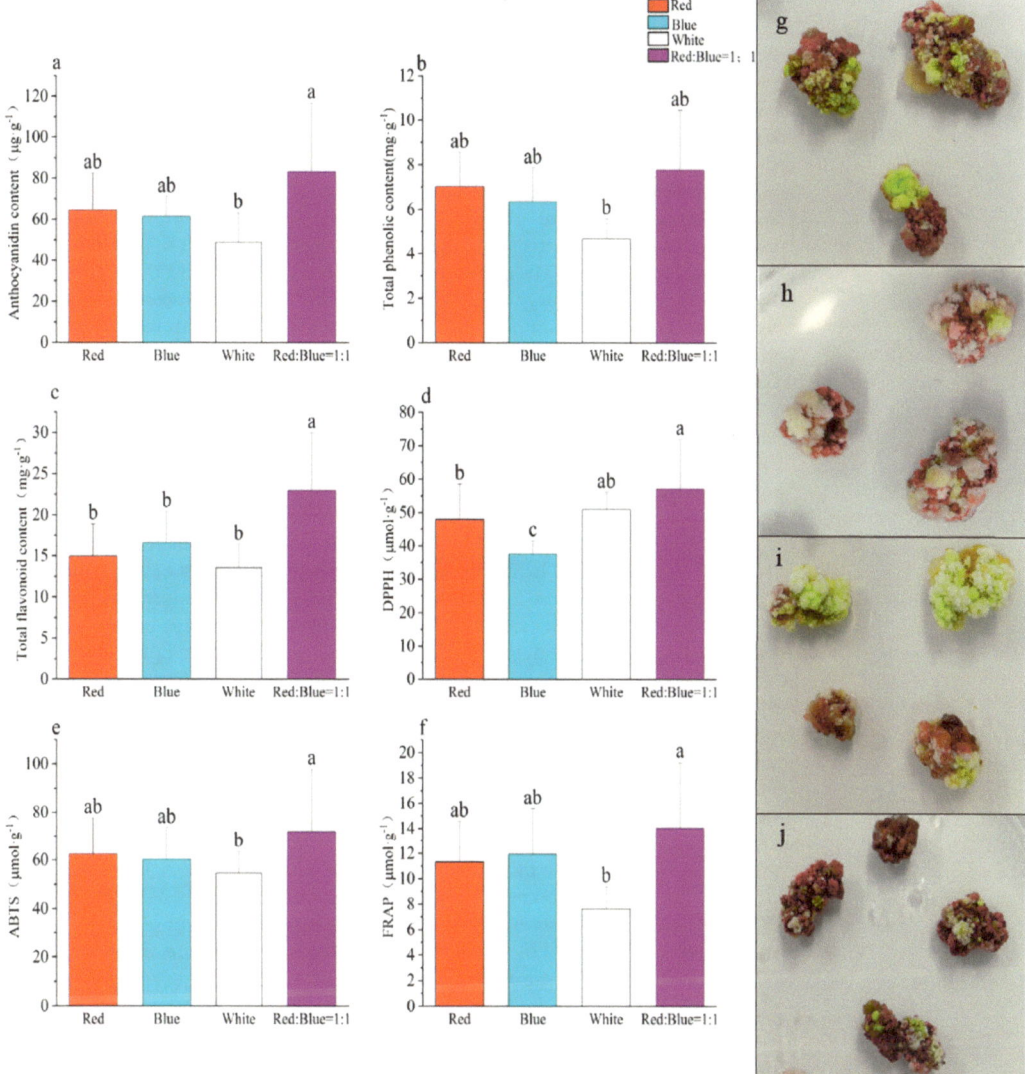

Figure 7. Effect of different light qualities on callus secondary metabolites ((**a**) anthocyanin content, (**b**) total phenolic content, (**c**) total flavonoid content, (**d**) DPPH, (**e**) ABTS, (**f**) FRAP, (**g**) red, (**h**) blue, (**i**) white, (**j**) red/blue = 1:1). Note: Normal letters in every column indicate significant differences at 0.05 level by Duncan's multiple range test.

Different wavelengths of light have varying effects on plant growth, and light is crucial for the formation of anthocyanins in most plants [57]. Several research studies have suggested that specific light sources have the capacity to directly stimulate the synthesis of critical secondary metabolites, such as flavonoids, caffeic acid derivatives, artemisinins, and anthocyanins [58–60]. Fazal et al. [61] reported that in prunella, while yellow light was found to be the optimal light source for maximum biomass accumulation in leaf explants, violet light was the most effective for promoting biomass accumulation in petiole explants. Additionally, blue light was found to be the ideal light condition for inducing the highest total phenolic content and total flavonoid content in callus growth. Stevia rebaudiana exhibited elevated levels of total phenolic content, total flavonoid content, and overall antioxidant capacity in callus cultures exposed to blue light, whereas green and red lights enhanced the reducing power (RPA) and DPPH free radical scavenging activity [62]. Blue and red lights increased the phenolic and flavonoid contents, respectively, in the callus of Ocimum; blue light increased rosemary acid and eugenol, while citric acid increased under continuous white light, and anthocyanins increased under red light [63]. Blueberry root callus exhibited the greatest concentration of anthocyanins when cultivated under red light, showing approximately a five-fold increase compared to conditions of darkness [64]. Purple basil demonstrated optimal growth, as well as increased levels of biomass, total phenolic content, total flavonoid content, and antioxidant activities (DPPH, FRAP, and ABTS) when exposed to blue light [65]. The application of red-blue light to Dendrobium callus led to an enhancement in both the dry weight and the levels of secondary metabolites, specifically phenolic and flavonoid compounds. Conversely, white light exposure stimulated the production of phenolphthalein [66]. Similarly to our findings, the use of red-blue light had a beneficial effect on callus growth and the accumulation of secondary metabolites [67].

The expressions of PAL and DFR were lower in the red light treatment than in the other treatments (Figure 8a,e). Under both red and blue light treatments, the expressions of ANS and CHI were comparatively lower than the other two treatments (Figure 8c,f). Nevertheless, the red light treatment resulted in an increased expression of MYB10 and CHS (Figure 8b,h). Conversely, both the blue and white light treatments notably decreased the expression of CHS and MYB10 compared to the other treatments (Figure 8b,h). On the other hand, the white light treatment led to the lowest expression of F3H, whereas the blue light treatment enhanced the expression of F3H. These findings suggest that the signaling pathways involved in anthocyanin regulation may vary between blue light and white light conditions (Figure 8d). Under the combined influence of red-blue light, PAL, CHS, CHI, ANS, UFGT, and MYB10 exhibited the highest levels of expression, confirming the positive effect of red-blue light on anthocyanin content promotion. In kiwifruit, it was observed that the presence of red light led to a reduction in the anthocyanin content, along with the inhibition of key genes involved in anthocyanin synthesis, including CHS, DFR, ANS, UFGT, and CRY. On the other hand, blue light effectively increased the accumulation of anthocyanin in callus cultures and resulted in enhanced expression levels of genes such as CHS, F3H, DFR, CHI, UFGT, CRY, and F3′H [68]. In eggplant, the exposure to white light containing UV light resulted in an increase in the anthocyanin levels, CHS, and DFR [69]. Similarly, blue light was observed to elevate the expression levels of MYB, DFR, ANS, and UFGT genes in purple pepper [70]. It is worth noting that different plant species exhibit varying pathways induced by different light qualities. For instance, in lettuce, the genes CHI, F3H, DFR, and ANS are also induced by red-blue light [67].

3.5. Effect of Different Temperatures on Callus under Light

The light treatment at 15 °C exhibited the highest anthocyanin content, while the differences in the dark treatments at 25 °C and 15 °C were not significant but significantly lower than the light treatment (Figure 9a). The dark treatment at 25 °C resulted in the highest rate of callus proliferation, which was significantly higher than the rate under the light treatment. However, the growth amount was inhibited by the low temperature (Figure A5). The 15 °C light treatment showed the highest total phenolic content, whereas the lowest content was observed in the dark 25 °C treatment (Figure 9b). There was no significant difference in the flavonoid content between the dark treatments at 25 °C and 15 °C, while the light treatment at 15 °C exhibited a higher flavonoid content than the light treatment at 25 °C (Figure 9c). The antioxidant capacity indicators demonstrated similar trends to that of anthocyanins, total phenols, and total flavonoids (Figure 9d–f). Based on these findings, it can be concluded that light effectively increases the content of secondary metabolites in callus, and a low temperature also contributes to the increase in the secondary metabolite content.

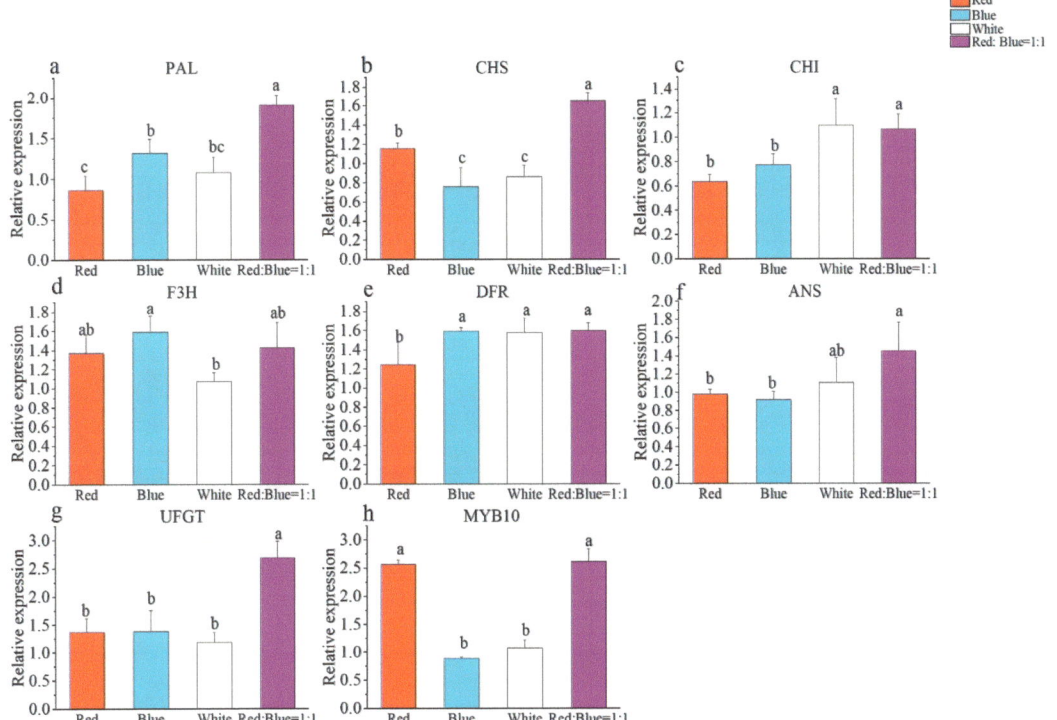

Figure 8. Effect of different light matter on anthocyanin synthesis genes ((**a**) PAL, (**b**) CHS, (**c**) CHI, (**d**) F3H, (**e**) DFR, (**f**) ANS, (**g**) UFGT, (**h**) MYB10). Note: Normal letters in every column indicate significant differences at 0.05 level by Duncan's multiple range test.

According to Simões et al. [51], the optimal temperature for anthocyanin production in rose callus was found to be 24 ± 2 °C, as higher temperatures led to the browning of the callus. Grass coral callus demonstrated the fastest growth at temperatures within the range of 23–28 °C, while severe damage occurred at both 15 °C and 32 °C. The highest production of flavonoids was observed at 26 °C [71]. In potato callus culture, the anthocyanin content was lower at 4 °C compared to 25 °C, indicating that low temperatures negatively affected callus growth [72]. Optimal temperatures for anthocyanin biosynthesis in carrot callus cultures were determined to be 30 °C (in solid medium) and 25 °C (in liquid medium) [73]. Light exerts a crucial influence on primary and secondary metabolism, as well as various developmental processes in plants [74]. It is a pivotal physical factor, and the quality of light significantly impacts photosynthesis and morphogenesis, ultimately modulating plant growth and development [75]. Temperature also plays a vital role in influencing the secondary metabolism yield, and specific temperature requirements must be carefully regulated at different stages of plant tissue culture based on the specific needs of the cultivated tissues [76].

Light increased the expression of PAL and UFGT (Figure 10a,g), whereas the CHS, CHI, F3H, and DFR expression levels were all higher at 15 °C than at 25 °C under the treatment (Figure 10b–e). The expressions of ANS, UFGT, and MYB10 were highest in light at 15 °C, while in light at 25 °C and in the dark at 15 °C, the expressions of ANS and MYB10 did not differ much but were still higher than in the dark at 25 °C (Figure 10f–h). The synergistic effect of light and a low temperature may lead to the upregulation of these genes, which, in turn, regulates anthocyanin synthesis. Previous studies have demonstrated that higher temperatures lead to the downregulation of anthocyanin biosynthesis genes, including

CHS, DFR, LDOX, UFGT, and MYB10. This downregulation is consistent with the increased expression of genes such as MYB15 [77]. In apple callus cultures, the interplay of light and temperature was observed. It was noted that under light conditions, low temperatures (16 °C) led to the upregulation of regulatory and structural genes, including CHI, F3H, CHS, DFR, UFGT, and MYB10. On the other hand, high temperatures (32 °C) induced the expression of MYB16, which had a detrimental impact on anthocyanin biosynthesis [76]. Previous studies have shown that the exposure to low temperatures can significantly stimulate the expression of CHS in Arabidopsis and in red orange [78,79]. Similarly, African chrysanthemums exhibited higher levels of CHS, F3H, and ANS expression at all developmental stages when subjected to a temperature of 6 °C compared to 22 °C [80]. These findings collectively emphasize the beneficial role of light and low temperature in regulating the production of anthocyanins.

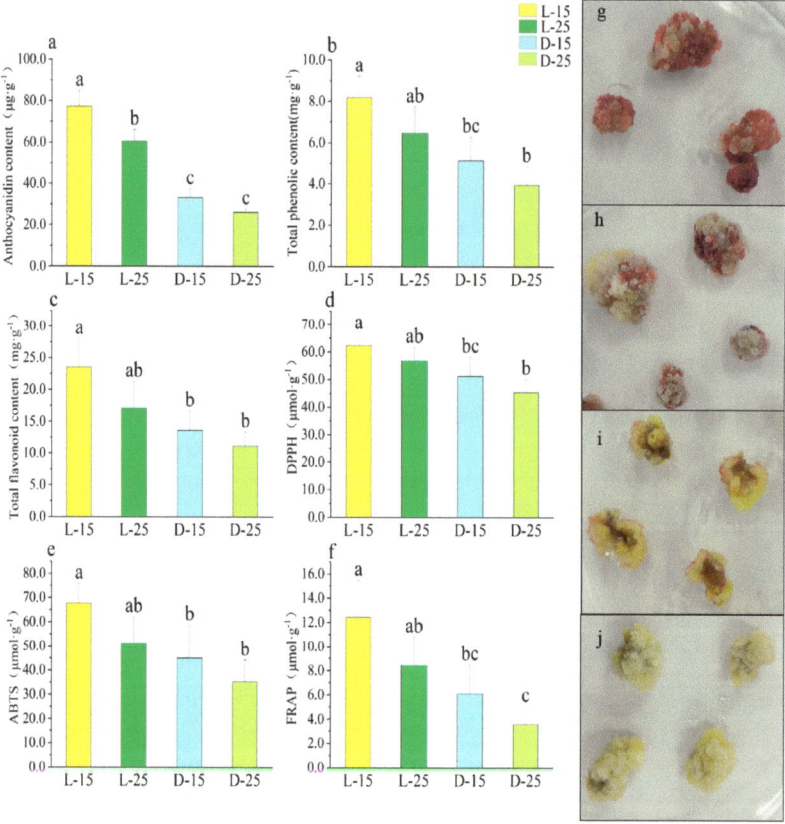

Figure 9. Effect of different light temperatures on callus secondary metabolites ((**a**) anthocyanin content, (**b**) total phenolic content, (**c**) total flavonoid content, (**d**) DPPH, (**e**) ABTS, (**f**) FRAP, (**g**) light 15 °C, (**h**) light 25 °C, (**i**) dark 15 °C, (**j**) dark 25 °C). Note: Normal letters in every column indicate significant differences at 0.05 level by Duncan's multiple range test.

3.6. Correlation Analysis of Different Treatments

From Table 2, it can be observed that there is a positive correlation among anthocyanins, total phenols, total flavonoids, and antioxidant index in all treatments. In some treatments, the positive correlation is even statistically significant. This indicates that despite the variations in the experimental conditions, the regulatory mechanisms and interactions among these substances remain relatively consistent. Interestingly, the treatments with different sugar sources showed a strong positive correlation between the callus accumulation rate and anthocyanin production. However, most of the other treatments exhibited negative correlations. This suggests that, in most cases, as the anthocyanin content increases, the growth rate of the callus tissue decreases. Moreover, genes involved in anthocyanin synthesis exhibited different correlations under various conditions. In the treatments with different sugar sources, it was found that CHS, F3H, DFR, and ANS showed a significant positive correlation with the anthocyanin content. In the treatments with different sugar concentrations, PAL, CHS, CHI, F3H, DFR, and ANS were significantly correlated with the anthocyanin content. However, in the treatments with different MS medium concentrations, only PAL showed a significant positive correlation. Under different light qualities, only F3H exhibited a positive correlation with the anthocyanin content. Similarly, in the treatments with different temperature and lighting conditions, only UFGT showed a significant positive correlation. These results indicate that a wide range of genes or regulatory pathways are involved in influencing anthocyanin synthesis. Different factors may utilize different genes to regulate this process, which warrants further investigation.

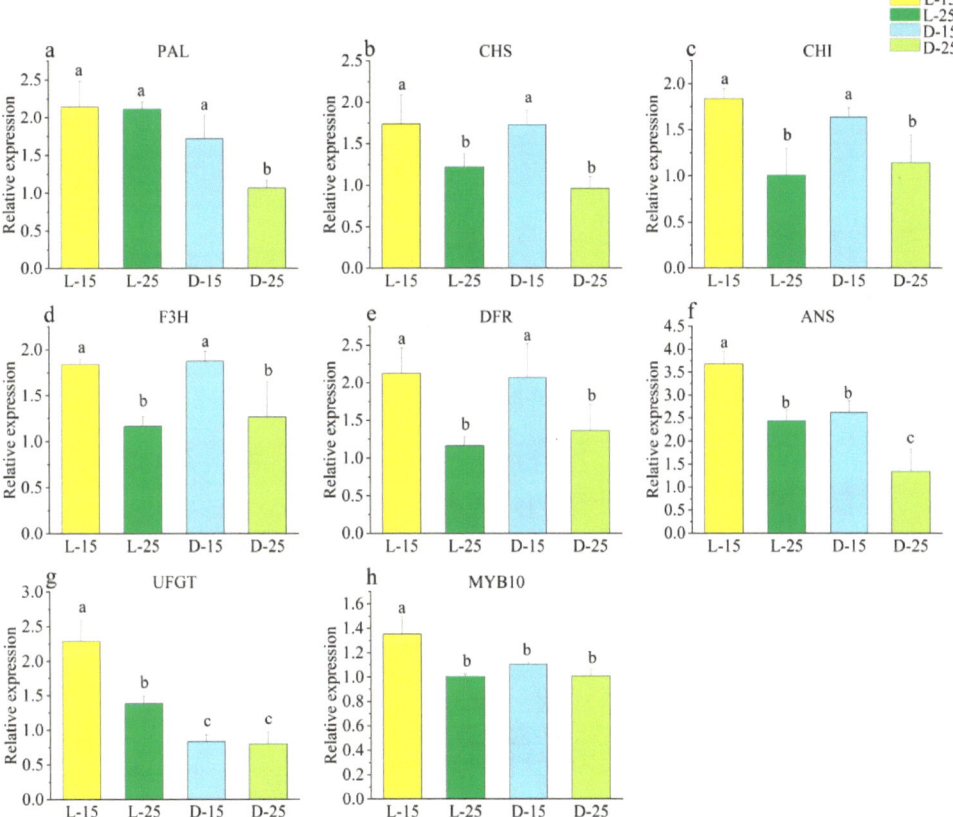

Figure 10. Effect of different light temperatures on anthocyanin synthesis genes ((**a**) PAL, (**b**) CHS, (**c**) CHI, (**d**) F3H, (**e**) DFR, (**f**) ANS, (**g**) UFGT, (**h**) MYB10). Note: Normal letters in every column indicate significant differences at 0.05 level by Duncan's multiple range test.

Table 2. Correlation analysis of different treatments.

Different sugar source treatment	Total phenolic content (mg·g⁻¹)	Total flavonoid content (mg·g⁻¹)	DPPH (μmol·g⁻¹)	ABTS (μmol·g⁻¹)	FRAP (μmol·g⁻¹)	Callus proliferation rate	PAL	CHS	CHI	F3H	DFR	ANS	UFGT	MYB10
Anthocyanidin content (μg·g⁻¹)	0.889 *	0.573	0.953 *	0.987 **	0.946 *	0.953 *	−0.658	0.932 *	0.117	0.959 **	0.903 *	0.973 **	−0.586	−0.756
Different sucrose concentration treatments	Total phenolic content (mg·g⁻¹)	Total flavonoid content (mg·g⁻¹)	DPPH (μmol·g⁻¹)	ABTS (μmol·g⁻¹)	FRAP (μmol·g⁻¹)	Callus proliferation rate	PAL	CHS	CHI	F3H	DFR	ANS	UFGT	MYB10
Anthocyanidin content (μg·g⁻¹)	0.931 *	0.889 *	0.831	0.858	0.913 *	−0.521	0.888 *	0.953 *	0.917 *	0.988 **	0.935 *	0.964 **	0.719	0.642
Different MS concentration treatments	Total phenolic content (mg·g⁻¹)	Total flavonoid content (mg·g⁻¹)	DPPH (μmol·g⁻¹)	ABTS (μmol·g⁻¹)	FRAP (μmol·g⁻¹)	Callus proliferation rate	PAL	CHS	CHI	F3H	DFR	ANS	UFGT	MYB10
Anthocyanidin content (μg·g⁻¹)	0.922 *	0.911 *	0.739	0.765	0.946 *	−0.948 *	0.935 *	0.823	0.597	0.458	0.447	0.641	0.344	−0.312
Different light quality treatment	Total phenolic content (mg·g⁻¹)	Total flavonoid content (mg·g⁻¹)	DPPH (μmol·g⁻¹)	ABTS (μmol·g⁻¹)	FRAP (μmol·g⁻¹)	Callus proliferation rate	PAL	CHS	CHI	F3H	DFR	ANS	UFGT	MYB10
Anthocyanidin content (μg·g⁻¹)	0.938	0.944	0.457	0.999 **	0.938	0.038	0.602	0.645	−0.524	0.968 *	−0.25	0.766	0.76	0.665
Different temperature light treatment	Total phenolic content (mg·g⁻¹)	Total flavonoid content (mg·g⁻¹)	DPPH (μmol·g⁻¹)	ABTS (μmol·g⁻¹)	FRAP (μmol·g⁻¹)	Callus proliferation rate	PAL	CHS	CHI	F3H	DFR	ANS	UFGT	MYB10
Anthocyanidin content (μg·g⁻¹)	0.983 *	0.974 *	0.977 *	0.957 *	0.973 *	−0.739	0.875	0.446	0.351	0.193	0.227	0.836	0.960 *	0.688

Note: ** At the 0.01 level (two-tailed), the correlation is significant. * At the 0.05 level (two-tailed), the correlation is significant.

4. Conclusions

Plants serve as a significant source of secondary metabolites, which possess diverse roles in human welfare, including therapeutic implications. However, the yield of secondary metabolites obtained from natural populations of plants is insufficient to meet commercial demands due to their limited accumulation [81]. Various in vitro culture techniques, including callus, hairy root, shoot, and suspension cultures, can be utilized to enhance the biosynthesis of specific metabolites in commercially important plants [82]. Among these techniques, callus culture is regarded as a promising approach for the biosynthesis of bioactive compounds in endangered species of most medicinal plants [83]. The secondary metabolites derived from Dendrobium officinale exhibit pharmacological efficacy in alleviating acute alcoholic liver injury [84]. The secondary metabolites of plants, such as terpenoids, lignans, polyphenols, phenolic acids, alkaloids, lactones, and flavonoids, exhibit anti-HBV activity. These natural anti-HBV products can be considered as potential lead compounds or candidate drugs [85]. Fruit trees typically have a prolonged growth cycle, which, combined with the arduous and time-intensive process of inducing their metabolites and validating their gene functions, can present a substantial challenge. Nonetheless, tissue culture presents a viable solution to address this problem, offering a more efficient and effective means of analysis. The present study aims to establish an efficient approach for boosting the production of anthocyanins in the callus of pear fruit by subjecting it to various treatments. The results demonstrate that the composition and concentration of sugar, MS content, light quality, and temperature have a significant impact on the synthesis of secondary metabolites, including anthocyanins in red pear callus. Importantly, these phenolics were found to be closely associated with antioxidant capacity. Further, the analysis of genes related to anthocyanin synthesis was carried out, unveiling the intrinsic factors influencing the process under varying conditions. Overall, these findings provide a roadmap for natural anthocyanin production in red pear fruit, along with their potential applications in agricultural production, such as the promotion of fruit color and genetic variation studies in fruit color.

Author Contributions: Visualization, W.Y., Y.W., Y.L. (Ya Luo) and X.W.; writing—original draf, W.Y.; Validation, D.L.; investigation, X.Z., H.W., Y.Z. (Yunting Zhang), M.L. and W.H.; methodology, X.Z. and J.L.; data curation, H.W.; formal analysis, Y.L. (Yuanxiu Lin); software, Q.C.; project administration, H.T.; writing—review and editing, Y.Z. (Yong Zhang); conceptualization, Y.Z. (Yong Zhang); funding acquisition, Y.Z. (Yong Zhang). All authors have read and agreed to the published version of the manuscript.

Funding: This work was supported by grants from the Science and Technology Plan Project of Sichuan Province (Grant No. 2021YFYZ0023-03) and the Double Support Project of Discipline Construction of Sichuan Agricultural University (03573134).

Data Availability Statement: Data is contained within the article. The data presented in this study are available in Appendix A.

Conflicts of Interest: All authors declare that they have no conflict of interest.

Appendix A

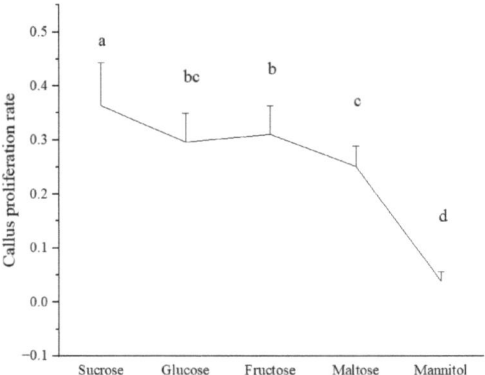

Figure A1. Effect of different sugar sources on the proliferation rate of callus. Note: Normal letters in every column indicate significant differences at 0.05 level by Duncan's multiple range test.

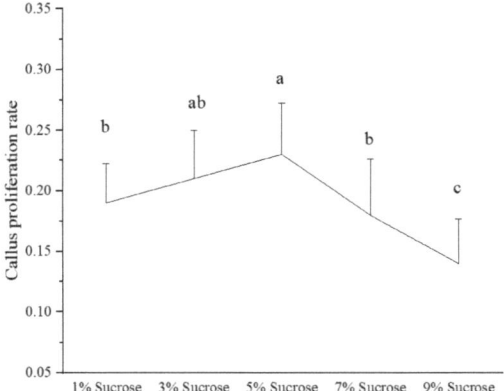

Figure A2. Effect of different sucrose concentrations on the proliferation rate of the callus. Note: Normal letters in every column indicate significant differences at 0.05 level by Duncan's multiple range test.

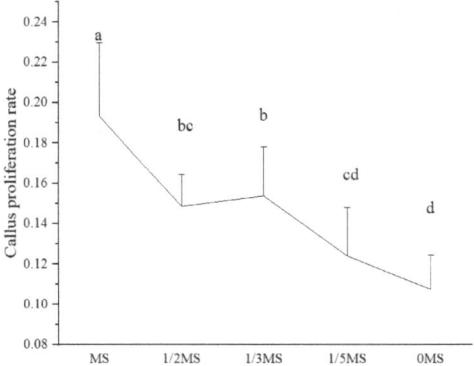

Figure A3. Effect of different MS concentrations on the proliferation rate of the callus. Note: Normal letters in every column indicate significant differences at 0.05 level by Duncan's multiple range test.

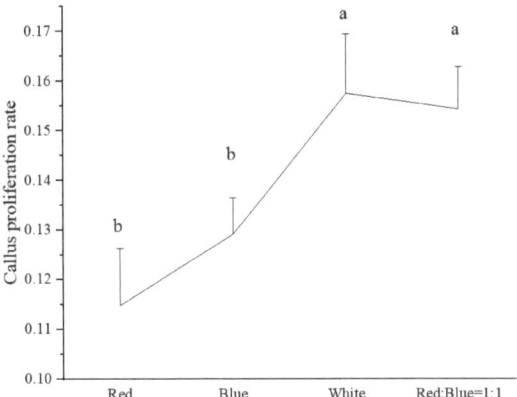

Figure A4. Effect of different light quality on the proliferation rate of the callus. Note: Normal letters in every column indicate significant differences at 0.05 level by Duncan's multiple range test.

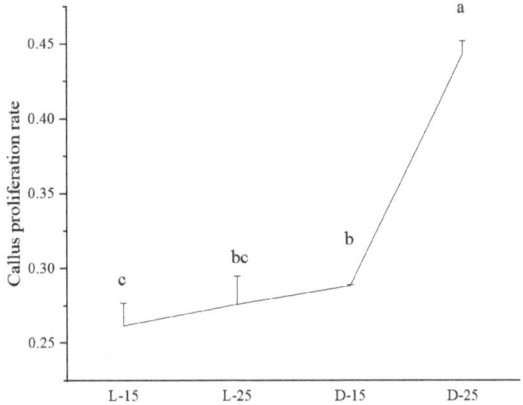

Figure A5. Effect of different temperature illumination on the proliferation rate of callus. Note: Normal letters in every column indicate significant differences at 0.05 level by Duncan's multiple range test.

Table A1. Reagent information.

Reagent	Catalog Number	Company	Country
NAA	N600	Phyto Tech	United States
2,4-D	D299	Phyto Tech	United States
6-BA	B800	Phyto Tech	United States
MS Medium	M519	Phyto Tech	United States

References

1. Kapoor, L.; Simkin, A.J.; George Priya Doss, C.; Siva, R. Fruit ripening: Dynamics and integrated analysis of carotenoids and anthocyanins. *BMC Plant Biol.* **2022**, *22*, 27. [CrossRef] [PubMed]
2. Wang, L.; Yang, S.; Ni, J.; Teng, Y.; Bai, S. Advances of anthocyanin synthesis regulated by plant growth regulators in fruit trees. *Sci. Hortic.* **2023**, *307*, 111476. [CrossRef]
3. Sun, L.P.; Huo, J.T.; Liu, J.Y.; Yu, J.Y.; Zhou, J.L.; Sun, C.D.; Wang, Y.; Leng, F. Anthocyanins distribution, transcriptional regulation, epigenetic and post-translational modification in fruits. *Food Chem.* **2023**, *411*, 135540. [CrossRef] [PubMed]
4. Sheng, J.; Chen, X.; Song, B.; Liu, H.; Li, J.; Wang, R.; Wu, J. Genome-wide identification of the MATE gene family and functional characterization of PbrMATE9 related to anthocyanin in pear. *Hortic. Plant J.* **2023**. [CrossRef]

5. Qu, S.S.; Li, M.M.; Wang, G.; Yu, W.T.; Zhu, S.J. Transcriptomic, proteomic and LC-MS analyses reveal anthocyanin biosynthesis during litchi pericarp browning. *Sci. Hortic.* **2021**, *289*, 110443. [CrossRef]
6. Jo, K.; Bae, G.Y.; Cho, K.; Park, S.S.; Suh, H.J.; Hong, K.B. An anthocyanin-enriched extract from Vaccinium uliginosum improves signs of skin aging in UVB-Induced photodamage. *Antioxidants* **2020**, *9*, 844. [CrossRef]
7. Xu, L.; Tian, Z.; Chen, H.; Zhao, Y.; Yang, Y. Anthocyanins, Anthocyanin-Rich Berries, and Cardiovascular Risks: Systematic Review and Meta-Analysis of 44 Randomized Controlled Trials and 15 Prospective Cohort Studies. Frontiers in Nutrition, An Anthocyanin-Enriched Extract from Vaccinium Uliginosum Improves Signs of Skin Aging in UVB-Induced Photodamage. *Antioxidants* **2021**, *8*, 747884. Available online: https://www.frontiersin.org/articles/10.3389/fnut.2021.747884 (accessed on 1 January 2023).
8. Li, B.; Wang, L.; Bai, W.; Chen, W.; Chen, F.; Shu, C. Biological Activity of Anthocyanins. In *Anthocyanins*; Springer: Singapore, 2021. [CrossRef]
9. Nomi, Y.; Iwasaki-Kurashige, K.; Matsumoto, H. Therapeutic effects of anthocyanins for vision and eye health. *Molecules* **2019**, *24*, 3311. [CrossRef]
10. Silva, S.; Costa, E.M.; Mendes, M.; Morais, R.; Calhau, C.; Pintado, M. Antimicrobial, antiadhesive and antibiofilm activity of an ethanolic, anthocyanin-rich blueberry extract purified by solid phase extraction. *J. Appl. Microbiol.* **2016**, *121*, 693–703. [CrossRef]
11. Cao, X.Y.; Sun, H.L.; Wang, X.Y.; Li, W.X.; Wang, X.Q. ABA signaling mediates 5-aminolevulinic acid-induced anthocyanin biosynthesis in red pear fruits. *Sci. Hortic.* **2022**, *304*, 111290. [CrossRef]
12. Du, Y.; Qu, B.; Li, R. Research progress on red pear resources and fruit coloring mechanism in China. *J. Agric. Yanbian Univ.* **2021**, *43*, 98–107. (In Chinese) [CrossRef]
13. Tao, R.; Yu, W.; Gao, Y.; Ni, J.; Yin, L.; Zhang, X.; Li, H.; Wang, D.; Bai, S.; Teng, Y. Light-Induced Basic/Helix-Loop-Helix64 Enhances Anthocyanin Biosynthesis and Undergoes Constitutively Photomorphogenic1-Mediated Degradation in Pear. *Plant Physiol.* **2020**, *4*, 1684–1701. [CrossRef] [PubMed]
14. Jia, J.X. *Catalogue of Fruit Germplasm Resources (First 1)*; Agricultural Press: Beijing, China, 1993; pp. 22–49. (In Chinese)
15. Steyn, W.J.; Holcroft, D.M.; Wand, J.E.S.; Jacobs, G. Anthocyanin degradation in detached pome fruit with reference to preharvest red color loss and pigmentation patterns of blushed and fully red pears. *J. Am. Soc. Hortic. Sci.* **2004**, *129*, 13–19. [CrossRef]
16. Kong, J.M.; Chia, L.S.; Goh, N.K.; Chia, T.F.; Brouillard, R. Analysis and biological activities of anthocyanins. *Phytochemistry* **2003**, *64*, 923–993. [CrossRef]
17. Wang, C.; Gan, Q.; Meng, F.; Yang, J.; Yan, H.; Jiang, X. Active Components and Antioxidant Activities of Four Kinds Small Berry Juices. *Sci. Technol. Food Ind.* **2019**, *40*, 71–76. [CrossRef]
18. Quattrocchio, F.; Wing, J.F.; Leppen, H.; Mol, J.; Koes, R.E. Regulatory Genes Controlling Anthocyanin Pigmentation Are Functionally Conserved among Plant Species and Have Distinct Sets of Target Genes. *Plant Cell* **1993**, *5*, 1497–1512. [CrossRef]
19. Winkelshirley, B. Flavonoid Biosynthesis. A Colorful Model for Genetics, Biochemistry, Cell Biology, and Biotechnology. *Plant Physiol.* **2001**, *126*, 485–493. [CrossRef] [PubMed]
20. Wu, X.; Liu, Z.; Liu, Y.; Wang, E.; Zhang, D.; Huang, S.; Li, C.; Zhang, Y.; Chen, Z.; Zhang, Y. SlPHL1 is involved in low phosphate stress promoting anthocyanin biosynthesis by directly upregulation of genes SlF3H, SlF3′H, and SlLDOX in tomato. *Plant Physiol. Biochem.* **2023**, *200*, 107801. [CrossRef]
21. Belwal, T.; Singh, G.; Jeandet, P.; Pandey, A.; Giri, L.; Ramola, S.; Bhatt, I.D.; Venskutonis, P.R.; Georgiev, M.I.; Clément, C.; et al. Anthocyanins, multi-functional natural products of industrial relevance: Recent biotechnological advances. *Biotechnol. Adv.* **2020**, *43*, 107600. [CrossRef]
22. María, M.; Paola, P.; Dirk, P.; Christian, G.; Lorena, J.; Perla, F.; Katy, P.; Rolando, C. Kinetics and modeling of cell growth for potential anthocyanin induction in cultures of Taraxacum officinale G.H. Weber ex Wiggers (Dandelion) in vitro. *Electron. J. Biotechnol.* **2018**, *36*, 15–23. [CrossRef]
23. Lai, C.; Fan, L.; He, X.; Xie, H. Callus induction and screening of high yield proanthocyanidin cell lines. *J. Plant Physiol.* **2014**, *50*, 1683–1691. (In Chinese) [CrossRef]
24. Singleton, V.L.; Rossi, J.A.J. Colorimetry to Total Phenolics with Phosphomolybdic Acid Reagents. *Am. J. Enol. Vitic.* **1965**, *16*, 144–158. Available online: https://10.5344/ajev.1965.16.3.144 (accessed on 3 June 2023). [CrossRef]
25. Kim, D.O.; Jeong, S.W.; Lee, C.Y. Antioxidant capacity of phenolic phytochemicals from various cultivars of plums. *Food Chem.* **2003**, *81*, 321–326. [CrossRef]
26. Barreca, D.; Bellocco, E.; Caristi, C.; Leuzzi, U.; Gattuso, C. Elucidation of the flavonoid and furocoumarin composition and radical-scavenging activity of green and ripe chinotto (*Citrus myrtifolia* Raf.) fruit tissues, leaves and seeds. *Food Chem.* **2011**, *129*, 1504–1512. [CrossRef]
27. Almeida, M.M.B.; Sousa, P.H.M.; Arriaga, Â.M.C.; Prado, G.M.; Magalhães, C.E.C.; Maia, G.A.; Lemos, T.L.G. Bioactive compounds and antioxidant activity of fresh exotic fruits from northeastern Brazil. *Food Res. Int.* **2011**, *44*, 2155–2159. [CrossRef]
28. Jang, H.D.; Chang, K.S.; Chang, T.C.; Hsu, C.L. Antioxidant potentials of buntan pumelo (*Citrus grandis* Osbeck) and its ethanolic and acetified fermentation products. *Food Chem.* **2010**, *118*, 554–558. [CrossRef]
29. Chen, Q.; Yu, H.W.; Wang, X.R.; Xie, X.L.; Yue, X.Y.; Tang, H.R. An alternative cetyltrimethylammonium bromide-based protocol for RNA isolation from blackberry (*Rubus* L.). *Genet. Mol. Res.* **2012**, *11*, 1773–1782. [CrossRef] [PubMed]
30. Khan, T.; Abbasi, H.B.; Zeb, A.; Ali, G.S. Carbohydrate-induced biomass accumulation and elicitation of secondary metabolites in callus cultures of Fagonia indica. *Ind. Crops Prod.* **2018**, *126*, 168–176. [CrossRef]

31. Wang, C. Factors affecting the production of secondary metabolites by plant tissue culture. *Rural Econ. Sci. Technol.* **2020**, *31*, 34+48. (In Chinese)
32. Bong, F.J.; Chear, N.J.Y.; Ramanathan, S.; Mohana-Kumaran, N.; Subramaniam, S.; Chew, B.L. The development of callus and cell suspension cultures of Sabah Snake Grass (*Clinacanthus nutans*) for the production of flavonoids and phenolics. *Biocatal. Agric. Biotechnol.* **2021**, *33*, 101977. [CrossRef]
33. Naczk, M.; Shahidi, F. Extraction and analysis of phenolics in food. *J. Chromatogr. A* **2004**, *1054*, 95–111. [CrossRef]
34. Gu, K.D.; Wang, C.K.; Hu, D.G.; Hao, Y.J. How do anthocyanins paint our horticultural products? *Sci. Hortic.* **2019**, *249*, 257–262. [CrossRef]
35. Kumar, G.P.; Sivakumar, S.; Govindarajan, S.; Sadasivam, V.; Manickam, V.; Mogilicherla, K.; Thiruppathi, S.K.; Narayanasamy, J. Evaluation of different carbon sources for high frequency callus culture with reduced phenolic secretion in cotton (*Gossypium hirsutum* L.) cv. SVPR-2. *Biotechnol. Rep.* **2015**, *7*, 72–80. [CrossRef] [PubMed]
36. Chen, J.; Chen, H.; Wang, H.; Zhan, J.; Yuan, X.; Cui, J.; Su, N. Selenium treatment promotes anthocyanin accumulation in radish sprouts (*Raphanus sativus* L.) by its regulation of photosynthesis and sucrose transport. *Food Res. Int.* **2023**, *165*, 112551. [CrossRef]
37. Durán-Soria, S.; Delphine, M.P.; Sonia, O.; José, G.V. Sugar Signaling During Fruit Ripening. *Front. Plant Sci.* **2020**, *11*, 564917. Available online: https://www.frontiersin.org/articles/10.3389/fpls.2020.564917 (accessed on 6 June 2023). [CrossRef] [PubMed]
38. Teng, S.; Keurentjes, J.; Bentsink, L.; Koornneef, M.; Smeekens, S. Sucrose-specific induction of anthocyanin biosynthesis in Arabidopsis requires the *MYB75/PAP1* gene. *Plant Physiol.* **2005**, *139*, 1840–1852. [CrossRef] [PubMed]
39. Dai, Z.W.; Meddar, M.; Renaud, C.; Merlin, I.; Hilbert, G.; Delrot, S. Long-term in vitro culture of grape berries and its application to assess the effects of sugar supply on anthocyanin accumulation. *J. Exp. Bot.* **2014**, *65*, 4665–4677. [CrossRef]
40. Sarmadi, M.; Karimi, N.; Palazón, J.; Ghassempour, A.; Mirjalili, M.H. The effects of salicylic acid and glucose on biochemical traits and taxane production in a Taxus baccata callus culture. *Plant Physiol. Biochem.* **2018**, *132*, 271–280. [CrossRef]
41. Julkunen-Tiitto, R. Defensive efforts of Salix myrsinifolia plantlets I photomixotrophic culture conditions: The effect of sucrose, nitrogen and pH on the phytomass and secondary phenolic accumulation. *Ecoscience* **1996**, *3*, 297–303. [CrossRef]
42. Pasqua, G.; Monacelli, B.; Mulinacci, N.; Rinaldi, S.; Giaccherini, C.; Innocenti, M.; Vinceri, F.F. The effect of growth regulators and sucrose on anthocyanin production in Camptotheca acuminata cell cultures. *Plant Physiol. Biochem.* **2005**, *43*, 293–298. [CrossRef]
43. Ram, M.; Prasad, K.V.; Kaur, C.; Singh, S.K.; Arora, A.; Kumar, S. Induction of anthocyanin pigments in callus cultures of Rosa hybrida L. in response to sucrose and ammonical nitrogen levels. *Plant Cell Tissue Organ Cult.* **2011**, *104*, 171–179. [CrossRef]
44. Kim, S.H.; Kim, Y.S.; Jo, Y.D.; Kang, S.Y.; Ahn, J.W.; Kang, B.C.; Kim, J.B. Sucrose and methyl jasmonate modulate the expression of anthocyanin biosynthesis genes and increase the frequency of flower-color mutants in chrysanthemum. *Sci. Hortic.* **2019**, *256*, 108602. [CrossRef]
45. Ai, T.N.; Naing, A.H.; Arun, M.; Lim, S.H.; Kim, C.K. Sucrose-induced anthocyanin accumulation in vegetative tissue of Petunia plants requires anthocyanin regulatory transcription factors. *Plant Sci.* **2016**, *252*, 144–150. [CrossRef] [PubMed]
46. Zheng, Y.; Tian, L.; Liu, H.; Pan, Q.; Zhan, J.; Huang, W. Sugars induce anthocyanin accumulation and flavanone 3-hydroxylase expression in grape berries. *Plant Growth Regul.* **2009**, *58*, 251–260. [CrossRef]
47. Gollop, R.; Even, S.; Colova-Tsolova, V.; Peri, A. Expression of the grape dihydroflavonol reductase gene and analysis of its promoter region. *J. Exp. Bot.* **2002**, *53*, 13971409. [CrossRef]
48. Gollop, R.; Farhi, S.; Peri, A. Regulation of the leucoanthocyanidin dioxygenase gene expression in Vitis vinifera. *Plant Sci.* **2001**, *161*, 579–588. [CrossRef]
49. Schiozer, A.L.; Barata, L.E.S. Stability of Natural Pigments and Dyes. *Rev. Fitos* **2007**, *3*, 6–24. Available online: https://revistafitos.far.fiocruz.br/index.php/revista-fitos/article/view/71 (accessed on 9 June 2023). [CrossRef]
50. Landi, M.; Tattini, M.; Gould, K.S. Multiple functional roles of anthocyanins in plant-environment interactions. *Environ. Exp. Bot.* **2015**, *119*, 4–17. [CrossRef]
51. Simões, C.; Bizarri, C.H.B.; da Silva Cordeiro, L.; Castro, T.C.; Coutada, L.C.M.; da Silva, A.J.R.; Albarello, N.; Mansur, E. Anthocyanin production in callus cultures of Cleome rosea: Modulation by culture conditions and characterization of pigments by means of HPLC-DAD/ESIMS. *Plant Physiol. Biochem.* **2009**, *47*, 895–903. [CrossRef]
52. Meng, J.X.; Gao, Y.; Han, M.L.; Liu, P.Y.; Yang, C.; Shen, T.; Li, H.H. In vitro Anthocyanin Induction and Metabolite Analysis in Malus spectabilis Leaves Under Low Nitrogen Conditions. *Hortic. Plant J.* **2020**, *6*, 284–292. [CrossRef]
53. Ji, X.H.; Wang, Y.T.; Zhang, R.; Wu, S.J.; An, M.M.; Li, M.; Wang, C.Z.; Chen, X.L.; Zhang, Y.M.; Chen, X.S. Effect of auxin, cytokinin and nitrogen on anthocyanin biosynthesis in callus cultures of red-fleshed apple (Malus sieversii f. niedzwetzkyana). *Plant Cell Tissue Organ Cult.* **2015**, *120*, 325–337. [CrossRef]
54. Li, H. *Modern Plant Physiology*, 3rd ed.; Life World: Beijing, China, 2012; p. 2. (In Chinese)
55. Li, H.; He, K.; Zhang, Z.Q.; Hu, Y. Molecular mechanism of phosphorous signaling inducing anthocyanin accumulation in Arabidopsis. *Plant Physiol. Biochem.* **2023**, *196*, 121–129. [CrossRef] [PubMed]
56. Huang, J.; Gu, M.; Lai, Z.; Fan, B.; Shi, K.; Zhou, Y.H.; Yu, J.Q.; Chen, Z. Functional analysis of the Arabidopsis PAL gene family in plant growth, development, and response to environmental stress. *Plant Physiol.* **2010**, *153*, 1526–1538. [CrossRef]
57. Abbasi, B.H.; Tian, C.L.; Murch, S.J.; Saxena, P.K.; Liu, C.Z. Light-enhanced caffeic acid derivatives biosynthesis in hairy root cultures of Echinacea purpurea. *Plant Cell Rep.* **2007**, *26*, 1367–1372. [CrossRef]
58. Kreuzaler, F.; Hahlbrock, K. Flavonoid glycosides from illuminated cell suspension cultures of Petroselinum hortense. *Phytochemistry* **1973**, *12*, 1149–1152. [CrossRef]

59. Zhong, J.J.; Seki, T.; Kinoshita, S.I.; Yoshida, T. Effect of light irradiation on anthocyanin production by suspended culture of Perilla frutescens. *Biotechnol. Bioeng.* **1991**, *38*, 653–658. [CrossRef]
60. Liu, C.Z.; Guo, C.; Wang, Y.C.; Ouyang, F. Effect of light irradiation on hairy root growth and artemisinin biosynthesis of *Artemisia annua* L. *Process. Biochem.* **2002**, *38*, 581–585. [CrossRef]
61. Fazal, H.; Abbasi, B.H.; Ahmad, N.; Ali, S.S.; Akbar, F.; Kanwal, F. Correlation of different spectral lights with biomass accumulation and production of antioxidant secondary metabolites in callus cultures of medicinally important *Prunella vulgaris* L. *J. Photochem. Photobiol. B Biol.* **2016**, *159*, 1–7. [CrossRef] [PubMed]
62. Ahmad, N.; Rab, A.; Ahmad, N. Light-induced biochemical variations in secondary metabolite production and antioxidant activity in callus cultures of Stevia rebaudiana (Bert). *J. Photochem. Photobiol. B Biol.* **2016**, *154*, 51–56. [CrossRef]
63. Nadeem, M.; Abbasi, B.H.; Younas, M.; Ahmad, W.; Zahir, A.; Hano, C. LED-enhanced biosynthesis of biologically active ingredients in callus cultures of Ocimum basilicum. *J. Photochem. Photobiol. B Biol.* **2019**, *190*, 172–178. [CrossRef]
64. Abou El-Dis, G.R.; Zavdetovna, K.L.; Nikolaevich, A.A.; Abdelazeez, W.M.A.; Arnoldovna, T.O. Influence of light on the accumulation of anthocyanins in callus culture of *Vaccinium corymbosum* L. cv. Sunt Blue Giant. *J. Photochem. Photobiol.* **2021**, *8*, 100058. [CrossRef]
65. Nazir, M.; Ullah, M.A.; Younas, M.; Siddiquah, A.; Shah, M.; Guivarc'h, G.; Hano, C.; Abbasi, B.H. Light-mediated biosynthesis of phenylpropanoid metabolites and antioxidant potential in callus cultures of purple basil (*Ocimum basilicum* L. var purpurascens). *Plant Cell Tissue Organ Cult.* **2020**, *142*, 107–120. [CrossRef]
66. Adil, M.; Ren, X.X.; Jeong, B.R. Light elicited growth, antioxidant enzymes activities and production of medicinal compounds in callus culture of Cnidium officinale Makino. *J. Photochem. Photobiol. B Biol.* **2019**, *196*, 111509. [CrossRef]
67. Sng, B.J.R.; Mun, B.; Mohanty, B.; Kim, M.; Phua, Z.W.; Yang, H.; Lee, D.Y.; Jang, I.C. Combination of red and blue light induces anthocyanin and other secondary metabolite biosynthesis pathways in an age-dependent manner in Batavia lettuce. *Plant Sci.* **2021**, *310*, 110977. [CrossRef]
68. Jie, X.; Liu, X.; Wang, W.; Bai, J.; Li, D.; Liu, Y. Effects of Light Quality on Callus Growth Rate and Anthocyanin Synthesis of Hongyang Kiwifruit. *Shanxi Agric. Sci.* **2021**, *49*, 1166–1172. (In Chinese)
69. Toguri, T.; Umemoto, N.; Kobayashi, O.; Ohtani, T. Activation of anthocyanin synthesis genes by white light in eggplant hypocotyl tissues, and identification of an inducible P-450 cDNA. *Plant Mol. Biol.* **1993**, *23*, 933–946. [CrossRef] [PubMed]
70. Liu, Y.; Schouten, R.E.; Tikunov, Y.; Liu, X.X.; Visser, R.G.F.; Tan, F.; Bovy, A.; Marcelis, L.F.M. Blue light increases anthocyanin content and delays fruit ripening in purple pepper fruit. *Postharvest Biol. Technol.* **2022**, *192*, 112024. [CrossRef]
71. Tu, Y.; Wang, B.; Jiang, H.; Xi, X.; Ding, J. Culture effect of different color light, temperature, and pH on grass coral callus. *Jiangxi Sci.* **1994**, *1994*, 85–89. (In Chinese)
72. Wei, C.; Niu, Z.; Dou, F.; Wang, Q. Effect of test tube microenvironment on anthocyanin content in potato 'GSAP-H' callus. *J. Gansu Agric. Univ.* **2016**, *51*, 47–53. (In Chinese) [CrossRef]
73. Narayan, M.S.; Thimmaraju, R.; Bhagyalakshmi, N. Interplay of growth regulators during solid-state and liquid-state batch cultivation of anthocyanin producing cell line of Daucus carota. *Process Biochem.* **2005**, *40*, 351–358. [CrossRef]
74. Liu, C.; Zhao, Y.; Wang, Y. Artemisinin: Current state and perspectives for biotechnological production of an antimalarial drug Appl. *Microb. Biotechnol.* **2006**, *72*, 11–20. [CrossRef] [PubMed]
75. Kim, S.J.; Hahn, E.J.; Heo, J.W.; Paek, K.Y. Effects of LEDs on net photosynthetic rate, growth and leaf stomata of chrysanthemum plantlets in vitro. *Sci. Hortic.* **2004**, *101*, 143–151. [CrossRef]
76. Wang, N.; Zhang, Z.; Jiang, S.; Xu, H.; Wang, Y.; Feng, S.; Chen, X. Synergistic effects of light and temperature on anthocyanin biosynthesis in callus cultures of red-fleshed apple (Malus sieversii f. niedzwetzkyana). *Plant Cell Tissue Organ Cult.* **2016**, *127*, 217–227. [CrossRef]
77. Rehman, R.N.U.; You, Y.; Yang, C.; Khan, A.R.; Li, P.; Ma, F. Characterization of phenolic compounds and active anthocyanin degradation in crabapple (*Malus orientalis*) flowers. *Hortic. Environ. Biotechnol.* **2017**, *58*, 324–333. [CrossRef]
78. Leyva, A.; Jarillo, J.A.; Salinas, J.; Martinez-Zapater, J.M. Low Temperature Induces the Accumulation of Phenylalanine Ammonia-Lyase and Chalcone Synthase mRNAs of Arabidopsis thaliana in a Light-Dependent Manner. *Plant Physiol.* **1995**, *108*, 39–46. [CrossRef] [PubMed]
79. Lo Piero, A.R.; Puglisi, I.; Rapisarda, P.; Petrone, G. Anthocyanins accumulation and related gene expression in red orange fruit induced by low-temperature storage. *J. Agric. Food Chem.* **2005**, *53*, 9083–9088. [CrossRef]
80. Naing, A.H.; Park, D.Y.; Park, K.I.; Kim, C.K. Differential expression of anthocyanin structural genes and transcription factors determines coloration patterns in gerbera flowers. *3 Biotech* **2018**, *8*, 393. [CrossRef]
81. Bagal, D.; Chowdhary, A.A.; Mehrotra, S.; Mishra, S.; Rathore, S.; Srivastava, V. Metabolic engineering in hairy roots: An outlook on production of plant secondary metabolites. *Plant Physiol. Biochem.* **2023**, *201*, 107847. [CrossRef] [PubMed]
82. Selwal, N.; Rahayu, F.; Herwati, A.; Latifah, E.; Supriyono; Suhara, C.; Suastika, I.B.K.; Mahayu, W.M.; Wani, A.K. Enhancing secondary metabolite production in plants: Exploring traditional and modern strategies. *J. Agric. Food Res.* **2023**, *14*, 100702. [CrossRef]
83. Koufan, M.; Belkoura, I.; Mazri, M.A.; Amarraque, A.; Essatte, A.; Elhorri, H.; Zaddoug, F.; Alaoui, T. Determination of antioxidant activity, total phenolics and fatty acids in essential oils and other extracts from callus culture, seeds and leaves of *Argania spinosa* (L.) Skeels. *Plant Cell Tiss Organ Cult.* **2020**, *141*, 217–227. [CrossRef]

84. Yang, M.; Zhang, Q.; Lu, A.; Yang, Z.; Tan, D.; Lu, Y.; Qin, L.; He, Y. The preventive effect of secondary metabolites of Dendrobium officinale on acute alcoholic liver injury in mice. *Arab. J. Chem.* **2023**, *16*, 105138. [CrossRef]
85. Liu, X.; Ma, C.; Liu, Z.; Kang, W. Natural Products: Review for Their Effects of Anti-HBV. *BioMed Res. Int.* **2020**, *9*, 3972390. [CrossRef]

Disclaimer/Publisher's Note: The statements, opinions and data contained in all publications are solely those of the individual author(s) and contributor(s) and not of MDPI and/or the editor(s). MDPI and/or the editor(s) disclaim responsibility for any injury to people or property resulting from any ideas, methods, instructions or products referred to in the content.

Article

Magnesium Oxide Nanoparticles: An Influential Element in Cowpea (*Vigna unguiculata* L. Walp.) Tissue Culture

Rabia Koçak [1,†], Melih Okcu [1,*,†], Kamil Haliloğlu [1,†], Aras Türkoğlu [2,†], Alireza Pour-Aboughadareh [3,*], Bita Jamshidi [4], Tibor Janda [5,*], Azize Alaylı [6] and Hayrunnisa Nadaroğlu [7,8]

[1] Department of Field Crops, Faculty of Agriculture, Ataturk University, 25240 Erzurum, Türkiye; rabiakocakrabia12@gmail.com (R.K.); kamilh@atauni.edu.tr (K.H.)
[2] Department of Field Crops, Faculty of Agriculture, Necmettin Erbakan University, 42310 Konya, Türkiye; aras.turkoglu@erbakan.edu.tr
[3] Seed and Plant Improvement Institute, Agricultural Research, Education and Extension Organization (AREEO), Karaj P.O. Box 3158854119, Iran
[4] Department of Food Security and Public Health, Khabat Technical Institute, Erbil Polytechnic University, Erbil 44001, Iraq; bita.alimer@epu.edu.iq
[5] Department of Plant Physiology and Metabolomics, Agricultural Institute, Centre for Agricultural Research, 2462 Martonvásár, Hungary
[6] Department of Nursing, Faculty of Health Sciences, Sakarya University of Applied Sciences, 54187 Sakarya, Türkiye; aalayli@subu.edu.tr
[7] Department of Food Technology, Vocational College of Technical Sciences, Ataturk University, 25240 Erzurum, Türkiye; hnisa25@atauni.edu.tr
[8] Department of Nano-Science and Nano-Engineering, Institute of Science, Ataturk University, 25240 Erzurum, Türkiye
* Correspondence: melihokcu@atauni.edu.tr (M.O.); a.poraboghadareh@edu.ikiu.ac.ir (A.P.-A.); janda.tibor@atk.hu (T.J.)
† These authors contributed equally to this work.

Citation: Koçak, R.; Okcu, M.; Haliloğlu, K.; Türkoğlu, A.; Pour-Aboughadareh, A.; Jamshidi, B.; Janda, T.; Alaylı, A.; Nadaroğlu, H. Magnesium Oxide Nanoparticles: An Influential Element in Cowpea (*Vigna unguiculata* L. Walp.) Tissue Culture. *Agronomy* 2023, *13*, 1646. https://doi.org/10.3390/agronomy13061646

Academic Editors: Justyna Lema-Rumińska, Danuta Kulpa and Alina Trejgell

Received: 10 May 2023
Revised: 14 June 2023
Accepted: 17 June 2023
Published: 19 June 2023

Copyright: © 2023 by the authors. Licensee MDPI, Basel, Switzerland. This article is an open access article distributed under the terms and conditions of the Creative Commons Attribution (CC BY) license (https://creativecommons.org/licenses/by/4.0/).

Abstract: Nanotechnology is a rapidly growing field of science and technology that deals with the development of new solutions by understanding and controlling matter at the nanoscale. Since the last decade, magnesium oxide nanoparticles (MgO-NPs) have gained tremendous attention because of their unique characteristics and diverse applications in materials sciences and because they are non-toxic and relatively cheaply available materials. MgO-NPs can improve plant growth and contribute to plant tolerance of heavy metal toxicity. The effects of MgO-NPs on cowpea (*Vigna unguiculata* L. Walp.) plants were surveyed under in vitro conditions to find the optimum combination for cowpea tissue culture. The MgO-NPs used in the study were synthesized using walnut shell extract by the green synthesis method. MgO nanoparticles with 35–40 nm size was used in this research. When the size distribution of the MgO-NPs' structure was examined, two peaks with 37.8 nm and 78.8 nm dimensions were obtained. The zeta potential of MgO-NPs dispersed in water was measured around −13.3 mV on average. The results showed that different doses of MgO-NPs applied to cowpea plant on all in vitro parameters significantly affected all measured parameters of cowpea plantlets under in vitro condition in a positive way. The best results in morphogenesis were MS medium supplemented with high MgO-NP applications (555 mg/L), resulting in a 25% increase in callus formation. The addition of Mg-NPs in the induction medium at concentrations at 370 mg/L increased shoot multiplication. The highest root length with 1.575 cm was obtained in MS medium containing 370 mg/L MgO. This study found that MgO-NPs greatly influenced the plantlets' growth parameters and other measured traits; in addition, our results indicate that the efficiency of tissue culture of cowpea could be improved by increased application of MgO in the form of nanoparticles. In conclusion, the present work highlights the possibility of using MgO-NPs in cowpea tissue culture.

Keywords: MgO-NPs; nano fertilizer; cowpea feed; regeneration

1. Introduction

Nanotechnology, which is a multidisciplinary science, uses nano-sized materials [1], which are defined as substances smaller than 100 nanometers in size, and controls them at the atomic level and makes them useful [2]. Nanotechnology covers many different fields, including pesticide distribution, nano sensors, pesticide degradation, use of micronutrients in agriculture, and plant protection and nutrition [3]. Nanotechnology offers effective methods to protect soil health and conditions by helping to minimize agricultural waste and environmental pollution [4,5]; therefore, it can greatly improve the functioning of precision agriculture [6]. Nanotechnology is useful in the agricultural sector, in the form of nano-pesticides and nano-fertilizers [7]. Among various methods of NP synthesis, the plant tract-mediated method is preferred due to its cost-effective nature [8]. MgO-NPs have applications in various fields and have earlier been synthesized using plant extract [9].

Magnesium plays an important role in plant growth and development, serves as a component of the chlorophyll molecules, and regulates the activity of key photosynthetic enzymes in the chloroplast [10]. Magnesium is a macronutrient that activates more enzymes than other nutrients [11] and has structural and regulatory functions related to nucleophilic ligands in plants [10,12]. It is one of the essential elements in the function and synthesis of nucleic acids and ATP [13]. Magnesium deficiency suppresses plant growth and decreases yield [14]. Magnesium oxide is an important inorganic material with a wide band range [15]. This material is used in many applications, such as catalysis, catalyst supports, toxic waste reclamation, refractory materials and adsorbents, additives in heavy fuel oils, reflective and anti-reflective coatings, substrate such as superconducting and ferroelectric thin films, superconductors, and lithium-ion batteries [16,17]. Nano MgO, on the other hand, has many special physical and chemical properties brought about by its nano size. With its size, nanoparticles can be used more by plant cells, induce plant growth, and have antimicrobial, antifungal, and antiviral effects against pests [18]. The small size of nanoparticles allows them to penetrate into the plant cell. Hence, we assessed the impact of MgO-NPs on the legume, cowpea plant.

Cowpea (*Vigna unguiculata* L. Walp.) is a very common annual plant, especially in Africa, South America, Asia, and the United States, and is one of the most important legumes worldwide [19]. It is a good pre-plant that has the ability to grow in poor soils and increases the yield of the next product with the help of nitrogen fixation [20,21]. Cowpea, which is considered as a green vegetable and dry grain in human nutrition and as a fodder in animal nutrition, belongs to the legume family and contains 2.0–4.3% protein in fresh beans and 4.5–5.0% in fresh grains. Protein content in dry cowpea grains that have reached maturity varies between 20.42 and 34.60%, depending on the variety and environmental conditions. In addition, its grains contain rich source of essential amino acids, except cysteine and methionine [22]. The protein in its seeds is rich in Lysine and Tryptophan amino acids compared to cereal seeds, and is insufficient in terms of Methionine and Cystine compared to animal proteins [23].

Nanoparticles, due to their special physical and chemical properties, may lead to unpredictable changes in the morphological characteristics of the plant [24]. The synthesized nanomaterials may provide protection that is effective in controlling pests and pathogens that significantly affect the yield of the plant [25]. Toxic effects of NPs for plants and animals have been reported, but there is no report documented until now that showed the harmful effects of NPs on tissue culture plants [26]. There have been many studies on the use of NPs in plant tissue culture systems [26–29]. Wide applications of NPs in plant tissue culture include the elimination of microbial contaminants from explants, callus induction, organogenesis, somatic embryogenesis, somaclonal variation, genetic transformation, and secondary metabolite development. Compounds can be developed by integrating the concept of nanotechnology into plant tissue culture techniques, synthesis, purification, and desired plant-derived yield [30]. There is limited information available in the literature regarding the effect of NPs on in vitro regeneration characters on cowpea plant, which has a great importance in human and animal nutrition. Therefore, the aim of this study is to determine the possible effects of callus formation on morphogenesis and plant regeneration by applying different doses of MgO-NPs of cowpea plant under plant tissue culture condition.

2. Materials and Methods

2.1. Synthesis of Mg Nanoparticles (MgO-NPs)

MgO-NPs were synthesized using walnut shell extract. The walnut shell extraction used for green synthesis was prepared with distilled water, and for this purpose, 25 g walnut shells were washed and crushed using the freeze–thaw technique in liquid nitrogen; 250 mL of distilled water was added and mixed in a magnetic stirrer for 1–2 h. The extract was obtained by first filtering through cheesecloth and then filter paper, before being kept at $-25\ °C$ until use. An amount of 0.1 M Mg $(NO_3)_2$ solution was actively used for conversation of MgO in plants. The formation of the synthesis was followed qualitatively and quantitatively by UV-Vis spectrophotometer (Epoch). After the method optimization, the obtained NPs were characterized. For this purpose, different morphological and molecular detection methods, such as scanning electron microscopy (SEM; Zeiss Sigma 300), Fourier-transform infrared spectroscopy (FT-IR; VERTEX 70 v FT-IR Spectrometer, Billerica, MA, USA), and X-ray diffraction (XRD; Malvern Panalytical B.V., Almelo, The Netherlands), were used. The obtained MgO-NPs were washed under vacuum using distilled water and ethanol and were used in the reaction after drying in an oven. In the application range, after MgO-NPs were weighed and homogenized in pure water with the help of an ultrasonicator, they were used in plant in vitro experiments [31,32].

2.2. Plant Material

The cowpea cultivar, registered as "Ulkem", used in this study was obtained from Ondokuz Mayıs University and is usually used for forage purposes. Forage cowpea seeds were washed with tap water and surface-sterilized with 70% ethanol for 5 min, treated for 25 min with solution containing 1% sodium hypochlorite with a few drops of Tween 20 with constant stirring, and rinsed three times with sterile distilled water thereafter. The seeds were imbibed in sterile water for 24 h in the dark. Plumula parts were aseptically dissected and used as explants in the experiment.

2.3. Tissue Culture Applications

MS medium, mineral salts, and vitamins [33] were used in the experiments. Magnesium and MgO-NPs in MS medium were exchanged with different concentrations. In the experiment, there were 5 treatments with different magnesium content: 0 mg/L (MS medium without $MgSO_4.7H_2O$ and MgO-NPs), $MgSO_4.7H_2O$, which is used commonly in MS medium 370 mg/L as macronutrient elements, and three treatment with removed $MgSO_4.7H_2O$ from the MS medium and replaced with a nanoparticle version of these elements: 1/2X (185 mg/L MgO-NPs), 1X (370 mg/L MgO-NPs), 2X (555 mg/L MgO-NPs) concentrations. Plumules were cultured on MS medium containing MgO-NPs previously prepared. The explants were kept in the dark at $24 \pm 1\ °C$. In shoot regeneration medium (MS salt and vitamins + 1.0 mg/L BAP) containing different Mg types and concentrations, explant were kept for 8 weeks, and they were then placed in root formation medium (MS salt and vitamins + 0.5 mg/L BAP) containing different MgO-NPs types and concentrations and kept under white fluorescent light (Preheat Daylight-42 µmol photons $m^{-2}s^{-1}$) for 4 weeks at $24 \pm 1\ °C$ in a 16-h light photo period. Morphogenesis and callus formation measurements were made after 30 days. The shoot formation rate, number of shoots, number of shoots per explant, shoot length, root formation rate, number of roots per explant, and root length were calculated after 60 days. This study was carried out in a complete randomized experimental design arrangement with four replications. Each petri dish was considered as an experimental unit, and 10 cowpea plumula were cultured in each petri dish. Analysis of variance and Duncan multiple comparison tests were computed with SPSS statistical analysis program (Version 20).

3. Results

3.1. Surface Morphological Characterization of MgO-NPs

Surface characterization of MgO-NPs was performed using SEM, FT-IR, and XRD analyzes, and the results are given in Figure 1. SEM analysis determined that the MgO-NPs obtained by green synthesis were well dispersed and cubic (Figure 1A), while the peaks at 39.2° (111), 62.53° (220), 77.8° (311), and 81.7° (222) 2θ° in the XRD graph in Figure 1B belong to Mg (OH)$_2$. FT-IR analysis is an effective technique used to identify possible peaks of MgO-NPs and extract used for the reduction of metal. FT-IR analysis of MgO-NPs obtained by green synthesis is given in Figure 1C. As seen from the FT-IR diagram, the wavelength between 400 and 4000 cm^{-1} was scanned. From the findings, it was determined that intense absorption peaks occurred at 3699, 3351, 2293, 1600, 1354, 1014, 763, and 519 cm^{-1}. While the peak observed at 3699 cm^{-1} belongs to the -OH band, the wide peak band observed at 3351 cm^{-1} indicates the presence of -NH$_2$ and -OH groups in the medium. The peaks seen around 1600 cm^{-1} indicate the presence of peaks defined as the primary amine group (N-H) overlapping with the amide and carboxylate group. The peak at 1354 cm^{-1} is matched with the Mg-OH group, while all bands between 400 and 736 confirm the presence of MgO-NPs. The size of MgO-NPs was determined to be 35–40 nm as a result of measurements and calculations. The obtained spectrum showed that the walnut shell extract had a high ability to reduce and stabilize MgO-NPs.

When the size distribution of the MgO NPs structure was examined, two peaks with 37.8 nm and 78.8 nm dimensions were obtained. These peaks showed that the MgO NP structure did not have much agglomeration, but there was a small amount of agglomeration. However, the structure was smaller than 100 nm, and the findings support the SEM image (Figure 2A). The zeta potential analysis is an important indicator of the surface charges of chemical compounds. The zeta potential of MgO NPs dispersed in water was measured around −13.3 mV on average. It shows a negative potential value (Figure 2B). The negative value obtained is due to the oxygen atoms in the MgO structure (10.1002/cbdv.201900608 (13 June 2023)).

Figure 1. *Cont.*

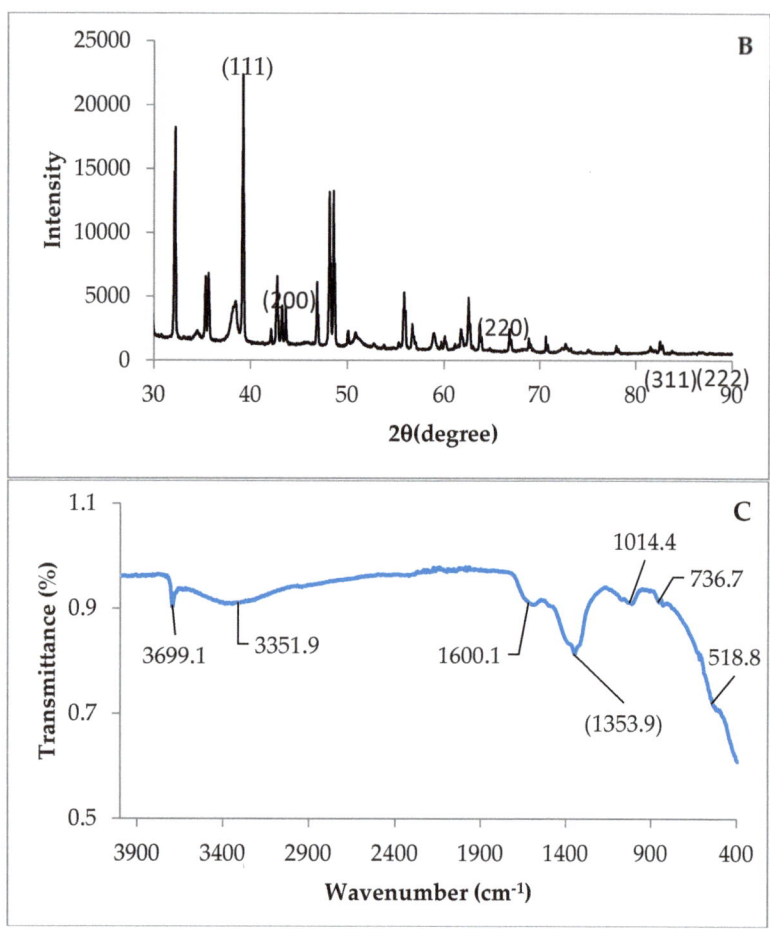

Figure 1. (**A**) SEM image of MgO-NPs synthesized by walnut shell extract. (**B**) XRD pattern of MgO nanoparticles. (**C**) The FT-IR spectrum of green synthesized MgO-NPs using walnut shell extract.

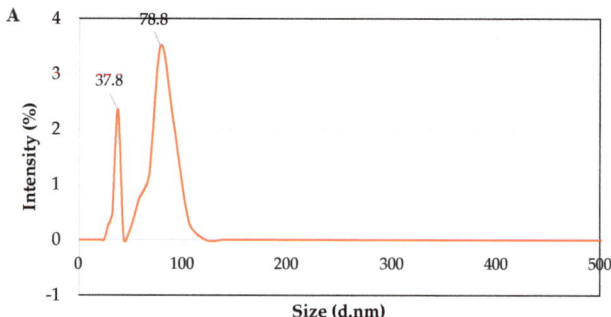

Figure 2. *Cont.*

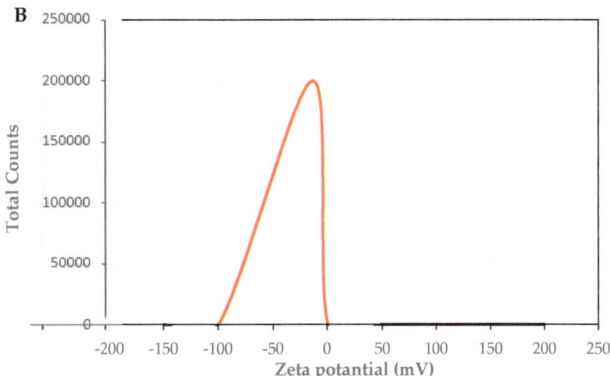

Figure 2. (**A**) Size distribution of MgO NPs. (**B**) Zeta potential analysis of MgO-NPs using walnut shell extract.

3.2. Morphogenesis

The averages of the characters determined by the treatments of different concentrations of MgO-NPs to the cowpea and the related variance analysis results are given in Table 1. In this study, morphogenesis basically refers to any changes in the explant, such as elongation, contraction, color, and structure changes, during the course of in vitro culture, except for the formation of callus, shoots, roots, or whole plants.

It has been observed that different MgO-NP applications have significant effects on morphogenesis ($p < 0.01$). These changes began to be observed after the first week of culture initiation. While the average number of explants showing morphogenesis was 9.25 from the original 10 explants at control (MS medium without MgO-NPs) and MS medium containing 370 mg/L MgO applications, this number was increased in Mg-NP applications (185 mg/L, 370 mg/L and 555 mg/L MgO-NPs). A significant 8.11% increase ($p < 0.05$) in the number of morphogenesis was observed in MS medium containing 185 mg/L, 370 mg/L, and 555 mg/L MgO-NP applications compared to the control. Morphological changes such as tissue growth and tissue swelling were observed in the cultured cowpea explants (Figure 3).

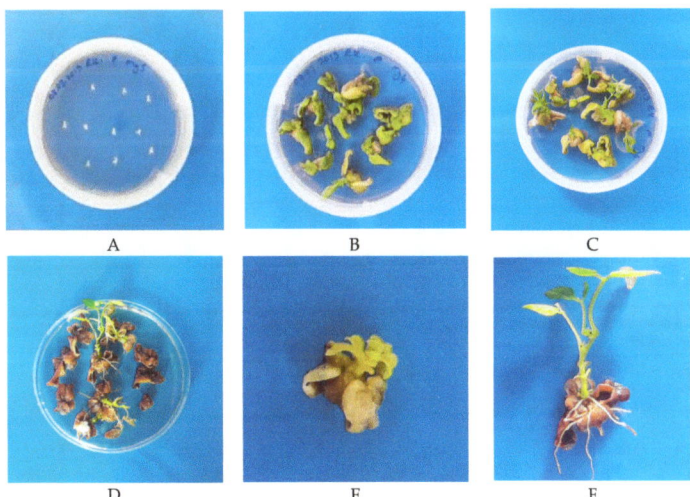

Figure 3. (**A**) Cultured explants. (**B**) Morphogenesis. (**C,D**) Shoot and root formation. (**E**) Callus formation and (**F**) regenerated cow pea plantlets in MS medium supplementary with 370 mg/L MgO-NPs.

Table 1. Average values and analysis of variance of the parameter ratios of magnesium nanoparticles at different doses examined in cowpea plant.

Mg	Morphogenesis		Callus Formation		Shoot Formation		Number of Shoots		Number of Shoots per Explant		Shoot Length		Root Formation Rate		Number of Roots per Explant		Root Length	
	Number	%[1]	Number	%	%	%	Number	%	Number	%	cm	%	%	%	Number	%	cm	%
Control	9.25 ab[2]	-	8.00 b	-	60.00 ab	-	21.75 b	-	6.50 bc	-	0.475 d	-	27.50 a	-	2.750 b	-	1.0750 b	-
370 mg/L MgSO$_4$·7H$_2$O	9.25 ab	-	9.00 ab	12.50	42.50 bc	−29.17	21.75 b	-	6.75 bc	3.85	1.200 c	152.63	27.50 a	-	6.750 a	145.45	1.5750 a	46.51
185 mg/L MgO-NPs	10.00 a	8.11	8.50 ab	6.25	30.00 c	−50	3.75 c	−82.76	1.25 c	−80.77	0.175 d	−63.16	19.75 b	−28.18	0.0009 c	−99.96	0.00015 c	−99.96
370 mg/L MgO-NPs	10.00 a	8.11	9.50 ab	18.75	72.50 a	20.83	61.25 a	181.61	17.50 a	169.23	2.075 a	336.84	22.50 ab	−18.18	0.7500 bc	−72.72	0.2750 bc	−74.41
555 mg/L MgO-NPs	10.00 a	8.11	10.00 a	25	82.50 a	37.50	36.25 ab	66.67	10.00 ab	53.85	1.450 b	205.26	10.00 c	−63.64	0.7500 bc	−72.72	0.2000 bc	−81.39
Variation Sources	1.350 *[3]		1837.5 **		3.459 *		5.675 **		88.901 **		1.905 *		18.030 **		2.679 *		20.402 **	
Error	15																	

[1] Percent compared to control groups. [2] a–d—Mean values with the same letter are not significantly different ($p < 0.05$). [3] ** and *: Significant at the 0.01 and 0.05 probability levels, respectively.

3.3. Callus Formation

It has been observed that different MgO-NP applications have significant effects on callus formation. Under the present experimental conditions, where 10 explants were originally started, the average callus formation depending on the MgO-NPs varied between 8 and 10. Callus formation reached the highest value with 10 in the application of MS medium containing 555 mg/L MgO-NPs, and this value was statistically significant from the control plants. This application was followed by 9.5 in MS medium containing 370 mg/L MgO-NPs and 9 in MS medium containing 370 mg/L $MgSO_4.7H_2O$ applications. Different Mg treatments significantly increased callus formation compared to control. The highest increase was observed with 25% in MS medium supplemented with 555 mg/L MgO-NP application, followed by 18.75% with MS medium including 370 mg/L MgO-NPs, 12.50% in MS medium comprising 370 mg/L $MgSO_4.7H_2O$ and 6.25% in MS medium comprising 185 mg/L MgO-NP applications (Table 1).

3.4. Shoot Formation

Different MgO (with or without NPs) treatments led to significant differences in shoot formation ($p < 0.01$). According to the effect of different doses of magnesium NPs applied to the cowpea, shoot formation rates varied between 20.83% and 37.50%. The highest value in shoot formation rate was obtained from MS medium containing 555 mg/L MgO-NP application with 82.50%, and the lowest value was obtained from MS medium containing 185 mg/L MgO-NP application with 30%. Shoot formation rate was 60% from the control application, 42.50% from the MS medium containing 370 mg/L $MgSO_4.7H_2O$ application, and 72.50% from the MS medium containing 370 mg/L MgO-NP application (Table 1).

3.5. Number of Shoots

The highest number of shoots was obtained from MS medium containing 370 mg/L MgO-NP application (61.25), followed by 36.25 in MS medium containing 555 mg/L MgO-NP application. Control and 370 mg/L $MgSO_4.7H_2O$ applications gave the same shoot number value as 21.75, while MS medium containing 185 mg/L MgO-NP application gave the lowest value with 3.75, but this did not differ significantly from the control (Table 1).

3.6. Number of Shoots per Explant

The mean values of the number of shoots per explant determined in the cowpea at different doses varied between 1.25 and 17.50. In terms of applications, the highest number of shoots per explant (17.50) was obtained from the application of MS medium comprising 370 mg/L MgO-NPs, and this value was statistically higher than that found in the control treatment (6.5). The lowest value (1.25) was obtained from the MS medium comprising 185 mg/L MgO-NP application. The number of shoots per explant obtained from other applications in our study was determined as 6.75 in the MS medium comprising 370 mg/L $MgSO_4.7H_2O$ application and 10 in the 555 mg/L MgO-NPs, but they were not statistically different from the control value (Table 1).

3.7. Shoot Length

In terms of shoot length, different MgO applications created significant differences. The longest shoot length was 2.07 cm in the MS medium containing 370 mg/L MgO-NP application, and the shortest shoot was 0.175 cm in the MS medium containing 185 mg/L MgO-NP application. MS medium supplemented with 370 mg/L MgO-NP application was followed by MS medium comprising 555 mg/L MgO-NPs (1.450 cm), MS medium comprising 370 mg/L $MgSO_4.7H_2O$ (1.200 cm), and control (0.475 cm) applications. Except for the MS medium comprising 185 mg/L MgO-NP application, the other MS medium comprising 370 mg/L $MgSO_4.7H_2O$, MS medium comprising 370 mg/L MgO-NPs, and MS medium comprising 555 mg/L MgO-NP applications showed an increasing effect compared to the control. The highest increase was obtained from the administration of MS medium supplemented with 370 mg/L MgO-NPs, with 336.84% (Table 1).

3.8. Root Formation Rate

Different Mg applications had significant effects on root formation rate. The average root formation rate of the applied magnesium NPs varied between 10 and 27.5%. In terms of root formation rate, the highest value among magnesium NPs was obtained from control and MS medium containing 370 mg/L MgSO$_4$.7H$_2$O applications with 27.50%; this application was followed by 22.50% in MS medium containing 370 mg/L MgO-NP application and 19.75% with MS medium containing 185 mg/L MgO-NP applications. The lowest value of 10% was obtained in MS medium containing 555 mg/L MgO-NP application. The highest value was recorded in the application of MS medium containing 555 mg/L MgO-NPs, with a 63.64% decrease in the variation of magnesium NPs compared to the control (Table 1).

3.9. Number of Roots per Explant

Different MgO-NP applications had significant effects in terms of root number per explant. The maximum number of roots per explant determined in cowpea was obtained as 6.75 in the application of MS medium containing 370 mg/L MgO-NPs; this application was followed by 2.75 in the control application. The applied magnesium nanoparticles at different doses caused significant reductions compared to the control, which was statistically significant at 185 mg/L MgO-NPs (Table 1).

3.10. Root Length

According to the analysis of variance, root length was significantly affected by certain MgO-NP applications (Table 1). It was determined that the root length values obtained from the cowpea with different doses of magnesium NPs varied between 0.015 cm (MS medium supplementary with 185 mg/L MgO-NPS) and 157.50 cm (MS medium supplementary with 370 mg/L MgO-NPS). These extreme values were significant from the control; the other treatments used in this experiment were not (Table 1).

4. Discussion

NPs have begun to be used extensively in plant tissue culture studies. It is known that the kind, type, concentration, and size of NPs can be effective in studies conducted within the scope of plant tissue culture. The present study clearly showed the beneficial effects of MgO-NPs on the in vitro parameters of cowpea. In this study, MgO-NPs were synthesized using walnut shell extract by the green synthesis method. The size of MgO nanoparticles was determined to be 35–40 nm as a result of measurements and calculations. SEM analysis determined that the MgO-NPs obtained by green synthesis were well dispersed and cubic. The diffraction peaks are points that represent cubic MgO-NPs at 42.76° (200)2θ and 62.6° (220)2θ [34]. FT-IR analysis is an effective technique used to identify possible peaks of MgO-NPs and the extract used for the reduction of metal. As seen from the FT-IR diagram, the wavelength between 400 and 4000 cm^{-1} was scanned. Similar findings were found in the literature in the FT-IR analyzes of MgO-NPs obtained using some plant extracts, and they support our study [35–41].

Morphological changes such as tissue growth and tissue swelling were observed in the cultured cowpea explants. Our findings showed that positive effects of MgO-NPs on callus induction, shoot regeneration, and explant growth were observed. The results showed that the highest value of callus formation was obtained in MS medium containing with 555 mg/L MgO-NPs. Our result showed that NPs not only overcome negative effects but also improve callus formation. It seems that MgO-NPs may play a role similar to plant hormones such as cytokinins and gibberellins owing to their ability to induce plant cell division and stimulate cellular expansion; however, the mechanism of its action in darkness is still unknown. A similar finding was obtained by Mandeh et al. [27]. Several studies have shown positive effects of NPs on callus induction. Different concentrations of silver or gold NPs alone or combined with naphthalene acetic acid (NAA) were evaluated for callus culture growth in Prunella vulgaris L. The silver (30 μg L^{-1}), silver and gold (1:2), and silver and gold (2:1) NPs in combination with NAA (2.0 mg L^{-1}) enhanced callus proliferation (100%) as compared to the control (95%) [42]. The number and size of calli increased when barley mature embryos were grown in MS medium supplemented with 20 mg L^{-1} 2,4-D and 60 mg mL^{-1} TiO$_2$-NPs [27].

The explants were cultured on shoot multiplication medium supplemented with MS medium supplementary with 370 mg/L Mg-NPs. These media were optimum for the formation and development of shoot. Our result clearly showed that 370 mg/L Mg-NPs had a stimulation effect on the growth of shoot parameters, whiles 185 mg/L decreased NS and NSE and the 555 mg/L Mg-NPs concentrations induced an inhibitory effect. One reason for this may be that Mg-NPs block ethylene signaling and trigger shoot growth. Similar results have been found with AgNPs in banana plants by Do et al. [43]. Sharma et al. [44] also reported that AgNPs increased plant growth processes such as shoot and root lengths, area of the leaf, and biochemical parameters, such as carbohydrate and protein contents of common bean and corn.

The explants were also cultured on rooting medium supplemented with MS medium supplementary with 370 mg/L MgSO$_4$.7H$_2$O. These media were optimum for the formation and development of roots. Our result suggested that 370 mg/L MgSO$_4$.7H$_2$O had a stimulation effect on the growth of root parameters, while the application of NPs induced an inhibitory effect for root formation, especially in higher concentrations of NPs; this is very clear and obvious. These results are in agreement with earlier findings. For example, Helaly et al. [45] reported that root lengths were increased when Zn and nano ZnO-NPs were added to the MS medium. Zhang et al. [46] stated that copper nanoparticles inhibit primary root elongation and enhance lateral root emergence. Auxins, especially indole-3-acetic acid (IAA), have an important role in root development in plants. Regulation of auxin levels in different cells of the root is involved in many root functions, including growth, lateral root elongation, and root hair formation. There is a sharp toxicity threshold because at higher doses the toxic effects of ion exposure are manifest, with such responses as root shortening [47]. The reduction in elongation in cowpea roots exposed to high doses of Mg-NPs may possibly be related to the abnormal distribution of auxin. These results are in conflict with those obtained in barley [27]. However, this is not surprising given that nanoparticles can explain their effects depending on the size and/or shape of the particles, the concentrations applied, the particular experimental conditions, plant species, and uptake mechanisms [48]. Sotoodehnia-Korani [49] emphasized the belief that nanomaterials such as MgO-NPs have the ability to improve the efficiency of tissue cultures in vitro and boost agricultural yield.

5. Conclusions

The effects of various metal and metal oxide NPs on plants are well documented in vivo. Such NPs can be used to promote or enhance the morphogenetic potential of explants obtained from different plant species. NPs have been used widely in plant tissue culture studies. The influence of different concentrations and combinations of NPs on different media (shoot induction, shoot propagation, and rooting media) should also be investigated to gain a clear understanding of the underlying mechanisms behind the role of NPs in plant tissue culture. Given the potential for future research, it is vital to understand the role of MgO-NPs in callogenesis performance, micropropagation, and cell culture elicitation. The MgO-NPs used in the study were synthesized using walnut shell extract by green synthesis method. When the size distribution of the MgO-NPs structure was examined, two peaks with 37.8 nm and 78.8 nm dimensions were obtained. The zeta potential of MgO-NPs dispersed in water was measured at approximately −13.3 mV on average. The present study provides the first evidence of Mg-NPs effects on the in vitro culture of cowpea, showing the possibility of using MgO-NPs in cowpea tissue culture. The results showed that different doses of MgO-NPs applied to cowpea plants on all in vitro parameters significantly affected all measured parameters of cowpea plantlets under in vitro condition in a positive way. The best results in morphogenesis were MS medium supplemented with 555 mg/L MgO-NP applications, resulting in a 25% increase in callus formation. The addition of Mg-NPs in the induction medium at concentrations of 370 mg/L increased shoot multiplication. The highest root length with 1.575 cm was obtained in MS medium containing 370 mg/L MgO. However, for a clear understanding of the mechanisms underlying the role of MgO-NPs in cowpea tissue culture, it is recommended to investigate in detail the actual mechanisms of the promoting or inhibitory effects of MgO-NPs on each parameter.

Author Contributions: Conceptualization, M.O. and K.H.; methodology, K.H., A.T. and M.O.; software, M.O.; validation, K.H., H.N., A.A. and A.T.; formal analysis, M.O., K.H., A.T. and R.K.; investigation, K.H. and A.T.; resources, K.H., M.O., A.A. and H.N.; data curation, K.H. and A.T.; writing—original draft preparation, M.O. and A.T.; writing—review and editing, K.H., A.T., A.A., H.N., A.P.-A., B.J. and T.J.; visualization, M.O., K.H. and A.T.; supervision, M.O.; project administration, K.H., A.A. and H.N.; funding acquisition, T.J. All authors have read and agreed to the published version of the manuscript.

Funding: This work was funded by a grant from the National Research, Development, and Innovation Office (grant No. K142899).

Data Availability Statement: All data supporting the conclusions of this article are included in this article.

Conflicts of Interest: The authors declare no conflict of interest.

References

1. Arnall, A.H. Future Technologies, Today's Choices-Nanotechnology, Artificial Intelligence and Robotics. A technical, political and institutional map of emerging technologies. *AHS* **2003**, *56*, 1329–1332.
2. Bergeson, L.L.; Cole, M.F. Regulatory implications of nanotechnology. *Biointeract. Nanomater.* **2014**, 315. [CrossRef]
3. Ghormade, V.; Deshpande, M.V.; Paknikar, K.M. Perspectives for nano-biotechnology enabled protection and nutrition of plants. *Biotechnol. Adv.* **2011**, *29*, 792–803. [CrossRef]
4. Duhan, J.S.; Kumar, R.; Kumar, N.; Kaur, P.; Nehra, K.; Duhan, S. Nanotechnology: The new perspective in precision agriculture. *Biotechnol. Rep.* **2017**, *15*, 11–23. [CrossRef]
5. Raliya, R.; Saharan, V.; Dimkpa, C.; Biswas, P. Nanofertilizer for precision and sustainable agriculture: Current state and future perspectives. *J. Agric. Food Chem.* **2017**, *66*, 6487–6503. [CrossRef]
6. Panpatte, D.G.; Jhala, Y.K. *Nanotechnology for Agriculture: Crop Production & Protection*; Springer: Berlin/Heidelberg, Germany, 2019.
7. Bratovcic, A.; Hikal, W.M.; Said-Al Ahl, H.A.; Tkachenko, K.G.; Baeshen, R.S.; Sabra, A.S.; Sany, H. Nanopesticides and nanofertilizers and agricultural development: Scopes, advances and applications. *Open J. Ecol.* **2021**, *11*, 301–316. [CrossRef]
8. Sharma, P.; Gautam, A.; Kumar, V.; Guleria, P. In vitro exposed magnesium oxide nanoparticles enhanced the growth of legume Macrotyloma uniflorum. *ESPR* **2022**, *29*, 13635–13645. [CrossRef] [PubMed]
9. Kumar, V.; Jain, A.; Wadhawan, S.; Mehta, S.K. Synthesis of biosurfactant-coated magnesium oxide nanoparticles for methylene blue removal and selective Pb^{2+} sensing. *IET Nanobiotechnol.* **2018**, *12*, 241–253. [CrossRef]
10. Shaul, O. Magnesium transport and function in plants: The tip of the iceberg. *Biometals* **2002**, *15*, 307–321. [CrossRef]
11. Epstein, E.; Bloom, A. *Mineral Nutrition of Plants: Principles and Perspectives*, 2nd ed.; Sinauer Associates Inc.: Sunderland, UK, 2005.
12. Cakmak, I.; Kirkby, E.A. Role of magnesium in carbon partitioning and alleviating photooxidative damage. *Physiol. Plant.* **2008**, *133*, 692–704. [CrossRef]
13. Igamberdiev, A.U.; Kleczkowski, L.A. Optimization of ATP synthase function in mitochondria and chloroplasts via the adenylate kinase equilibrium. *Front. Plant Sci.* **2015**, *6*, 10. [CrossRef]
14. Stagnari, F.; Pisante, M. The critical period for weed competition in French bean (*Phaseolus vulgaris* L.) in Mediterranean areas. *Crop. Prot.* **2011**, *30*, 179–184. [CrossRef]
15. Al-Gaashani, R.; Radiman, S.; Al-Douri, Y.; Tabet, N.; Daud, A.R. Investigation of the optical properties of $Mg(OH)_2$ and MgO nanostructures obtained by microwave-assisted methods. *J. Alloys Compd.* **2012**, *521*, 71–76. [CrossRef]
16. Mirzaei, H.; Davoodnia, A. Microwave assisted sol-gel synthesis of MgO nanoparticles and their catalytic activity in the synthesis of hantzsch 1,4-dihydropyridines. *Chin. J. Catal.* **2012**, *33*, 1502–1507. [CrossRef]
17. Ouraipryvan, P.; Sreethawong, T.; Chavadej, S. Synthesis of crystalline MgO nanoparticle with mesoporous-assembled structure via a surfactant-modified sol–gel process. *Mater. Lett.* **2009**, *63*, 1862–1865. [CrossRef]
18. Ramezani Farani, M.; Farsadrooh, M.; Zare, I.; Gholami, A.; Akhavan, O. Green Synthesis of Magnesium Oxide Nanoparticles and Nanocomposites for Photocatalytic Antimicrobial, Antibiofilm and Antifungal Applications. *Catalysts* **2023**, *13*, 642. [CrossRef]
19. Xiong, H.; Shi, A.; Mou, B.; Qin, J.; Motes, D.; Lu, W.; Ma, J.; Weng, Y.; Yang, W.; Wu, D. Genetic diversity and population structure of cowpea (*Vigna unguiculata* L. Walp.). *PLoS ONE* **2016**, *11*, e0160941. [CrossRef]

20. Miller, B.; Oplinger, E.; Rand, R.; Peters, J.; Weis, G. Effect of planting date and plant population on sunflower performance 1. *J. Agron.* **1984**, *76*, 511–515. [CrossRef]
21. Pemberton, I.; Smith, G.; Miller, J. Inheritance of ineffective nodulation in cowpea. *Crop Sci.* **1990**, *30*, 568–571. [CrossRef]
22. Affrifah, N.S.; Phillips, R.D.; Saalia, F.K. Cowpeas: Nutritional profile, processing methods and products—A review. *Legume Sci.* **2022**, *4*, e131. [CrossRef]
23. Davis, D.; Oelke, E.; Oplinger, E.; Doll, J.; Hanson, C.; Putnam, D. Cowpea. In *Alternative Field Crops Manual*; University of Wisconsin Cooperative or Extension Service: Madison, WI, USA, 1991.
24. Kumar, V.; Guleria, P.; Ranjan, S. Phytoresponse to nanoparticle exposure. *Nanotoxicol. Nanoecotoxicol.* **2021**, *1*, 251–286.
25. Prasad, R.; Bhattacharyya, A.; Nguyen, Q.D. Nanotechnology in sustainable agriculture: Recent developments, challenges, and perspectives. *Front. Microbiol.* **2017**, *8*, 1014. [CrossRef] [PubMed]
26. Zafar, H.; Ali, A.; Ali, J.S.; Haq, I.U.; Zia, M. Effect of ZnO nanoparticles on Brassica nigra seedlings and stem explants: Growth dynamics and antioxidative response. *Front. Plant Sci.* **2016**, *7*, 535. [CrossRef] [PubMed]
27. Mandeh, M.; Omidi, M.; Rahaie, M. In vitro influences of TiO_2 nanoparticles on barley (*Hordeum vulgare* L.) tissue culture. *Biol. Trace Elem. Res.* **2012**, *150*, 376–380. [CrossRef]
28. Nalci, O.B.; Nadaroglu, H.; Hosseinpour, A.; Gungor, A.A.; Haliloglu, K. Effects of ZnO, CuO and γ-Fe_3O_4 nanoparticles on mature embryo culture of wheat (*Triticum aestivum* L.). *PCTOC* **2019**, *136*, 269–277. [CrossRef]
29. Anwaar, S.; Maqbool, Q.; Jabeen, N.; Nazar, M.; Abbas, F.; Nawaz, B.; Hussain, T.; Hussain, S.Z. The effect of green synthesized CuO nanoparticles on callogenesis and regeneration of *Oryza sativa* L. *Front. Plant Sci.* **2016**, *7*, 1330. [CrossRef]
30. Kim, D.H.; Gopal, J.; Sivanesan, I. Nanomaterials in plant tissue culture: The disclosed and undisclosed. *RSC Adv.* **2017**, *7*, 36492–36505. [CrossRef]
31. Gultekin, D.D.; Nadaroglu, H.; Gungor, A.A.; Kishali, N.H. Biosynthesis and characterization of copper oxide nanoparticles using Cimin grape (*Vitis vinifera* cv.) extract. *IJSM* **2017**, *4*, 77–84. [CrossRef]
32. Nadaroglu, H.; Güngör, A.A.; Selvi, İ. Synthesis of nanoparticles by green synthesis method. *JIRR* **2017**, *1*, 6–9.
33. Murashige, T.; Skoog, F. A revised medium for rapid growth a bioassays with tobacco tissue cultures. *Physiol. Plant* **1962**, *15*, 473–497. [CrossRef]
34. Nguyen, D.T.C.; Dang, H.H.; Vo, D.-V.N.; Bach, L.G.; Nguyen, T.D.; Van Tran, T. Biogenic synthesis of MgO nanoparticles from different extracts (flower, bark, leaf) of *Tecoma stans* (L.) and their utilization in selected organic dyes treatment. *JHM Lett.* **2021**, *404*, 124146. [CrossRef] [PubMed]
35. Aasim, M.; Bakhsh, A.; Khawar, K.M.; Ozcan, S. Past, present and future of tissue culture and genetic transformation research on cowpea (*Vigna unguiculata* L.). *COBIOT* **2011**, *22*, S131. [CrossRef]
36. Asami, H.; Tokugawa, M.; Masaki, Y.; Ishiuchi, S.-I.; Gloaguen, E.; Seio, K.; Saigusa, H.; Fujii, M.; Sekine, M.; Mons, M. Effective strategy for conformer-selective detection of short-lived excited state species: Application to the IR spectroscopy of the N1H Keto tautomer of guanine. *J. Phys. Chem.* **2016**, *120*, 2179–2184. [CrossRef]
37. Dobrucka, R. Synthesis of MgO nanoparticles using Artemisia abrotanum herba extract and their antioxidant and photocatalytic properties. *Iran. J. Sci. Technol. Trans. A Sci.* **2018**, *42*, 547–555. [CrossRef]
38. Karimi, B.; Khorasani, M.; Vali, H.; Vargas, C.; Luque, R. Palladium nanoparticles supported in the nanospaces of imidazolium-based bifunctional PMOs: The role of plugs in selectivity changeover in aerobic oxidation of alcohols. *ACS Catal.* **2015**, *5*, 4189–4200. [CrossRef]
39. Saied, E.; Eid, A.M.; Hassan, S.E.-D.; Salem, S.S.; Radwan, A.A.; Halawa, M.; Saleh, F.M.; Saad, H.A.; Saied, E.M.; Fouda, A. The catalytic activity of biosynthesized magnesium oxide nanoparticles (MgO-NPs) for inhibiting the growth of pathogenic microbes, tanning effluent treatment, and chromium ion removal. *Catalysts* **2021**, *11*, 821. [CrossRef]
40. Somanathan, T.; Krishna, V.M.; Saravanan, V.; Kumar, R.; Kumar, R. MgO nanoparticles for effective uptake and release of doxorubicin drug: pH sensitive controlled drug release. *JNN* **2016**, *16*, 9421–9431. [CrossRef]
41. Suresh, J.; Pradheesh, G.; Alexramani, V.; Sundrarajan, M.; Hong, S.I. Green synthesis and characterization of hexagonal shaped MgO nanoparticles using insulin plant (*Costus pictus* D. Don) leave extract and its antimicrobial as well as anticancer activity. *Adv. Powder Technol.* **2018**, *29*, 1685–1694. [CrossRef]
42. Ewais, E.A.; Desouky, S.A.; Elshazly, E.H. Evaluation of callus responses of *Solanum nigrum* L. exposed to biologically synthesized silver nanoparticles. *J. Nanosci. Nanotechnol.* **2015**, *5*, 45–56.
43. Do, D.G.; Dang, T.K.T.; Nguyen, T.H.T.; Nguyen, T.D.; Tran, T.T.; Hieu, D.D. Effects of nano silver on the growth of banana (*Musa* spp.) cultured in vitro. *J. Vietnam. Environ.* **2018**, *10*, 92–98. [CrossRef]
44. Salama, H.M. Effects of silver nanoparticles in some crop plants, common bean (*Phaseolus vulgaris* L.) and corn (*Zea mays* L.). *Int. Res. J. Biotechnol.* **2012**, *3*, 190–197.
45. Helaly, M.N.; El-Metwally, M.A.; El-Hoseiny, H.; Omar, S.A.; El-Sheery, N.I. Effect of nanoparticles on biological contamination of 'in vitro' cultures and organogenic regeneration of banana. *Aust. J. Crop Sci.* **2014**, *8*, 612–624.
46. Zhang, Z.; Ke, M.; Qu, Q.; Peijnenburg, W.; Lu, T.; Zhang, Q.; Ye, Y.; Xu, P.; Du, B.; Sun, L. Impact of copper nanoparticles and ionic copper exposure on wheat (*Triticum aestivum* L.) root morphology and antioxidant response. *Environ. Pollut.* **2018**, *239*, 689–697. [CrossRef]
47. Pena, L.B.; Méndez, A.A.; Matayoshi, C.L.; Zawoznik, M.S.; Gallego, S.M. Early response of wheat seminal roots growing under copper excess. *Plant Physiol. Biochem.* **2015**, *87*, 115–123. [CrossRef] [PubMed]

48. Ruffini Castiglione, M.; Giorgetti, L.; Geri, C.; Cremonini, R. The effects of nano-TiO$_2$ on seed germination, development and mitosis of root tip cells of *Vicia narbonensis* L. and *Zea mays* L. *J. Nanopart Res.* **2011**, *13*, 2443–2449. [CrossRef]
49. Sotoodehnia-Korani, S.; Iranbakhsh, A.; Ebadi, M.; Majd, A.; Oraghi-Ardebili, Z. Efficacy of magnesium nanoparticles in-modifying growth, antioxidant activity, nitrogen status, and expression of WRKY1 And BZIP transcription factors in pepper (*Capsicum annuum* L.); an in vitro biological assessment. *Russ. J. Plant Physiol.* **2023**, *70*, 39. [CrossRef]

Disclaimer/Publisher's Note: The statements, opinions and data contained in all publications are solely those of the individual author(s) and contributor(s) and not of MDPI and/or the editor(s). MDPI and/or the editor(s) disclaim responsibility for any injury to people or property resulting from any ideas, methods, instructions or products referred to in the content.

Article

Elicitation and Enhancement of Phenolics Synthesis with Zinc Oxide Nanoparticles and LED Light in *Lilium candidum* L. Cultures In Vitro

Piotr Pałka [1,*], Bożena Muszyńska [2], Agnieszka Szewczyk [2] and Bożena Pawłowska [1]

1. Faculty of Biotechnology and Horticulture, Department of Ornamental Plants and Garden Art, University of Agriculture in Krakow, 29 Listopada 54, 31-425 Kraków, Poland; bozena.pawlowska@urk.edu.pl
2. Faculty of Pharmacy, Department of Pharmaceutical Botany, Jagiellonian University Medical College, ul. Medyczna 9, 30-688 Kraków, Poland; bozena.muszynska@uj.edu.pl (B.M.); agnieszka.szewczyk@uj.edu.pl (A.S.)
* Correspondence: piotr.palka@urk.edu.pl

Citation: Pałka, P.; Muszyńska, B.; Szewczyk, A.; Pawłowska, B. Elicitation and Enhancement of Phenolics Synthesis with Zinc Oxide Nanoparticles and LED Light in *Lilium candidum* L. Cultures In Vitro. *Agronomy* 2023, 13, 1437. https://doi.org/10.3390/agronomy13061437

Academic Editors: Justyna Lema-Rumińska, Danuta Kulpa and Alina Trejgell

Received: 17 April 2023
Revised: 19 May 2023
Accepted: 19 May 2023
Published: 23 May 2023

Copyright: © 2023 by the authors. Licensee MDPI, Basel, Switzerland. This article is an open access article distributed under the terms and conditions of the Creative Commons Attribution (CC BY) license (https://creativecommons.org/licenses/by/4.0/).

Abstract: In this study, we identified and determined the content of phenolic compounds in *Lilium candidum* adventitious bulbs formed in vitro. HPLC analysis revealed the presence of four phenolic acids: chlorogenic, caffeic, *p*-coumaric, and ferulic acid. Phenolic acid content was assessed in adventitious bulbs formed in vitro on media supplemented with zinc oxide nanoparticles (ZnO NPs at 25, 50, and 75 mg/L) under fluorescent light (FL) or in darkness (D). The second experiment analyzed the effects of light-emitting diodes (LEDs) of variable light spectra on the formation of adventitious bulbs and their contents of phenolic acids. Spectral compositions of red (R; 100%), blue (B; 100%), red and blue (RB; 70% and 30%, respectively), a mix of RB and green (RBG) in equal proportions (50%), and white light (WLED, 33.3% warm, neutral, and cool light, proportionately) were used in the study. FL and D conditions were used as controls for light spectra. Bulbs grown in soil served as control samples. The most abundant phenolic acid was *p*-coumaric acid. Treatment with LED light spectra, i.e., RB, RBG, WLED, and B, translated into the highest *p*-coumaric acid concentration as compared with other treatments. Moreover, all the bulbs formed in light, including those grown on the media supplemented with ZnO NPs and under FL light, contained more *p*-coumaric acid than the bulbscales of the control bulbs grown in soil. On the other hand, control bulbs grown in soil accumulated about two to three times higher amounts of chlorogenic acid than those formed in vitro. We also found that the levels of all examined phenolics decreased under FL, R, and D conditions, while the bulblets formed in vitro under RB light showed the highest phenolic content. The use of ZnO NPs increased the content of *p*-coumaric, chlorogenic, and caffeic acid in the bulblets formed under FL as compared with those grown in darkness.

Keywords: adventitious bulbs; phenolic acids; HPLC

1. Introduction

Lilium candidum L., commonly known as Madonna lily or white lily, is a bulb geophyte that occurs naturally in Mediterranean countries. The species produces pure white flowers with a strong and pleasant scent [1]. Flower and bulb extracts of Madonna lily have been used in folk medicine to treat ulcers, wounds, burns, and muscle pains [2–4]. These properties have been confirmed in studies on burn treatment [5]; moreover, previous experiments have also demonstrated anti-inflammatory, antioxidant, anticancer, antidiabetic, and hepatoprotective properties of these extracts [3,4]. These properties, as well as the significant ornamental potential of Madonna lily, have contributed to the spread of its cultivation in many countries. They have also encouraged the species' harvest in its natural sites, which leads to the depletion of the environment [1]. Moreover, the generative multiplication rate of *L. candidum* is low, due to which the species was taken under protection in the countries of its natural occurrence [6].

Little information is available in the literature on the micropropagation of Madonna lily [6–15]. Available studies mainly concern the elimination of contaminations that occur during in vitro culture initiation [9,10,12], the selection of mother plant explant [7,11,14], and growth regulators used during organogenesis [8,15]. Our previous work [6] investigated the effects of LED light on adventitious organogenesis in *L. candidum*. The use of LED lighting has numerous energy-saving advantages. Moreover, LED lamps allow for the use of light of a specific wavelength so that experiments can be conducted under strictly controlled conditions [16–18]. LED light spectrum quality was demonstrated to affect the direction and performance of organogenesis and metabolite production in in vitro cultures of white lily [6] and other species of this genus [19,20].

One of the methods of influencing plant metabolism is the use of elicitors, and nanoparticles (NPs) can serve this purpose [21]. Elicitors stimulate biosynthetic pathways of compounds responsible for defense against stress associated with pathogen and pest attacks [22–24]. Nanoparticles are defined as materials with a size range between 1 and 100 nm. Because of their physical and chemical characteristics in the nanoscale, NPs show properties different from the bulk material [21]. Living organisms may react differently to NPs than to their bulk counterparts [25]. Zinc (Zn) is involved in numerous enzymatic reactions and physiological processes, which makes it an essential micronutrient [26]. This element is a component of enzymes, and it is involved in the synthesis of chlorophyll, proteins, carbohydrates, and nucleic acids, as well as the metabolism of these compounds [27]. Zinc oxide nanoparticles (ZnO NPs) have the form of a white, inorganic powder and can be chemically synthesized or obtained from plant extracts [28–31]. NPs can be used as a fertilizer [32], and they also exhibit antibacterial and antifungal properties [30,33,34]. To date, only a few studies have reported on the use of ZnO NPs in in vitro cultures [35–42], and some researchers have used ZnO NPs in ex vitro conditions [43,44].

Phenolic compounds are a widespread group of compounds found in living organisms, mainly in plants. Altogether, more than 8000 of these compounds have been distinguished, which greatly vary in their chemical structure. All phenols share a common feature, which is the presence of at least one aromatic ring with one hydroxyl group. Among the most important phenolic compounds are phenolic acids that contain a single phenyl group in their structure substituted by one carboxylic group and at least one hydroxy group [45]. In plants, phenols play a crucial role in the regulation of their growth and development, with antioxidant, protective, signaling, and structural functions [46]. In in vitro cultures, phenolic content has been demonstrated to be highly dependent on species, organ, and culture conditions, including light [47–49].

Chlorogenic acid is a phenolic compound commonly detected in plant tissues, and it is also an important component of the human diet [50]. To date, its presence has been demonstrated in numerous traditionally used medicinal plants [51]. Human consumption of foods containing chlorogenic acid may have health benefits related to its antioxidant properties [50]. This compound can also act as a free radical scavenger. Moreover, it shows a wide array of other functions, inter alia, and acts antivirally, antimicrobially, and antipyretically; it is a cardio- and hepatoprotective chemical, as well as a stimulator of the central nervous system [52]. It has been demonstrated to be effective against fungal pathogens of plants [53] and insect herbivores [54].

Caffeic acid is also widespread in the plant kingdom and is therefore often found in food and medicinal products of plant origin [55]. This compound has anticancer [56,57], antioxidant, and antibacterial properties [58], as well as the potential to prevent the development of cardiovascular diseases [59]. Studies have shown that caffeic acid exhibits even greater antioxidant potential in many lipid systems in combination with other phenolic acids, such as chlorogenic acid [55].

Another phenolic acid widely distributed in plants is *p*-coumaric acid, which occurs either in free form or conjugated with other chemicals, such as amines, alcohols, lignin, and mono- and oligosaccharides. Conjugates of *p*-coumaric acid exhibit particularly wide biological activity and are the subject of intense study. Moreover, they occur in plants in higher concentration than the free form of the compound [60]. Its antioxidant [61,62], antibacterial [63,64], anticancer [65–67], wound-healing [68,69], and skin discoloration leveling [70] effects have also been proven.

Ferulic acid occurs in numerous plant species used in traditional medicine and in vegetables and fruits used for food. The compound is rarely found in its free form, but it forms conjugates with other chemicals [71–73]. It has been shown to have antioxidant [72,74], anti-inflammatory [75,76], anticancer [77], and antidiabetic [78] effects. Ferulic acid is easily assimilated and remains in the blood longer than other compounds with antioxidant activity [72]. Because of its negligible toxicity and strong antioxidant properties, ferulic acid is approved as a food and cosmetic additive [71], and as it exhibits protective effects on the skin (i.e., inhibits melanin production and accelerates wound healing), it is used in cosmetics, including sunscreens [74].

Our study aimed to identify phenolic compounds in *L. candidum* adventitious bulbs formed in vitro. We also assessed the effect of ZnO NPs added to the culture media on the content of phenolic acids in adventitious bulbs of *L. candidum* formed under either dark or light conditions. Our second experiment analyzed the effects of different LED light spectra used during bulb formation on phenolic acid content in the bulbs and the intensity of adventitious organogenesis.

2. Materials and Methods

2.1. Plant Material

Adventitious bulbs of *L. candidum* L. from the in vitro collection of the Department of Ornamental Plants and Garden Art, University of Agriculture in Kraków, were used as the experimental material. The cultures were formed on bulbscales of lilies grown in the field collection. The in vitro-formed adventitious bulbs were stored at 4 °C for 12 months. Bulbs with 11 to 12 individual bulbscales were used as explants.

2.2. Experimental Conditions

Individual bulbscales were placed on Petri dishes with Murashige and Skoog (MS) [79] medium containing 3% sucrose, pH 5.7, solidified with 0.5% BioAgar (BIOCORP, Warszawa, Poland).

In the first experiment, ZnO NPs (≤40 nm average particle size) (Sigma-Aldrich, St. Louis, MO, USA) suspended in distilled water were added to the medium at three concentrations: 25 mg/L (Zn25), 50 mg/L (Zn50), and 75 mg/L (Zn75). The nanoparticles were added before sterilization of the medium. To disperse the NPs, they were placed in a Sonic 3 ultrasonic stirrer (Polsonic, Poland) for one hour. The cultures were maintained either under florescent light (FL Zn) (OSRAM LUMILUX Cool White L 36W/840) or in darkness (D Zn). A total of six factor combinations were tested: D Zn25 and FL Zn25, D Zn50 and FL Zn50, and D Zn75 and FL Zn75.

In the second experiment, the medium did not contain ZnO NPs. The cultures were maintained under six combinations of LED light quality (i.e., different wavelengths) [80]: 100% red at 670 nm (R); 100% blue at 430 nm (B); a mix of 70% red and 30% blue (RB); 50% RB and 50% green at 528 nm (RBG); 33.3% warm white (2700 K), 33.3% neutral white (4500 K), and 33.3% cool white (5700 K) (WLED); as well as fluorescent lamp light (FL) and darkness (D). A total of seven combinations were tested in this experiment.

The cultures were maintained in a culture room at 23/21 °C (day/night), 80% relative humidity, and 16 h photoperiod (16 h day/8 h night).

2.3. Data Collection

After eight weeks of culture, biometric observations of the formed bulbs and roots (i.e., number of bulbs, bulb diameter, bulb weight, and number of roots) were performed. The phenolic compound content was analyzed in the obtained bulblets. For each combination, three weighed amounts of three grams each were prepared. The material was then frozen at −80 °C until further analyses. Next, the material was lyophilized (Freezone 4.5, Labconco, Kansas City, MO, USA). From the lyophilized material, weighed amounts of one gram each were prepared and homogenized in an agate mortar and extracted with methanol in an ultrasonic bath at 49 kHz (Sonic-2, Polsonic, Warszawa, Poland). The obtained extracts were evaporated. The whole procedure was carried out a total of four times. The resulting filtrate was left to evaporate at room temperature to obtain dry extracts. The dry extracts were washed with an appropriate amount of HPLC-grade methanol. The samples were brought to final volumes, then filtered through syringe filters (Millex, Millipore Corporation, Burlington, MA, USA) into HPLC vials. Bulbscales obtained from plants growing in soil in the university collection served as control samples.

2.4. High-Performance Liquid Chromatography Analysis (HPLC) of Phenolic Compounds

Determination of phenolic compounds was carried out by RP-HPLC using an HPLC VWR Hitachi-Merck instrument with an L2200 autosampler, an L-2130 pump, an RP-18e LiChrospher column (4 mm × 250 mm, 5 μm) thermostated at 25 °C, an L-2350 column oven, and an L-2455 diode array detector operating in the UV wavelength range of 200–400 nm. The mobile phase consisted of solvent A (a mixture of methanol and 0.5% acetic acid solution by volume (1:4)) and solvent B (methanol). The gradient was as follows: 100:0 for 0–25 min, 70:30 for 35 min, 50:50 for 45 min, 0:100 for 50–55 min, and 100:0 for 57–67 min. Comparison of UV spectra and retention times with standard compounds enabled identification of phenolic compounds present in the analytical samples. Quantitative analysis of free phenolic acids was carried out using a calibration curve, assuming a linear relationship of the area under the curve to the concentration of the standard. Results were expressed in mg/100 g dry weight (dw).

2.5. Statistical Analysis

All the study findings were analyzed statistically (ANOVA) using Statistica 13.3 software (StatSoft, TIBCO Software Inc., Palo Alto, CA, USA). A post hoc multiple range Duncan test was used. Significantly different means were separated at $p \leq 0.05$.

3. Results and Discussion

3.1. Adventitious Bulblet and Root Formation

The experiment yielded correctly formed bulblets and adventitious roots (Figure 1). The organs differed in their size and greenness, depending on the factor combinations. The bulblets obtained under red LED (R) (Figure 1g) and in darkness (D and D Zn) (Figure 1d,e,h) were white due to their low chlorophyll content, as confirmed by literature data [6,81–86].

ZnO NP medium concentration above 25 mg/L inhibited the formation of adventitious bulbs. It was also the only concentration that allowed for obtaining more than two bulblets, both in the light and in the dark (Table 1). Chamani et al. [35] obtained the greatest number of *Lilium ledebourii* bulblets from a culture maintained under a fluorescent lamp on a medium supplemented with 50 mg/L ZnO NPs, but higher concentrations of NPs decreased the regeneration efficiency.

In the second experiment, individual explants maintained under the analyzed light combinations yielded 1.3 to 2.2 bulblets. The greatest number of bulblets was formed under blue (B), red–blue (RB), and white LED light (WLED) (Table 2). These findings are similar to those from our previous study [6].

The diameter of all the resulting bulblets was greater than 4 mm. The largest bulblets (above 5 mm) were formed on Zn50 medium under fluorescent light (Table 1). Bulblets of this size were also formed on the bulbscales placed on ZnO NPs-free medium under RBG light (Table 2), which confirmed our previous results [6]. The mixture of red and blue (RB) light increased the bulblet weight, which was even two times greater than that under the remaining light combinations. The bulblets produced under RBG light were also characterized by considerable weight (Table 2). Under fluorescent light, increasing concentrations of ZnO NPs in the medium enhanced the bulblet weight from 0.10 to 0.22 g. Such a trend was not observed in the dark (Table 1). Mosavat et al. [36] reported improved callus growth in species of the *Thymus* genus at higher concentrations of ZnO NPs in the culture medium. In contrast, in a study by Garcia-Lopez et al. [87], the dry weight of *Capsicum annuum* seedlings germinated ex vitro did not depend on the applied concentration of ZnO NPs. Nanoparticles may lower plant biomass due to their phytotoxicity [88]. This happened in our study in the cultures maintained in the light.

Figure 1. Adventitious organogenesis on *Lilium candidum* bulbscales in vitro depending on elicitation factor: on medium suplemented with zinc oxide nanoparticles (ZnO NP) (Zinc oxide nanoparticle elicitation): in darkness with (**d**) 25 (D Zn25); (**e**) 75 mg/L (D Zn75) and under fluorescent lamp with (**a**) 25 (Fl Zn25); (**b**) 50 (Fl Zn50); (**c**) 75 mg/L (Fl Zn75). Under different light quality in vitro (Light elicitation): (**h**) darkness (D); (**j**) fluorescent lamp (Fl); under LED light (%): (**f**) 100 blue (B); (**g**) 100 red (R); (**k**) 35 R + 15 B + 50 green (RBG) and (**i**) white: 33.3 warm, 33.3 neutral + 33.3 cool (Wled). Bar = 1 cm.

Table 1. Adventitious organogenesis on bulbscales of *Lilium candidum* under different light and zinc oxide nanoparticle treatments in vitro.

Culture Condition	Bulblets per Regenerating Bulbsacale	Bulblet Diameter (mm)	Bulblet Weight (g)	Root per Bulblet
D [a]	1.70 ± 0.1 a–c [b]	4.21 ± 0.3 a	0.14 ± 0.0 ab	0.92 ± 0.1 ab
FL	1.97 ± 0.1 bc	4.23 ± 0.3 a	0.14 ± 0.0 ab	0.59 ± 0.1 a
D Zn25	2.07 ± 0.2 c	4.29 ± 0.2 a	0.22 ± 0.0 c	0.60 ± 0.1 a
D Zn50	1.33 ± 0.1 a	4.26 ± 0.3 a	0.18 ± 0.0 bc	0.84 ± 0.1 ab
D Zn75	1.59 ± 0.1 ab	4.32 ± 0.2 a	0.20 ± 0.0 c	0.91 ± 0.1 ab
FL Zn25	2.01 ± 0.1 c	4.18 ± 0.3 a	0.10 ± 0.0 a	0.51 ± 0.1 a
FL Zn50	1.74 ± 0.1 bc	5.28 ± 0.4 a	0.19 ± 0.0 bc	1.02 ± 0.2 b
FL Zn75	1.78 ± 0.2 bc	4.68 ± 0.4 a	0.22 ± 0.0 c	0.88 ± 0.1 ab
		Source of variation		
Culture condition	***	n.s.	***	n.s.

Significant effect: *** $p \leq 0.001$; n.s. not significant. [a] Culture in darkness (D) and under fluorescent lamp (FL) on medium supplemented with: 25 mg/L (D Zn25 and FL Zn25); 50 mg/L (D Zn50 and FL Zn50) and 75 mg/L (D Zn75 and FL Zn75) zinc oxide nanoparticles. [b] Means ± standard deviations within a column followed by the same letter are not significantly different according to Duncan's multiple range test at $p \leq 0.05$.

Table 2. Adventitious organogenesis on bulbscales of *Lilium candidum* under different light quality conditions in vitro.

Light	Bulblets per Regenerating Bulbsacale	Bulblet Diameter (mm)	Bulblet Weight (g)	Root per Bulblet
D [a]	1.70 ± 0.1 ab [b]	4.21 ± 0.3 a	0.14 ± 0.0 ab	0.92 ± 0.1 bc
FL	1.97 ± 0.1 bc	4.23 ± 0.3 a	0.14 ± 0.0 ab	0.59 ± 0.1 ab
R	1.34 ± 0.1 a	4.20 ± 0.4 a	0.16 ± 0.0 bc	0.81 ± 0.1 bc
B	2.17 ± 0.1 bc	4.06 ± 0.2 a	0.11 ± 0.0 a	0.39 ± 0.1 a
RB	2.14 ± 0.2 bc	4.63 ± 0.3 a	0.24 ± 0.0 d	0.98 ± 0.1 c
RBG	1.99 ± 0.1 bc	5.71 ± 0.2 b	0.19 ± 0.0 c	0.89 ± 0.1 bc
WLED	2.34 ± 0.3 c	4.20 ± 0.2 a	0.11 ± 0.0 a	0.67 ± 0.1 a-c
		Source of variation		
Light	***	***	***	***

Significant effect: *** $p \leq 0.001$; n.s. not significant. [a] Darkness (D); fluorescent lamp (FL); LED lights (%): 100 red (R); 100 blue (B); 70 red + 30 blue (RB); 35 R + 15 B + 50 green (RBG) and white: 33.3 warm, 33.3 neutral + 33.3 cool (WLED). [b] Means ± standard deviations within a column followed by the same letter are not significantly different according to Duncan's multiple range test at $p \leq 0.05$.

We found that increasing concentrations of ZnO NPs were associated with a tendency to form more numerous adventitious roots (Table 1). A greater number of roots, along with increasing concentration of ZnO NPs in the medium, was also reported in the cultures of *Phoenix dactylifera* [42]. *Lilium ledebourii* produced the longest roots in the medium supplemented with 50 mg/L ZnO NPs, but higher concentrations of NPs inhibited root growth [35]. The effect of Zn on root development may be associated with its role in the biosynthesis of tryptophan, which is an indispensable component in the biosynthesis of IAA responsible for adventitious root formation [89].

The bulblets formed in our second experiment under the light of different quality also produced adventitious roots, and their number was the highest under RB, RBG, and R light and in the dark (D) (Table 2), which again confirmed our previous results [6].

3.2. Identification of Phenolic Acids in the Bulblets

Chromatographic analysis revealed the peaks typical of four phenolic acids: chlorogenic, caffeic, *p*-coumaric, and ferulic acid. Example chromatograms are shown in Figure 2.

3.3. Effect of Zinc Oxide Nanoparticles on Phenolic Acid Content

Chlorogenic, caffeic, *p*-coumaric, and ferulic acids were detected in the adventitious bulblets maintained on the media containing ZnO NPs. One of the acids, namely *p*-coumaric acid, occurred at the highest concentrations, usually exceeding 15 mg/100 g dw (except for Zn combination in the dark) (Figure 3a).

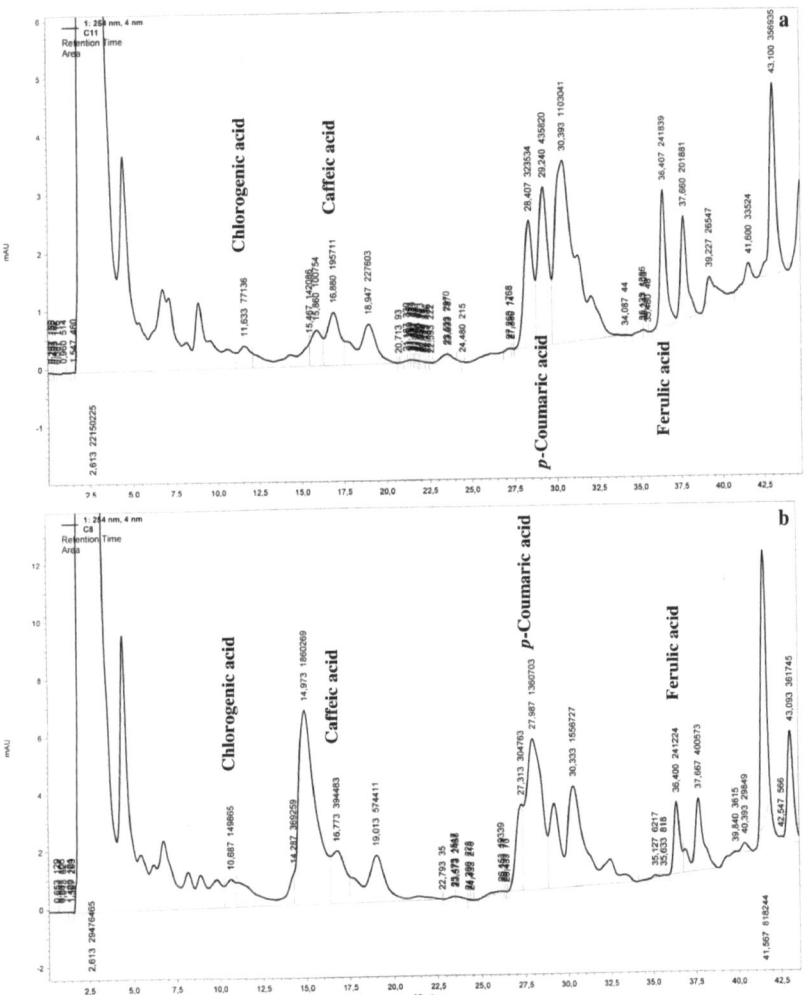

Figure 2. HPLC chromatographic separation of phenolic acids from *Lilium candidum* L. bulblets in vitro on medium supplemented with 25 mg/L zinc oxide nanoparticles in (**a**) darkness and (**b**) under a fluorescent lamp.

The content of *p*-coumaric acid was about two times higher (19.92 to 25.14 mg/100 g dw) in the bulblets formed in the light than in the dark, and the presence of 50 mg/L ZnO NPs promoted its accumulation under both conditions (D Zn50 and FL Zn50). The concentration of *p*-coumaric acid in the bulbscales of the control bulbs growing in soil reached about 15 mg/100 g dw (Figure 3a).

The content of chlorogenic acid in the bulblets formed on the media supplemented with ZnO NPs was the lowest in the dark, but the highest concentration of NPs (D Zn75) increased its accumulation up to 1.35 mg/100 g dw. A similar response was observed in the bulblets formed in the light, where the content of this acid was the highest for the FL Zn75 variant among all bulblets grown in vitro (Figure 4a) but three times lower than in the bulbscales of the control bulbs (C) that contained 6.22 mg/100 g dw chlorogenic acid (Figure 4a).

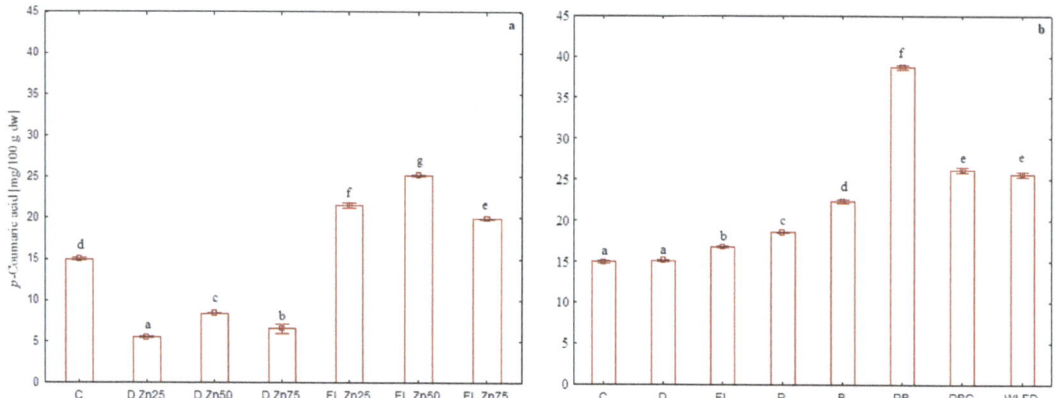

Figure 3. *p*-Coumaric acid content in *Lilium candidum* L. bulblets in vitro: (**a**) on medium with different concentration of zinc oxide nanoparticles: 25 mg/L (D Zn25 and FL Zn25); 50 mg/L (D Zn50 and FL Zn50) and 75 mg/L (D Zn75 and FL Zn75) in the darkness (D) and under fluorescent lamp (FL) and (**b**) under different light quality: in the darkness (D); under fluorescence lamp (FL); under LED light (%): 100 red of 670 nm (R); 100 blue of 430 nm (B); mix of 70 red and 30 blue (RB); 50 RB and 50 green of 528 nm (RBG); 33.3 warm white (2700 K), 33.3 neutral white (4500 K), and 33.3 cool white (5700 K) (WLED). Data are presented as means ± standard deviations. Different letters indicate significant differences between values according to Duncan's multiple range test at $p \leq 0.05$. Statistical analysis was performed for each experiment separately.

The highest concentrations of caffeic acid (2.74–2.78 mg/100 g dw) were found in the bulblets formed in the light, and they were independent of ZnO NP content in the medium. The bulblets grown in the dark accumulated two times less caffeic acid, and its content dropped with increasing concentration of ZnO NPs. The control bulbs growing in the field had an average content of this acid at a level of 2.3 mg/100 g dw (Figure 4b).

The content of ferulic acid in the bulblets grown in vitro was four times lower than that in the control bulbs, where it reached 8.43 mg/100 g dw. No differences were found between the cultures maintained in the dark and in the light. For all investigated combinations, the content of ferulic acid was similar (1.64–2.09 mg/100 g dw), except for FL Zn50, where it reached 3.33 mg/100 g dw (Figure 4c).

Statistical analysis examining the influence of light conditions (D, FL) and the presence of ZnO NPs in the medium (regardless of their concentration) revealed that the bulbs formed in the light on the medium with nanoparticles usually had the highest content of the investigated acids. The addition of NPs to the cultures carried out in the dark (D Zn) had an inhibitory effect on the synthesis of phenolic acids. For example, the content of *p*-coumaric acid was two times lower in the bulblets treated with NPs and growing in the dark, while in those growing under a fluorescent lamp (FL Zn), it increased by 30% as compared with the control. For chlorogenic acid, the most significant drop was found in the bulblets formed in the dark (D Zn), and for caffeic acid, the most significant change was a rise in its content in the bulblets grown in the light and in the presence of NPs (FL Zn). Ferulic acid was the only acid whose concentration was the highest in the bulblets formed in the dark (D) as compared with those grown under FL. Medium supplementation with ZnO NPs did not affect its content in the bulblets formed in the dark (D Zn), while those grown under FL (FL Zn) were the most abundant in ferulic acid (2.33 mg/100 g dw) (Table 3).

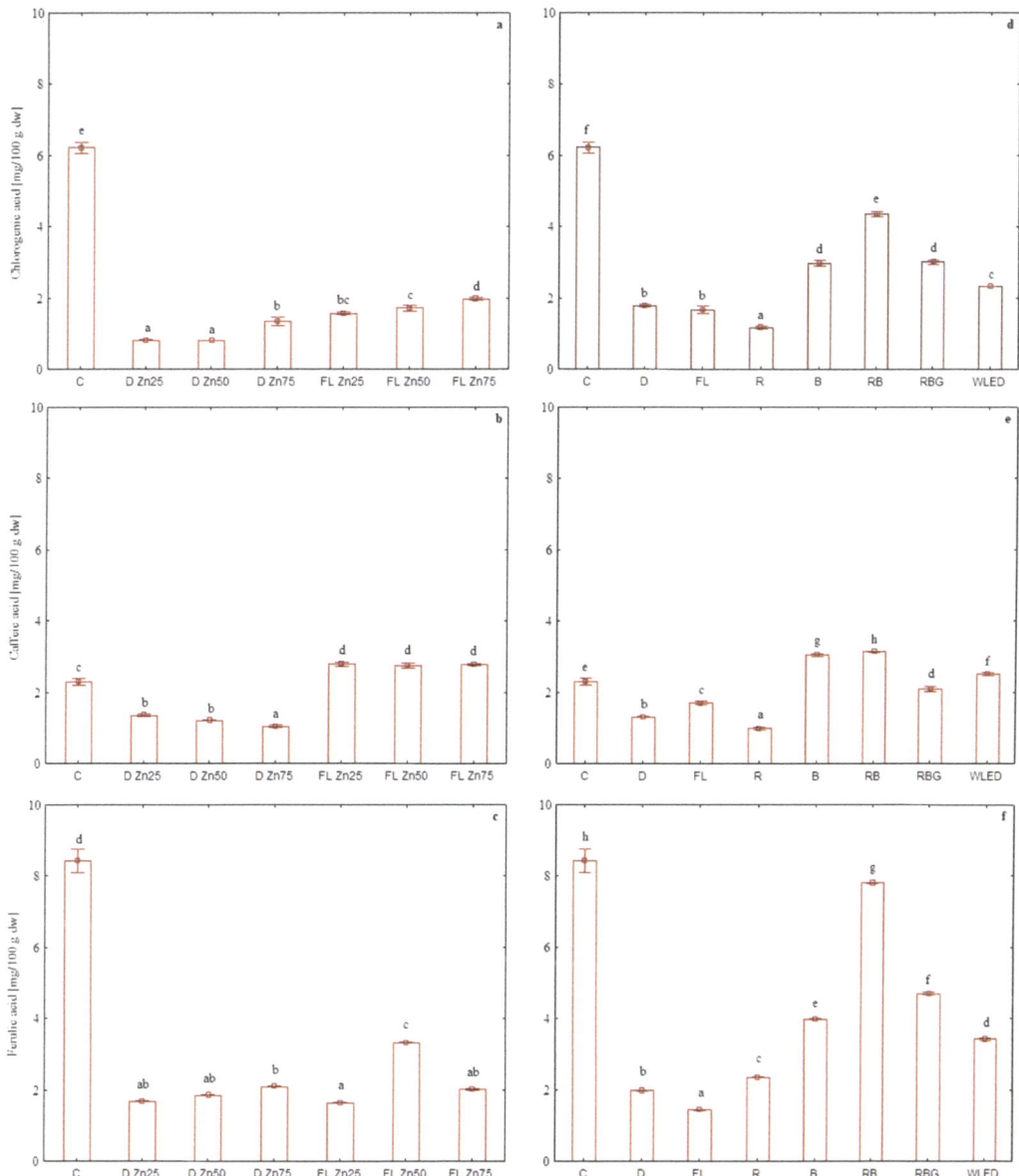

Figure 4. Chlorogenic acid, caffeic acid, and ferulic acid contents in *Lilium candidum* L. bulblets in vitro: (**a–c**) on medium with different concentrations of zinc oxide nanoparticles: 25 mg/L (D Zn25 and FL Zn25); 50 mg/L (D Zn50 and FL Zn50) and 75 mg/L (D Zn75 and FL Zn75) in the darkness (D) and under fluorescent lamp (FL) and (**d–f**) under different light quality: in the darkness (D); under fluorescence lamp (FL); under LED light (%): 100 red of 670 nm (R); 100 blue of 430 nm (B); mix of 70 red and 30 blue (RB); 50 RB and 50 green of 528 nm (RBG); 33.3 warm white (2700 K), 33.3 neutral white (4500 K), and 33.3 cool white (5700 K) (WLED). Data are presented as means ± standard deviations. Different letters indicate significant differences between values according to Duncan's multiple range test at $p \leq 0.05$. Statistical analysis was performed for each experiment separately.

Table 3. Phenolic acid contents (mg/100 g dw) in *Lilium candidum* bulblets depending on light and the presence of zinc oxide nanoparticles, regardless of the concentration of nanoparticles.

Culture Condition	Chlorogenic Acid	Caffeic Acid	p-Coumaric Acid	Ferulic Acid
D [a]	1.79 ± 0.1 b [b]	1.31 ± 0.0 a	15.40 ± 0.2 b	1.98 ± 0.0 ab
FL	1.67 ± 0.2 b	1.70 ± 0.1 b	16.85 ± 0.1 b	1.44 ± 0.0 a
D Zn	0.99 ± 0.3 a	1.25 ± 0.1 a	6.82 ± 1.4 a	1.88 ± 0.2 ab
FL Zn	1.75 ± 0.2 b	2.77 ± 0.1 c	22.19 ± 2.3 c	2.33 ± 0.8 b
	Source of variation			
Culture condition	***	n.s.	***	n.s.

Significant effect: *** $p \leq 0.001$; n.s. not significant. [a] Culture in darkness (D) and under fluorescent lamp (FL) on medium without zinc oxide nanoparticles and in darkness (D Zn) and under fluorescent lamp (FL Zn) on medium supplemented with zinc oxide nanoparticles. [b] Means ± standard deviations within a column followed by the same letter are not significantly different according to Duncan's multiple range test at $p \leq 0.05$.

The literature lacks data on the analysis of phenolic acids investigated in this work, typical for tissues formed in the presence of nanoparticles. The available information focuses solely on total content of phenols. Callus cultures of *Fagonia indica* showed increased total phenolic content following elicitation with iron-doped zinc oxide nanoparticles. This increase depended on the concentration of the nanoparticles and the duration of the culture [90]. The content of thymol and carvacrol in *Thymus* tissues rose a few times as compared with the control after application of ZnO NPs at 150 mg/L [36].

The effect of the application of ZnO NPs on the concentration of phenolic compounds ex vitro depended on plant species. For example, in the cultures of *Brassica nigra* [91] and Solanum tuberosum [92], ZnO NPs enhanced total phenolic content. In contrast, their contents in *Capsicum annuum* seedlings generally dropped in the presence of ZnO NPs, except for radicle, where they were clearly boosted [87]. While the effect of Zn NPs on plant physiology and biochemistry has been proven, their influence on secondary metabolism [93], especially of phenolic compounds, is still unclear [94].

3.4. Effect of LED Light on Phenolic Acid Content

The second experiment revealed that fluorescent light (FL), red LED light (R), and darkness (D) decreased the content of four of the investigated phenolic acids in the bulblets formed on the media not supplemented with ZnO NPs (Figures 3b and 4d–f). Similarly, as on the media supplemented with ZnO NPs, the most abundant phenolic acid was p-coumaric acid. The bulblets formed under RB light had the highest content of p-coumaric acid (38.73 mg/100 g dw). Its concentration was lower in the bulblets exposed to RBG and WLED, followed by B light. The bulblets subjected to the other combinations featured a lower content of p-coumaric acid, which was least abundant in the dark, where its content matched that of the control bulbs from field cultivation (Figure 3b).

The highest content of chlorogenic acid was found in the bulblets growing under RB LED (4.35 mg/100 g dw), but it was still lower than that in the control bulbs (by about two units). This acid was also abundant in the bulblets exposed to B and RBG light, but its content under red LED light (R) was almost three times lower (1.16 mg/100 g dw), which was the lowest result of all tested combinations (Figure 4d). Chen et al. [95] also reported on variable content of chlorogenic acid in the tissues of *Peucedanum japonicum* exposed to LED light of different quality. In this species, the content of chlorogenic acid was the highest in the plants acclimatized to ex vitro conditions (it was also 17 times higher than in the herbal raw material). The callus grown in vitro under a mixture of blue, red, and far-red LED light also featured high (three times higher than in the raw herbal material) content of chlorogenic acid.

The highest content of caffeic acid, also in comparison with control bulbscales of the field-cultivated bulbs, was detected in the bulblets formed under RB (3.14 mg/100 g dw), B (3.05 mg/100 g dw), and WLED light (2.51 mg/100 g dw). In the bulblets exposed to R light, the content of caffeic acid was below 1 mg/100 g dw (Figure 4e). The tissues of

Protea cynaroides exposed to a fluorescent lamp accumulated nearly two times more caffeic acid (15.9 mg/g) than those exposed to LED light. Red and blue light triggered similar accumulation of this acid (8.4–8.0 mg/g), and a mixture of red and blue LED light slightly increased the acid content (9.0 mg/g) [96]. There are no literature data on the effects of LED light on the content of individual phenolic acids, but there is information on its effect on the accumulation of total phenolics. Many researchers have observed the inhibiting effect of red light and the stimulating effect of blue light on the synthesis of phenolics [97–99]. This was also reported in our previous paper in *L. candidum* [6].

Our experiment revealed high contents (7.80 mg/100 g dw) of ferulic acid in the bulblets maintained under RB LED light. This value was only by 0.6 mg/100 g dw lower than that in the control bulbs (C). Values half as high as that in the control were achieved under blue (B) and white LED light (WLED), as well as when the spectrum also contained green light (RBG). The lowest content of ferulic acid (below 1.5 mg/100 g dw) was detected under the fluorescent lamp (FL) (Figure 4f). Wu and Lin [96] reported the highest content of this acid (9.7 mg/g) in the tissues of *P. cynaroides* exposed to fluorescent light. Ferulic acid concentration was lower under blue light (8.2 mg/g) and a mixture of red and blue LED light (8.7 mg/g) and the lowest under red light (7.4 mg/g). This confirmed a previously described relationship between total content of phenolics and the presence of blue and red light. The addition of blue light to the spectrum increases the content of phenolic metabolites, which is in accordance with literature data [100–104]. One should, however, keep in mind that the response to different stimuli, including light, is strongly species-dependent, even in in vitro cultures [105]. Photoreceptors and auxin-responsive factors indirectly affect gene expression and cellular responses, but the exact mechanisms have not been fully explained yet [106].

4. Conclusions

The bulblets of *Lilium candidum* formed during adventitious organogenesis contained the following phenolic acids: *p*-coumaric, chlorogenic, caffeic, and ferulic acid. Their contents depended on the concentration of ZnO NPs in the medium and the light quality during bulblet formation. *p*-Coumaric acid was the most abundant acid, especially in the samples exposed to LED light (RB, RBG, WLED, and B), and on the media without ZnO NPs. All bulbs formed in the light, including those maintained on the media supplemented with ZnO NPs, contained more *p*-coumaric acid than the bulbscales of control bulbs grown in soil.

The control bulbs always accumulated about two to three times higher amounts of chlorogenic acid than the bulblets formed in vitro. The bulblets exposed to R light accumulated the highest amounts of this acid, although 30% lower than in the control.

We found that FL, R, and darkness decreased the levels of all examined phenolics, the contents of which were the highest in the bulblets formed in vitro under RB light.

The use of ZnO NPs increased the contents of *p*-coumaric, chlorogenic, and caffeic acid in the bulblets formed under FL as compared with those grown in darkness.

The reported presence of phenolic acids in the bulblets confirmed the medicinal properties of *Lilium candidum* and the use of this species described in folk medicine sources. Moreover, the compounds we investigated act synergistically with each other, which makes Madonna lily a valuable object of future research. The methods of in vitro elicitation described in this work yield natural compounds of chemical and microbiological purity that can be produced with high efficiency.

Author Contributions: Conceptualization, P.P. and B.P.; methodology, B.P.; validation, P.P. and B.M.; formal analysis, P.P.; investigation, P.P.; resources, B.M., A.S. and B.P.; data curation, A.S.; writing—original draft preparation, P.P.; writing—review and editing, B.P.; visualization, P.P.; supervision, B.M. and B.P.; project administration, B.P. All authors have read and agreed to the published version of the manuscript.

Funding: This project was supported by the Polish Ministry of Science and Higher Education.

Data Availability Statement: The data presented in this study are available upon request from the corresponding author.

Acknowledgments: This project was supported by the Polish Ministry of Science and Higher Education.

Conflicts of Interest: The authors declare no conflict of interest.

Abbreviations

B: 100% blue LED light; D: darkness; D Zn: culture medium supplemented with zinc oxide nanoparticles in darkness; D Zn25: culture medium supplemented with 25 mg/L zinc oxide nanoparticles in darkness; D Zn50: culture medium supplemented with 50 mg/L zinc oxide nanoparticles in darkness; D Zn75: culture medium supplemented with 75 mg/L zinc oxide nanoparticles in darkness; FL: fluorescent lamp; FL Zn: culture medium supplemented with zinc oxide nanoparticles under fluorescent lamp; FL Zn25: culture medium supplemented with 25 mg/L zinc oxide nanoparticles under fluorescent lamp; FL Zn50: culture medium supplemented with 50 mg/L zinc oxide nanoparticles under fluorescent lamp; FL Zn75: culture medium supplemented with 75 mg/L zinc oxide nanoparticles under fluorescent lamp; g dw: gram of dry weight; LED: light-emitting diode; MS: Murashige and Skoog culture medium; R: 100% red LED; RB: 70% red + 30% blue LED; RBG: 35% red + 15% blue + 50% green LED; WLED: 33.3% warm + 33.3% neutral + 33.3% cool white LED.

References

1. Özen, F.; Temeltaş, H.; Aksoy, Ö. The anatomy and morphology of the medicinal plant, *Lilium candidum* L. (Liliaceae) distributed in Marmara region of Turkey. *Pak. J. Bot.* **2012**, *44*, 1185–1192.
2. Pieroni, A. Medicinal plants and food medicines in the folk traditions of the upper Lucca Province, Italy. *J. Ethnopharmacol.* **2000**, *70*, 235–273. [CrossRef] [PubMed]
3. Patocka, J.; Navratilova, Z. Bioactivity of *Lilium candidum* L.: A mini review. *Biomed. J. Sci. Technol. Res.* **2019**, *8*, 13859–13862. [CrossRef]
4. Zaccai, M.; Yarmolinsky, L.; Khalfin, B.; Budovsky, A.; Gorelick, J.; Dahan, A.; Ben-Shabat, S. Medicinal properties of *Lilium candidum* L. and its phytochemicals. *Plants* **2020**, *9*, 959. [CrossRef]
5. Momtaz, S.; Dibaj, M.; Abdollahi, A.; Amin, G.; Bahramsoltani, R.; Abdollahi, M.; Mahdaviani, P.; Abdolghaffari, A.H. Wound healing activity of the flowers of *Lilium candidum* L. in burn wound model in rats. *J. Med. Plants* **2020**, *19*, 109–118. [CrossRef]
6. Pałka, P.; Cioć, M.; Hura, K.; Szewczyk-Taranek, B.; Pawłowska, B. Adventitious organogenesis and phytochemical composition of Madonna lily (*Lilium candidum* L.) in vitro modeled by different light quality. *Plant Cell Tissue Organ Cult. (PCTOC)* **2023**, *152*, 99–114. [CrossRef]
7. Khawar, K.M.; Cocu, S.; Parmaksiz, I.; Sarihan, E.O.; Özcan, S. Mass proliferation of Madonna Lily (*Lilium candidum* L.) under in vitro conditions. *Pak. J. Bot.* **2005**, *37*, 243–248.
8. Sevimay, C.S.; Khawar, K.M.; Parmaksız, I.; Cocu, S.; Sancak, C.; Sarihan, E.; Özcan, S. Prolific in vitro bulblet formation from bulb scales of meadow lily (*Lilium candidum* L.). *Period. Biol.* **2005**, *107*, 107–111.
9. Altan, F.; Bürün, B.; Şahin, N. Fungal contaminants observed during micropropagation of *Lilium candidum* L. and the effect of chemotherapeutic substances applied after sterilization. *Afr. J. Biotechnol.* **2010**, *9*, 991–995. [CrossRef]
10. Burun, B.; Sahin, O. Micropropagation of *Lilium candidum* L.: A rare and native bulbous flower of Turkey. *Bangladesh J. Bot.* **2013**, *42*, 185–187. [CrossRef]
11. Saadon, S.; Zaccai, M. *Lilium candidum* bulblet and meristem development. *Vitr. Cell. Dev. Biol. Plant* **2013**, *49*, 313–319. [CrossRef]
12. Altan, F.; Bürün, B. The effect of some antibiotic and fungucide applications on the micropropagation of *Lilium candidum* L. *Mugla J. Sci. Technol.* **2017**, *3*, 86–91. [CrossRef]
13. Daneshvar Royandazagh, S. Efficient approaches to in vitro multiplication of *Lilium candidum* L. with consistent and safe access throughout year and acclimatization of plant under hot-summer mediterranean (Csa Type) climate. *Not. Bot. Horti Agrobot.* **2019**, *47*, 734–742. [CrossRef]
14. Tokgoz, H.B.; Altan, F. Callus induction and micropropagation of *Lilium candidum* L. using stem bulbils and confirmation of genetic stability via SSR-PCR. *Int. J. Second. Metab.* **2020**, *7*, 286–296. [CrossRef]
15. Akshay, M.P.; Pooja, P.G.; Sonali, D. In vitro micropropagation of *Lilium candidum* bulb by application of multiple hormone concentrations using plant tissue culture technique. *Int. J. Res. Appl. Sci. Biotechnol.* **2021**, *8*, 244–253. [CrossRef]
16. Bornwaßer, T.; Tantau, H.J. Evaluation of LED lighting systems in in vitro cultures. *Acta Hort.* **2012**, *956*, 555–562. [CrossRef]
17. Gupta, S.D.; Jatothu, B. Fundamentals and applications of lightemitting diodes (LEDs) in in vitro plant growth and morphogenesis. *Plant Biotechnol. Rep.* **2013**, *7*, 211–220. [CrossRef]

18. Bantis, F.; Smirnakou, S.; Ouzounis, T.; Koukounaras, A.; Ntagkas, N.; Radoglou, K. Current status and recent achievements in the field of horticulture with the use of light-emitting diodes (LEDs). *Sci. Hortic.* **2018**, *235*, 437–451. [CrossRef]
19. Lian, M.L.; Murthy, H.N.; Paek, K.Y. Effects of light emitting diodes (LED) on the in vitro induction and growth of bulblets of *Lilium* oriental hybrid 'Pesaro'. *Sci. Hortic.* **2002**, *94*, 365–370. [CrossRef]
20. Prokopiuk, B.; Ćioć, M.; Maślanka, M.; Pawłowska, B. Effects of light spectra and benzyl adenine on in vitro adventitious bulb and shoot formation of *Lilium regale* E. H. Wilson. *Propag. Ornam. Plants* **2018**, *18*, 12–18.
21. Rivero-Montejo, S.J.; Vargas-Hernandez, M.; Torres-Pacheco, I. Nanoparticles as novel elicitors to improve bioactive compounds in plants. *Agriculture* **2021**, *11*, 134. [CrossRef]
22. Namdeo, A.G. Plant cell elicitation for production of secondary metabolites: A review. *Pharmacogn. Rev.* **2007**, *1*, 69–79.
23. Shitan, N. Secondary metabolites in plants: Transport and self-tolerance mechanisms. *Biosci. Biotechnol. Biochem.* **2016**, *80*, 1283–1293. [CrossRef] [PubMed]
24. Jafari, S.M.; McClements, D.J. Chapter One—Nanotechnology approaches for increasing nutrient bioavailability. In *Advances in Food and Nutrition Research*, 1st ed.; Toldrá, F., Ed.; Academic Press: London, UK, 2017; Volume 81, pp. 1–30. [CrossRef]
25. Iranbakhsh, A.; Oraghi Ardebili, Z.; Oraghi Ardebili, N. Synthesis and characterization of zinc oxide nanoparticles and their impact on plants. In *Plant Responses to Nanomaterials*; Singh, V.P., Singh, S., Tripathi, D.K., Prasad, S.M., Chauhan, D.K., Eds.; Nanotechnology in the Life Sciences; Springer: Cham, Switzerland, 2021; pp. 33–93. [CrossRef]
26. Misra, A.; Srivastava, A.K.; Srivastava, N.K.; Khan, A. Zn-acquisition and its role in growth, photosynthesis, photosynthetic pigments, and biochemical changes in essential monoterpene oil(s) of *Pelargonium graveolens*. *Photosynthetica* **2005**, *43*, 153–155. [CrossRef]
27. Eisvand, H.R.; Kamaei, H.; Nazarian, F. Chlorophyll fluorescence, yield and yield components of bread wheat affected by phosphatebio-fertilizer, zinc and boron under late-season heat stress. *Photosynthetica* **2018**, *56*, 1287–1296. [CrossRef]
28. Raliya, R.; Tarafdar, J.C. ZnO nanoparticle biosynthesis and its effect on phosphorus-mobilizing enzyme secretion and gum contents in clusterbean (*Cyamopsis tetragonoloba* L.). *Agric. Res.* **2013**, *2*, 48–57. [CrossRef]
29. Sabir, S.; Arshad, M.; Chaudhari, S.K. Zinc oxide nanoparticles for revolutionizing agriculture: Synthesis and applications. *Sci. World J.* **2014**, *2014*, 925494. [CrossRef]
30. Singh, A.; Singh, N.B.; Afzal, S.; Singh, T.; Hussain, I. Zinc oxide nanoparticles: A review of their biological synthesis, antimicrobial activity, uptake, translocation and biotransformation in plants. *J. Mater. Sci.* **2018**, *53*, 185–201. [CrossRef]
31. Wojnarowicz, J.; Chudoba, T.; Lojkowski, W. A Review of Microwave Synthesis of Zinc Oxide Nanomaterials: Reactants, Process Parameters and Morphologies. *Nanomaterials* **2020**, *10*, 1086. [CrossRef]
32. Prasad, T.N.V.K.V.; Sudhakar, P.; Sreenivasulu, Y.; Latha, P.; Munaswamy, V.; Raja Reddy, K.; Sreeprasad, T.S.; Sajanlal, P.R.; Pradeep, T. Effect of nanoscale zinc oxide particles on the germination, growth and yield of peanut. *J. Plant Nutr.* **2012**, *35*, 905–927. [CrossRef]
33. Helaly, M.N.; El-Metwally, M.A.; El-Hoseiny, H.; Omar, S.A.; El-Sheery, N.I. Effect of nanoparticles on biological contamination of in vitro cultures and organogenic regeneration of banana. *Aust. J. Crop Sci.* **2014**, *8*, 612–624.
34. Raskar, S.V.; Laware, S.L. Effect of zinc oxide nanoparicles on cytology and seed germination in onion. *Int. J. Curr. Microbiol. Appl. Sci.* **2014**, *3*, 467–473.
35. Chamani, E.; Ghalehtaki, S.K.; Mohebodini, M.; Ghanbari, A. The effect of zinc oxide nano particles and humic acid on morphological characters and secondary metabolite production in *Lilium ledebourii* Bioss. *Iran. J. Genet. Plant Breed.* **2015**, *4*, 11–19.
36. Mosavat, N.; Golkar, P.; Yousefifard, M.; Javed, R. Modulation of callus growth and secondary metabolites in different *Thymus* species and *Zataria multiflora* micropropagated under ZnO nanoparticles stress. *Biotechnol. Appl. Biochem.* **2019**, *66*, 316–322. [CrossRef]
37. Ahmad, M.A.; Javed, R.; Adeel, M.; Rizwan, M.; Ao, Q.; Yang, Y. Engineered ZnO and CuO nanoparticles ameliorate morphological and biochemical response in tissue culture regenerants of candyleaf (*Stevia rebaudiana*). *Molecules* **2020**, *25*, 1356. [CrossRef] [PubMed]
38. El-Mahdy, M.T.; Elazab, D. Impact of zinc oxide nanoparticles on pomegranate growth under in vitro conditions. *Russ. J. Plant Physiol.* **2020**, *67*, 162–167. [CrossRef]
39. Mazaheri-Tirani, M.; Dayani, S. In vitro effect of zinc oxide nanoparticles on *Nicotiana tabacum* callus compared to ZnO micro particles and zinc sulfate ($ZnSO_4$). *Plant Cell Tissue Organ Cult. (PCTOC)* **2020**, *140*, 279–289. [CrossRef]
40. Tymoszuk, A.; Wojnarowicz, J. Zinc oxide and zinc oxide nanoparticles impact on in vitro germination and seedling growth in *Allium cepa* L. *Materials* **2020**, *13*, 2784. [CrossRef]
41. Zaeem, A.; Drouet, S.; Anjum, S.; Khurshid, R.; Younas, M.; Blondeau, J.P.; Tungmunnithum, D.; Giglioli-Guivarc'h, N.; Hano, C.; Abbasi, B.H. Effects of biogenic zinc oxide nanoparticles on growth and oxidative stress response in flax seedlings vs. in vitro cultures: A comparative analysis. *Biomolecules* **2020**, *10*, 918. [CrossRef] [PubMed]
42. Al-Mayahi, A.M.W. The effect of humic acid (HA) and zinc oxide nanoparticles (ZnO-NPS) on in vitro regeneration of date palm (*Phoenix dactylifera* L.) cv. Quntar. *Plant Cell Tissue Organ Cult.* **2021**, *145*, 445–456. [CrossRef]
43. Hezaveh, T.A.; Rahmani, F.; Alipour, H.; Pourakbar, L. Effects of foliar application of ZnO nanoparticles on secondary metabolite and micro-elements of camelina (*Camelina sativa* L.) under salinity stress. *J. Stress Physiol. Biochem.* **2020**, *16*, 54–69.
44. Sharifi-Rad, R.; Bahabadi, S.E.; Samzadeh-Kermani, A.; Gholami, M. The effect of non-biological elicitors on physiological and biochemical properties of medicinal plant *Momordica charantia* L. *Iran. J. Sci. Technol. Trans. A Sci.* **2020**, *44*, 1315–1326. [CrossRef]

45. Rosa, L.A.; Moreno-Escamilla, J.O.; Rodrigo-García, J.; Alvarez-Parrilla1, E. Chapter 12 Phenolic compounds. In *Postharvest Physiology and Biochemistry of Fruits and Vegetables*; Yahia, E.M., Ed.; Woodhead Publishing: Sawton, UK, 2018; pp. 253–271. [CrossRef]
46. Babenko, L.M.; Smirnov, O.E.; Romanenko, K.O.; Trunova, O.K.; Kosakivska, I.V. Phenolic compounds in plants: Biogenesis and functions. *Ukr. Biochem. J.* **2019**, *91*, 5–18. [CrossRef]
47. Ghasemzadeh, A.; Jaafar, H.Z.E.; Rahmat, A.; Wahab, P.E.M.; Halim, M.R.A. Effect of different light intensities on total phenolics and flavonoids synthesis and anti-oxidant activities in young ginger varieties (*Zingiber officinale* Roscoe). *Int. J. Mol. Sci.* **2010**, *11*, 3885–3897. [CrossRef]
48. Chang, H.P.; Kim, N.S.; Park, J.S.; Lee, S.Y.; Lee, J.W.; Park, S.U. Effects of light-emitting diodes on the accumulation of glucosinolates and phenolic compounds in sprouting canola (*Brassica napus* L.). *Foods* **2019**, *8*, 76. [CrossRef]
49. Hsie, B.S.; Bueno, A.I.S.; Bertolucci, S.K.V.; Carvalho, A.A.; Cunha, S.H.B.; Martins, E.R.; Pinto, J.E.B.P. Study of the influence of wavelengths and intensities of LEDs on the growth, photosynthetic pigment, and volatile compounds production of *Lippia rotundifolia* Cham in vitro. *J. Photochem. Photobiol. B Biol.* **2019**, *198*, 111577. [CrossRef]
50. Upadhyay, R.; Rao, L.J.M. An outlook on chlorogenic acids—Occurrence, chemistry, technology, and biological activities. *Crit. Rev. Food Sci. Nutr.* **2013**, *53*, 968–984. [CrossRef]
51. Marques, V.; Farah, A. Chlorogenic acids and related compounds in medicinal plants and infusions. *Food Chem.* **2009**, *113*, 1370–1376. [CrossRef]
52. Naveed, M.; Hejazi, V.; Abbas, M.; Kamboh, A.A.; Khan, G.J.; Shumzaid, M.; Ahmad, F.; Babazadeh, D.; Xia, F.; Modarresi-Ghazani, F.; et al. Chlorogenic acid (CGA): A pharmacological review and call for further research. *Biomed. Pharmacother.* **2018**, *97*, 67–74. [CrossRef]
53. Martínez, G.; Regente, M.; Jacobi, S.; Del Rio, M.; Pinedo, M.; de la Canal, L. Chlorogenic acid is a fungicide active against phytopathogenic fungi. *Pestic. Biochem. Physiol.* **2017**, *140*, 30–35. [CrossRef]
54. Kundu, A.; Vadassery, J. Chlorogenic acid-mediated chemical defence of plants against insect herbivores. *Plant Biol.* **2019**, *21*, 185–189. [CrossRef] [PubMed]
55. Magnani, C.; Isaac, V.L.B.; Correa, M.A.; Salgado, H.R.N. Caffeic Acid: A review of its potential use for medications and cosmetics. *Anal. Methods* **2014**, *6*, 3203–3210. [CrossRef]
56. Greenwald, P. Clinical trials in cancer prevention: Current results and perspectives for the future. *J. Nutr.* **2004**, *134*, 3507S–3512S. [CrossRef] [PubMed]
57. Bouzaiene, N.N.; Jaziri, S.K.; Kovacic, H.; Chekir-Ghedira, L.; Ghedira, K.; Luis, J. The effects of caffeic, coumaric and ferulic acids on proliferation, superoxide production, adhesion and migration of human tumor cells in vitro. *Eur. J. Pharmacol.* **2015**, *766*, 99–105. [CrossRef]
58. Sanchez-Moreno, C.; Jimenez-Escrig, A.; Saura-Calixto, F. Study of low-density lipoprotein oxidizability indexes to measure the antioxidant activity of dietary polyphenol. *Nutr. Res.* **2000**, *20*, 941–953. [CrossRef]
59. Vinson, J.A.; Teufel, K.; Wu, N. Red wine, dealcoholized red wine, and especially grape juice, inhibit atherosclerosis in a hamster model. *Atherosclerosis* **2001**, *156*, 67–72. [CrossRef]
60. Pei, K.; Ou, J.; Huang, J.; Ou, S. p-Coumaric acid and its conjugates: Dietary sources, pharmacokinetic properties and biological activities. *J. Sci. Food Agric.* **2016**, *96*, 2952–2962. [CrossRef]
61. Zang, L.Y.; Cosma, G.; Gardner, H.; Shi, X.; Castranova, V.; Vallyathan, V. Effect of antioxidant protection by p-coumaric acid on low-density lipoprotein cholesterol oxidation. *Am. J. Physiol. Cell Physiol.* **2000**, *279*, C954–C960. [CrossRef]
62. Kiliç, I.; Yeşiloğlu, Y. Spectroscopic studies on the antioxidant activity of p-coumaric acid. *Spectrochim. Acta A Mol. Biomol. Spectrosc.* **2013**, *115*, 719–724. [CrossRef]
63. Lou, Z.; Wang, H.; Rao, S.; Sun, J.; Ma, C.; Li, J. p-Coumaric acid kills bacteria through dual damage mechanisms. *Food Control* **2012**, *25*, 550–554. [CrossRef]
64. Boz, H. p-Coumaric acid in cereals: Presence, antioxidant and antimicrobial effects. *Int. J. Food Sci. Technol.* **2015**, *50*, 2323–2328. [CrossRef]
65. Janicke, B.; Hegardt, C.; Krogh, M.; Onning, G.; Akesson, B.; Cirenajwis, H.M.; Oredsson, S.M. The antiproliferative effect of dietary fiber phenolic compounds ferulic acid and p-coumaric acid on the cell cycle of Caco-2 cells. *Nutr. Cancer.* **2011**, *63*, 611–622. [CrossRef] [PubMed]
66. Jaganathan, S.K.; Supriyanto, E.; Mandal, M. Events associated with apoptotic effect of p-Coumaric acid in HCT-15 colon cancer cells. *World J. Gastroenterol.* **2013**, *19*, 7726–7734. [CrossRef] [PubMed]
67. Roy, N.; Narayanankutty, A.; Nazeem, P.A.; Valsalan, R.; Babu, T.D.; Mathew, D. Plant phenolics ferulic acid and p-coumaric acid inhibit colorectal cancer cell proliferation through EGFR down-regulation. *Asian Pac. J. Cancer Prev.* **2016**, *17*, 4019–4023. [PubMed]
68. Contardi, M.; Alfaro-Pulido, A.; Picone, P.; Guzman-Puyol, S.; Goldoni, L.; Benítez, J.J.; Heredia, A.; Barthel, M.J.; Ceseracciu, L.; Cusimano, G.; et al. Low molecular weight epsilon-caprolactone-p-coumaric acid copolymers as potential biomaterials for skin regeneration applications. *PLoS ONE* **2019**, *14*, e0214956. [CrossRef]
69. Contardi, M.; Heredia-Guerrero, J.A.; Guzman-Puyol, S.; Summa, M.; Benítez, J.J.; Goldoni, L.; Caputo, G.; Cusimano, G.; Picone, P.; Di Carlo, M.; et al. Combining dietary phenolic antioxidants with polyvinylpyrrolidone: Transparent biopolymer films based on p-coumaric acid for controlled release. *J. Mater. Chem. B* **2019**, *7*, 1384–1396. [CrossRef]

70. Boo, Y.C. *p*-Coumaric acid as an active ingredient in cosmetics: A review focusing on its antimelanogenic effects. *Antioxidants* **2019**, *8*, 275. [CrossRef]
71. Ou, S.; Kwok, K. Ferulic acid: Pharmaceutical functions, preparation and applications in foods. *J. Sci. Food Agric.* **2004**, *84*, 1261–1269. [CrossRef]
72. Srinivasan, M.; Sudheer, A.R.; Menon, V.P. Ferulic acid: Therapeutic potential through its antioxidant property. *J. Clin. Biochem. Nutr.* **2007**, *40*, 92–100. [CrossRef]
73. Li, D.; Rui, Y.; Guo, S.; Luan, F.; Liu, R.; Zeng, N. Ferulic acid: A review of its pharmacology, pharmacokinetics and derivatives. *Life Sci.* **2021**, *284*, 119921. [CrossRef]
74. Zduńska, K.; Dana, A.; Kolodziejczak, A.; Rotsztejn, H. Antioxidant properties of ferulic acid and its possible application. *Skin Pharmacol. Physiol.* **2018**, *31*, 332–336. [CrossRef] [PubMed]
75. Sakai, S.; Ochiai, H.; Nakajima, K.; Terasawa, K. Inhibitory effect of ferulic acid on macrophage inflammatory protein-2 production in a murine macrophage cell line, RAW264.7. *Cytokine* **1997**, *9*, 242–248. [CrossRef] [PubMed]
76. Ou, L.; Kong, L.; Zhang, X.; Niwa, M. Oxidation of ferulic acid by *Momordica charantia* peroxidase and related anti-inflammation activity changes. *Biol. Pharm. Bull.* **2003**, *26*, 1511–1516. [CrossRef] [PubMed]
77. Kawabata, K.; Yamamoto, T.; Hara, A.; Shimizu, M.; Yamada, Y.; Matsunaga, K.; Tanaka, T.; Mori, H. Modifying effects of ferulic acid on azoxymethane-induced colon carcinogenesis in F344 rats. *Cancer Lett.* **2000**, *157*, 15–21. [CrossRef]
78. Balasubashini, M.S.; Rukkumani, R.; Viswanathan, P.; Menon, V.P. Ferulic acid alleviates lipid peroxidation in diabetic rats. *Phytother. Res.* **2004**, *18*, 310–314. [CrossRef]
79. Murashige, T.; Skoog, F. A revised medium for rapid growth and bioassays with tobacco tissue cultures. *Physiol. Plant.* **1962**, *15*, 473–497. [CrossRef]
80. Pawłowska, B.; Żupnik, M.; Szewczyk-Taranek, B.; Cioć, M. Impact of LED light sources on morphogenesis and levels of photosynthetic pigments in *Gerbera jamesonii* grown in vitro. *Hortic. Environ. Biotechnol.* **2018**, *59*, 115–123. [CrossRef]
81. Li, H.; Xu, Z.; Tang, C. Effect of light-emitting diodes on growth and morphogenesis of upland cotton (*Gossypium hirsutum* L.) plantlets in vitro. *Plant Cell Tissue Organ Cult.* **2010**, *103*, 155–163. [CrossRef]
82. Fan, X.X.; Zang, J.; Xu, Z.G.; Guo, S.R.; Jiao, X.L.; Liu, X.Y.; Gao, Y. Effects of different light quality on growth, chlorophyll concentration and chlorophyll biosynthesis precursors of non-heading Chinese cabbage (*Brassica campestris* L.). *Acta Physiol. Plant.* **2013**, *35*, 2721–2726. [CrossRef]
83. Habiba, A.U.; Kazuhiko, S.; Ahasan, M.; Alam, M. Effects of different light quality on growth and development of protocorm-like bodies (PLBs) in *Dendrobium kingianum* cultured in vitro. *Bangladesh Res. Publ. J.* **2014**, *10*, 223–227.
84. Bello-Bello, J.J.; Martínez-Estrada, E.; Caamal-Velázquez, J.H.; Morales-Ramos, V. Effect of LED light quality on in vitro shoot proliferation and growth of vanilla (*Vanilla planifolia* Andrews). *Afr. J. Biotechnol.* **2016**, *15*, 272–277. [CrossRef]
85. Coelho, A.D.; Souza, C.K.; Bertolucci, S.K.V.; Carvalho, A.A.; Santos, G.C.; Oliveira, T.; Marques, E.A.; Salimena, J.P.; Pinto, J.E.B.P. Wavelength and light intensity enhance growth, phytochemical contents and antioxidant activity in micropropagated plantlets of *Urtica dioica* L. *Plant Cell Tissue Organ Cult. (PCTOC)* **2021**, *145*, 59–74. [CrossRef]
86. Klimek-Szczykutowicz, M.; Prokopiuk, B.; Dziurka, K.; Pawłowska, B.; Ekiert, H.; Szopa, A. The influence of different wavelengths of LED light on the production of glucosinolates and phenolic compounds and the antioxidant potential in in vitro cultures of *Nasturtium ofcinale* (watercress). *Plant Cell Tissue Organ Cult. (PCTOC)* **2022**, *149*, 113–122. [CrossRef]
87. García-López, J.I.; Zavala-García, F.; Olivares-Sáenz, E.; Lira-Saldívar, R.H.; Barriga-Castro, E.D.; Ruiz-Torres, N.A.; Ramos-Cortez, E.; Vázquez-Alvarado, R.; Niño-Medina, G. Zinc oxide nanoparticles boosts phenolic compounds and antioxidant activity of *Capsicum annuum* L. during germination. *Agronomy* **2018**, *8*, 215. [CrossRef]
88. Kim, D.H.; Gopal, J.; Sivanesan, I. Nanomaterials in plant tissue culture: The disclosed and undisclosed. *RSC Adv.* **2017**, *7*, 6492–36505. [CrossRef]
89. Garcia-Lópes, J.; Nino-Medina, G.; Olivares-Sàenz, E.; LiraSaldivar, R.; Barriga-Costro, E.; Vàzques-Alvarado, R.; Rodriguez-Salinas, P.; Zavala-Garcia, F. Foliar application of zinc oxide nanoparticles and zinc sulfate boosts the content of bioactive compound in Habanero peppers. *Plants* **2019**, *8*, 254. [CrossRef]
90. Khan, A.U.; Khan, T.; Khan, M.A.; Nadhman, A.; Aasim, M.; Khan, N.Z.; Ali, W.; Nazir, N.; Zahoor, M. Iron-doped zinc oxide nanoparticles-triggered elicitation of important phenolic compounds in cell cultures of *Fagonia indica*. *Plant Cell Tissue Organ Cult. (PCTOC)* **2021**, *147*, 287–296. [CrossRef]
91. Zafar, H.; Alli, A.; Ali, J.S.; Haq, I.U.; Zia, M. Effect of ZnO nanoparticles on *Brassica nigra* seedlings and stem explants: Growth dynamics and antioxidative response. *Front. Plant Sci.* **2016**, *7*, 535. [CrossRef]
92. Raigond, P.; Raigond, B.; Kaundal, B.; Singh, B.; Joshi, A.; Dutt, S. Effect of zinc nanoparticles on antioxidative system of potato plants. *J. Environ. Biol.* **2017**, *38*, 435–439. [CrossRef]
93. Marslin, G.; Sheeba, C.J.; Franklin, G. Nanoparticles alter secondary metabolism in plants via ROS burst. *Front. Plant Sci.* **2017**, *8*, 832. [CrossRef]
94. Michalak, A. Phenolic compounds and their antioxidant activity in plants growing under heavy metal stress. *Pol. J. Environ.* **2006**, *15*, 523–530.
95. Chen, C.; Agrawal, D.C.; Lee, M.; Lee, R.; Kuo, C.; Wu, C.; Tsay, H.; Chang, H. Influence of LED light spectra on in vitro somatic embryogenesis and LC–MS analysis of chlorogenic acid and rutin in *Peucedanum japonicum* Thunb.: A medicinal herb. *Bot. Stud.* **2016**, *57*, 9. [CrossRef] [PubMed]

96. Wu, H.; Lin, C. Red light-emitting diode light irradiation improves root and leaf formation in difficult-to-propagate *Protea cynaroides* L. plantlets in vitro. *HortScience* **2012**, *47*, 1490–1494. [CrossRef]
97. Guo, B.; Liu, Y.; Yan, Q.; Liu, C. Spectral composition of irradiation regulates the cell growth and flavonoids biosynthesis in callus cultures of *Saussurea medusa* Maxim. *Plant Growth Regulat.* **2007**, *52*, 259–263. [CrossRef]
98. Urbonaviciute, A.; Samuoliene, G.; Brazaityte, A.; Duchovskis, P.; Ruzgas, V.; Zukauskas, A. The effect of variety and lighting quality on wheatgrass antioxidant properties. *ZemdirbysteAgriculture* **2009**, *96*, 119–128.
99. Leal-Costa, M.V.; dos Santos Nascimento, L.B.; dos Santos Moreira, N.; Reinert, F.; Costa, S.; Lage, C.L.S.; Tavares, E.S. Influence of blue light on the leaf morphoanatomy of in vitro *Kalanchoe pinnata* (Lamarck) Persson (Crassulaceae). *Microsc. Microanal.* **2010**, *16*, 576–582. [CrossRef]
100. Szopa, A.; Ekiert, H. The importance of applied light quality on the production of lignans and phenolic acids in *Schisandra chinensis* (Turcz.) Baill. cultures in vitro. *Plant Cell Tissue Organ Cult.* **2016**, *127*, 115–121. [CrossRef]
101. Kawka, B.; Kwiecień, I.; Ekiert, H. Infuence of culture medium composition and light conditions on the accumulation of bioactive compounds in shoot cultures of *Scutellaria laterifora* L. (American Skullcap) grown in vitro. *Appl. Biochem. Biotechnol.* **2017**, *183*, 1414–1425. [CrossRef]
102. Kubica, P.; Szopa, A.; Ekiert, H. Production of verbascoside and phenolic acids in biomass of *Verbena ofcinalis* L. (vervain) cultured under diferent in vitro conditions. *Nat. Prod. Res.* **2017**, *31*, 1663–1668. [CrossRef]
103. Szopa, A.; Kokotkiewicz, A.; Bednarz, M.; Luczkiewicz, M.; Ekiert, H. Studies on the accumulation of phenolic acids and favonoids in diferent in vitro culture systems of *Schisandra chinensis* (Turcz.) Baill. using a DAD-HPLC method. *Phytochem. Lett.* **2017**, *20*, 462–469. [CrossRef]
104. Szopa, A.; Starzec, A.; Ekiert, H. The importance of monochromatic lights in the production of phenolic acids and favonoids in shoot cultures of *Aronia melanocarpa, Aronia arbutifolia* and *Aronia* × *prunifolia*. *J. Photochem. Photobiol. B* **2018**, *179*, 91–97. [CrossRef] [PubMed]
105. Samuoliene, G.; Brazaityte, A.; Vaštakaitė-Kairienė, V. Light-emitting diodes (LEDs) for improved nutritional quality. In *Light-Emitting Diodes for Agriculture*; Dutta Gupta, S., Ed.; Springer: New York, NY, USA, 2017; pp. 149–190. [CrossRef]
106. Zielińska, S.; Piątczak, E.; Kozłowska, W.; Bohater, A.; Jezierska-Domaradzka, A.; Kolniak-Ostek, J.; Matkowski, A. LED illumination and plant growth regulators' efects on growth and phenolic acids accumulation in *Moluccella laevis* L. in vitro cultures. *Acta Physiol. Plant.* **2020**, *42*, 72. [CrossRef]

Disclaimer/Publisher's Note: The statements, opinions and data contained in all publications are solely those of the individual author(s) and contributor(s) and not of MDPI and/or the editor(s). MDPI and/or the editor(s) disclaim responsibility for any injury to people or property resulting from any ideas, methods, instructions or products referred to in the content.

MDPI AG
Grosspeteranlage 5
4052 Basel
Switzerland
Tel.: +41 61 683 77 34

Agronomy Editorial Office
E-mail: agronomy@mdpi.com
www.mdpi.com/journal/agronomy

Disclaimer/Publisher's Note: The title and front matter of this reprint are at the discretion of the Guest Editors. The publisher is not responsible for their content or any associated concerns. The statements, opinions and data contained in all individual articles are solely those of the individual Editors and contributors and not of MDPI. MDPI disclaims responsibility for any injury to people or property resulting from any ideas, methods, instructions or products referred to in the content.

www.ingramcontent.com/pod-product-compliance
Lightning Source LLC
LaVergne TN
LVHW072351090526
838202LV00019B/2523